T0280851

Beyond Conventional Quantization

This text describes novel treatments of quantum problems using enhanced quantization procedures.

When treated conventionally, certain systems yield trivial and unacceptable results. This book describes enhanced procedures, generally involving extended correspondence rules for the association of a classical and a quantum theory, which, when applied to such systems, yield nontrivial and acceptable results. The text begins with a review of classical mechanics, Hilbert space, quantum mechanics, and scalar quantum field theory. Next, analytical skills are further developed, a special class of models is studied, and a discussion of continuous and discontinuous perturbations is presented. Later chapters cover two further classes of models, both of which entail discontinuous perturbations. The final chapter offers a brief summary, concluding with a conjecture regarding interacting covariant scalar quantum field theories. Symmetry is repeatedly used as a tool to help develop solutions for simple and complex problems alike. Challenging exercises and detailed references are included.

Requiring only a modest prior knowledge of quantum mechanics and quantum field theory, this book will be of interest to graduate students and researchers in theoretical physics, mathematical physics, and mathematics.

JOHN KLAUDER was born on January 24th 1932 and received his Ph.D. from Princeton University, where he was a student of John A. Wheeler. A former head of the Theoretical Physics and Solid State Spectroscopy Departments of Bell Telephone Laboratories, he has been a visiting professor at Rutgers University, Syracuse University and the University of Bern. Since 1988 Professor Klauder has been a Professor of Physics and Mathematics at the University of Florida. He has also served on the Physics Advisory Panel of the National Science Foundation and been Editor of the Journal of Mathematical Physics, President of the International Association of Mathematical Physics, Associate Secretary-General of the International Union of Pure and Applied Physics, and has authored over 200 articles published in international journals.

Beyond
Conventional Quantization

JOHN R. KLAUDER

University of Florida

CAMBRIDGE
UNIVERSITY PRESS

CAMBRIDGE UNIVERSITY PRESS
Cambridge, New York, Melbourne, Madrid, Cape Town, Singapore, São Paulo

Cambridge University Press
The Edinburgh Building, Cambridge CB2 2RU, UK

Published in the United States of America by Cambridge University Press, New York

www.cambridge.org
Information on this title: www.cambridge.org/9780521258845

© Cambridge University Press 2000

This book is in copyright. Subject to statutory exception
and to the provisions of relevant collective licensing agreements,
no reproduction of any part may take place without
the written permission of Cambridge University Press.

First published 2000
This digitally printed first paperback version 2005

A catalogue record for this publication is available from the British Library

Library of Congress Cataloguing in Publication data
Klauder, John R.
Beyond conventional quantization / John R. Klauder.
p. cm.
Includes bibliographical references and index.
ISBN 0 521 25884 7 (hb)
1. Quantum field theory. I. Title.
QC174.45.K53   1999
530.14′3–dc21   99–10007   CIP

ISBN-13 978-0-521-25884-5 hardback
ISBN-10 0-521-25884-7 hardback

ISBN-13 978-0-521-67548-2 paperback
ISBN-10 0-521-67548-0 paperback

To my family

Contents

Preface

Some of our theories are missing! With all its successes, the quantum theory of fields as presently understood and applied has run into apparently insoluble difficulties related to classes of problems for which the conventional and time-tested methods of standard formulations utterly fail. We have in mind such theories as ϕ_4^4, a scalar field with a quartic self-interaction in four-dimensional space time, the results of which although relatively specific cannot be regarded as acceptable. Likewise for a host of other theories, such as ϕ_n^p, where $p \geq 2n/(n-2)$, and possibly even gravity. The theories in question are either strictly renormalizable and not asymptotically free or they are nonrenormalizable altogether. One aspect of the unacceptability for such quantizations is the fact that the classical limit of the quantized theory is manifestly unequal to the classical theory that has been quantized. In the same spirit as the opening sentence of this paragraph, what follows in this text is a kind of detective story. We shall need to review the historical evidence that has led us to this point, to develop and sharpen our skills in analyzing clues, and most particularly to carefully examine analogous cases that have been successfully resolved. Only then will we be in a position to put forth a conjecture on how and where the missing theories may be found.

Over a number of years the author has studied a handful of model quantum field theories each of which fails to fit into the conventional mold in one way or another. These models are streamlined to eliminate inessential bells and whistles and, most importantly, they each have been endowed with an enormous degree of symmetry that ultimately permits their solution to be found. These soluble models provide examples of quantum field theories that are "extragalactic", i.e., disconnected from the galaxy of conventional models and the body of knowledge that has been created to understand them. The study of these several models has led to a store of information as to why in each case the conventional approach has failed, and, taken in total, it constitutes a reservoir of unconventional approaches to a variety of problems. An appreciation of this body of knowledge gives one courage to seek solutions to the Major Problems outside the restrictions of conventional formulations.

One could, of course, propose the unconventional solutions we advocate straight away, but, as experience has shown, there is simply too big a gap between the conventional approaches and the unconventional approaches to persuade others of the validity of an alternative viewpoint. It is better in this case to follow a more gradual path, essentially the path the author himself has taken, complete with the incremental discoveries and understanding that each model affords. Our style of presentation is intended to be pedagogical, and we make no apology for the fact that this text is more about discovering principles than it is about developing technical proficiency. In so doing we have eschewed a strictly rigorous and potentially narrow approach in favor of a more discursive and conversational style; however, we have endeavored to be careful when care is called for. Generally, in this regard, we have adopted a maxim attributed to Mark Kac: *"For a reasonable person, a demonstration suffices; only a stubborn one requires a proof."*

There are two rather different ways in which this book may be approached.

On the one hand, the material in this book may be used as a supplement to standard courses in quantum mechanics and/or quantum field theory. For instance, we have in mind the enriched discussion of Hilbert space techniques (Chapters 3 and 6), the novel analysis of the harmonic oscillator (Chapter 4), and a more careful and thorough examination of path-integral techniques (Chapter 4). These treatments offer insights into quantum mechanics that are normally unavailable elsewhere. Likewise, several of the model quantum field theories (Chapters 7, 9, and 10) offer separate and complete mini-examples of how the usual tenants of quantum field theory need to be occasionally relaxed to provide room for nontrivial yet classically consistent quantum solutions as well as offering simplified and explicit examples of some features of general quantum theory. This approach uses the material for its case studies only.

On the other hand, we observe that this text is as much about an overall point of view as it is about any individual set of specific details, and hence to obtain its maximum impact, it should be read as a whole for only then will the evidence for the proposals for nontrivial scalar field theories offered in the final chapter become as convincing and compelling as they possibly can be. In an effort to facilitate reading as a whole, very little reference is made to earlier material within this book itself, and even when it is done it is only as a guide and not as a necessary diversion. Prolific numbering of display equations, which books of this type often entail, tends to foster a reliance on frequent cross referencing to prior discussions, and so as an effort to make that unnecessary we have dispensed with any equation numbers whatsoever! This approach has made it necessary for an occasional repeating of an argument here and there, but we feel that this repetition is better than interrupting the flow of the reader. Besides, repetition can be a useful technique for reinforcing an important concept. (All this also helps the reader who only wishes to study a few cases just as well!)

Exercises appear at the end of most chapters and are intended to complement the argument of the main text. Even if the reader does not work out the exercises, he or she is at least urged to read them so as to assimilate the additional clues they provide. By the time the reader reaches the conclusion it is hoped that he or she will share the author's enthusiasm for an eventually successful resolution for such theories – and indeed every interested reader is encouraged to take up the challenge in their own right.

A number of people have contributed in various ways to this effort, some recently, others long ago. Influence has come by an occasional comment, by the philosophical approach adopted in works of certain authors, in several cases acquired through direct collaboration with specific individuals, and/or comments made on earlier drafts of this text. The list is extensive, but particular mention should be made of Huzihiro Araki, Klaus Baumann, Bernhard Bodmann, Arthur Broyles, Detlev Buchholz, Gerard Emch, Hiroshi Ezawa, Helmut Gausterer, Gerhard Hegerfeldt, James McKenna, Heidi Narnhofer, Larry Shepp, Ludwig Streit, and Chengjun Zhu. Many of the ideas presented in this text have been developed while the author was employed at Bell Telephone Laboratories, as this institution was known for many years. In the past several years, lectures and student feedback at the University of Florida have been useful in refocusing the author's interest and in stimulating further developments in the material set forth between these covers. Thanks are expressed to Peter V. Landshoff for his encouragement to prepare this book many years ago. Gentle and repeated inquiries over the intervening years by Rufus Neal, Senior Commissioning Editor, Cambridge University Press, have proved remarkably effective. Recent sabbatical visits to the I.H.E.S., Bures-sur-Yvette, France, and the I.T.P., Bern University, Bern, Switzerland, have provided much needed time to focus on text preparation. And thanks are expressed to Mme Marie-Claude Vergne of the I.H.E.S. for her excellent rendering of the figures.

Gainesville   John R. Klauder

1

Introduction and Overview

1.1 Another book on quantum field theory?

There is absolutely no question that there are many excellent texts on the general subject of Quantum Field Theory. The field is vast and there are good reasons for such a large number of books. Generally, each book adopts a point of view favored by the author, and this is how it should be (indeed, this book will be no exception). Despite the fact that there are so many different important topics to address, it is no exaggeration to state, generally speaking, that most of these books cover common ground. There are chapters on canonical quantization techniques, free boson and fermion fields, Abelian gauge fields, interactions with external sources, coupled fields, perturbation theory with an emphasis on Feynman graphs, dealings with divergences: regularization and renormalization, path integral and functional integration formulations, non-Abelian gauge fields, and more modern topics such as the renormalization group and dimensional regularization. These and related tools and techniques are well honed and serve magnificently for studying a wide variety of interactions that describe the real world. But perhaps less appreciated is the fact that these methods do not provide a satisfactory quantization scheme for all classical field theories that one would like to deal with. At present there is, in the author's opinion, no satisfactory quantum theory of self-interacting relativistic scalar fields in four and more space-time dimensions, and while there is no lack of attempts there is no consensus on a satisfactory theory of the quantum gravitational field. These problems are significant and need to be faced. It has been clear to some workers that the methods and confines of "conventional quantization" techniques are inadequate to resolve the very real irreconcilable conflicts that arise in the usual approaches, and, consequently, an enlarged framework is called for. The material in this text is designed to offer one view, and a rather conservative one at that, of just such an enlarged viewpoint on how quantum field theory may be formulated with an eye to dealing with hitherto insoluble problems. The approach adopted in this text is that of learning by examples, in particular, the study of examples of an

1

ever-increasing difficulty, culminating with a head-on study of self-interacting relativistic scalar fields, albeit very brief. The advantage of this approach is that certain threshold levels of inevitable logic may be introduced that are easier for the general reader to accept than would be a single, large logical jump all by itself.

The underlying philosophy

This is a book about bosonic degrees of freedom, either a finite number of them or a countably infinite number of them. The latter case is alternatively described by a scalar field theory. There is no restriction that the field theory satisfies relativistic covariance, although this is an important class of problems and is occasionally discussed. Our interest mainly lies in discussing (some of) the *general principles* of quantum field theories rather than analyzing intricate details brought about by specific kinematical requirements. Of course, such issues are very important, but they are adequately treated in conventional textbooks. It is our conviction that approaching problems with an infinite number of degrees of freedom through a sequence of cutoff problems, each of which has a finite number of degrees of freedom, while natural and acceptable in the classical realm, is fraught with potential danger in the quantum realm. Indeed, we will present several model field theory problems, which in some way form the core of the present book, that illustrate how a conventional approach to field theory through a sequence of cutoff models need *not* give rise to an acceptable answer, often in fact leading to a trivial result. We can make this claim because the models in question have, by design, a great deal of symmetry that permits us to derive an acceptable answer outside the confines of the conventional approaches. The lessons learned from these model studies strongly suggest an alternative and unconventional approach to relativistic theories, such as ϕ_n^4, $n \geq 4$, which when studied by conventional means have so far led only to trivial solutions. Although we have not succeeded in proving that these alternative ideas work for such asymptotically nonfree and even nonrenormalizable relativistic models, they nonetheless do offer the possibility of a solution outside the conventional schemes. How can this be true?

After a great many years, it is surely a fact that the analytical methods used in quantum field theory have been battle tested and are capable of attacking virtually any model problem. Operator techniques, functional techniques, renormalization group methods, etc., are all available to take on any problem. What is left open and is simply outside the scope of any of these methods is the choice of "Equation (1)", namely the starting point. No solution technique, no matter how powerful, can derive a result that is not already implicit in the starting equation. It is not unfair to say in our study of model problems that we critically reexamine the starting point, and in so doing we are led to an alternative form for "Equation (1)", which can then be studied by the very array of powerful technical tools already available. There are so many different possible initial equations, how is one to select from among them all so as to arrive at a useful alternative starting point? This is where *symmetry* plays an indispensable role. Each of the model problems successfully treated by starting from an unconventional standpoint has a vast symmetry group that narrows down the possible starting points to a limited few.

In order to offer alternative starting points acceptable to a skeptical reader, it is incumbent upon us to present a modest, but nevertheless fair, assessment of the conventional approach to scalar model field theories. In that discourse we will also develop the usual approaches in a more-or-less conventional fashion as a kind of standard with which to compare and contrast our unconventional approach. In point of fact, our approach is not so unconventional as that phrase may suggest, and generally it is based simply on exploiting the symmetry inherent in each model along with the assumption of a few general principles which should be readily acceptable (e.g., existence of a Hamiltonian operator with an appropriate spectrum, existence of a ground state invariant under the group of symmetry transformations). The solution to which this alternative starting point leads is then tested for acceptability, generally by seeing that it – and not the solution obtained by conventional means – exhibits an acceptable classical limit as the parameter $\hbar \rightarrow 0$ in an appropriate way.

What this book is not about

This book touches on only a very "small" corner of the topics that generally make up the subject of quantum field theory. Broadly viewed, we focus on scalar fields. Therefore, there is no discussion of vector or tensor fields, no discussion of spinor fields, essentially no discussion of massless fields, no discussion of gauge theories, no discussion of detailed calculational techniques for perturbation analysis, no discussion of scattering theory, and of course, no discussion of such modern topics as supersymmetry, (super)string theory, quantum gravity, etc. We make no excuse for this vast omission of topics; there is, in our opinion, still an extensive array of interesting and unanswered questions surrounding scalar fields alone. Of course, it is conceivable that any insight won into how scalar fields may receive alternative and successful quantum treatments might very well spill over into other aspects of quantum field theory; indeed, we hope this may be the case. However, such possibilities are not pursued here.

1.2 The principal dilemma

Without any attempt at the present moment to explain notation or justify formulas, let us try to state "in a nutshell" the principal concern that motivates the present text. In one mode of formulation, the starting expression for a specific quantum field theory (here the covariant model ϕ_n^4) is chosen as a functional integral conventionally – and formally – given by

$$S\{h\} = \mathcal{N} \int \exp(\int \{h(x)\phi(x) - \tfrac{1}{2}[\nabla\phi(x)]^2 - \tfrac{1}{2}m^2\phi(x)^2 - g\phi(x)^4\}\, d^n x)\, \mathcal{D}\phi\, .$$

The expression in the exponent within the integrand derives from the classical theory. Unfortunately, as it stands, the given expression is quite without any mathematical meaning, and in order to give it meaning one must step back and define it as the limit of a sequence of well-defined, "regularized" integrals. One of the currently favored regularizations interprets the expression given above as $S\{h\} = \lim_{a\to 0} S_a(h)$, which is the continuum limit of the sequence of multiple integrals

$$S_a(h) = N_a \int \exp\{\Sigma[h_k\phi_k$$
$$- \tfrac{1}{2}(\phi_{k^*} - \phi_k)^2 a^{-2} - \tfrac{1}{2}m_o^2\phi_k^2 - g_o\phi_k^4]a^n\}\, \Pi d\phi_k\, ,$$

which is arrived at essentially by replacing the integral in the exponent of the previous expression by a straightforward Riemann sum approximation. The former expression may be called "Equation (0)", while the latter expression is what we typically mean by "Equation (1)". The interpretation of Eq. (0) in terms of Eq. (1) works well in a low number of space-time dimensions ($n \leq 3$), and it is an eminently reasonable starting point in higher dimensions ($n \geq 4$). However, all evidence suggests that the result in higher dimensions is "trivial", more specifically, Gaussian in the unintegrated probe field h [Ai 82, Fr 82, Ca 88]. We do not question this conclusion. What we do question is the uncritical acceptance that this particular regularized expression necessarily belongs to the "physically correct" class of regularizations. One source of doubt on this point arises because the (trivial) classical limit of the resultant (trivial) quantum theory does not coincide with the (*nontrivial*) classical theory with which we started [Re 76]. Instead, we believe that there is an alternative regularized starting equation that leads to a nontrivial quantum theory in the sense (at the very least) that the classical limit of the alternative quantum theory agrees with the starting classical theory.*

It is one thing to suggest that alternative starting points may be appropriate, but it is quite another to actually find acceptable ones. Searching for clues through the study of model problems and ultimately offering our best guess for an alternative starting point for a relativistic theory constitutes the unifying thread of the detective story told in this book. Whether or not our conjecture resolves the mystery must, in turn, wait for additional evidence to be gathered.

---

* Additional arguments for considering nonrenormalizable quantum field theories are offered by Weinberg [We 95].

At any rate, it is time to open the case and lay it out for the jury's consideration. As is generally appropriate, we urge the jury to remain open minded until the bulk of the evidence has been presented. Let us start with an opening statement.

1.3 Outline

This text is informally divided into four parts. Part I, which covers Chapters 2-5, constitutes an introduction and review of the conventional approach to classical and quantum theory, for systems with a few degrees of freedom as well as for fields. In this text we confine our attention to scalar fields for the simple reason that many problems still exist in these kinematically simplified models and there is still much we have not understood about this limited class of theories. Part II, which covers Chapters 6–8, begins with a sharpening of some analytical tools useful in treating additional model field theories. In addition, in Part II, we undertake our first excursion beyond conventional quantization with the study of a specific class of models. The last chapter in Part II deals with continuous and discontinuous perturbations, and with some general consequences of highly singular perturbations. From the insight gained by such a general study we learn that even the one-dimensional harmonic oscillator is not immune to the disease of divergences that are known to plague quantum field theory. Part III, which consists of Chapters 9–10, deals with two types of quantum field models, each of which is more singular than corresponding relativistic models with the same nonlinear interaction. Viewed conventionally, these models are trivial. However, when considered outside the confines of the usual framework, perfectly satisfactory and nontrivial solutions are found. Although the several models we carefully analyze in Parts II and III of this text are not relativistically covariant, their solution nevertheless serves to illuminate some unconventional avenues to the successful quantum analysis of systems with an infinite number of degrees of freedom. In a certain sense, these models form the heart of the present text. Part IV consists only of Chapter 11. In this chapter we embark on some speculation, trying to put to further use some of the lessons we have learned in the earlier chapters. Although this chapter does not offer a confirmed solution for a nontrivial quantization of covariant self-interacting quantum scalar fields in four and more space-time dimensions, there is a relatively concrete proposal which is suggested, and, at the very least, some food for thought.

It is a special feature of this text that analytical techniques based on the use of coherent states are introduced early and used repeatedly.

2

Classical Mechanics

In this chapter we introduce the subject of Lagrangian and Hamiltonian formulations, following descriptions that focus on action functionals and their stationary variation to arrive at the relevant equations of motion. Initially we deal with one degree of freedom, and then with finitely many degrees of freedom. These studies are standard and introductory to the case of an infinite number of degrees of freedom, namely the case of fields. In the classical theory, the infinite number of variables case can be approached as a limit of the finite number of variables cases as the number of those variables tends to infinity. This seems completely reasonable, and, more or less, may be taken as an unwritten rule of how to treat classical systems with an infinite number of variables. Unfortunately, as we shall learn much later, this unwritten rule will not always carry over to the quantum theory, at least in any obvious and straightforward fashion.

2.1 Lagrangian classical mechanics

Without undue exaggeration, the goal of classical dynamics may be said to be the introduction of dynamical equations of motion for point particles, or for entities that may be idealized as point-like particles, and the analysis of properties of the solution to such equations, i.e., the time-dependent path $x(t)$ describing the classical motion.* In physical contexts these equations of motion are generally second-order differential equations, such as arise in Newton's equations of motion, and it is extremely useful to establish a uniform set of procedures to derive and develop such equations of motion. In such circumstances, the dynamical equations of motion for classical mechanics follow most naturally from the

---

* For a contemporary account of classical mechanics see [SuM 74].

6

stationary variation of a particular functional of the path known as the *action*. Let us base our initial discussion on simple, one-dimensional systems. If $x(t)$ denotes the one-dimensional coordinate of a particle as a function of the time t, $t' \le t \le t''$, $t' < t''$, and $\dot{x}(t) \equiv dx(t)/dt$, then a wide class of action functionals consists of expressions of the form

$$I = \int_{t'}^{t''} L(\dot{x}(t), x(t), t)\, dt \, ,$$

where the function L is called the Lagrangian. For generality, we have assumed that L may also depend explicitly on time. Consider as well the action evaluated for a neighboring path $x(t) + \delta x(t)$, where $\delta x(t)$ is constrained to vanish at the end points, i.e., $\delta x(t') = 0 = \delta x(t'')$. With the integration limits fixed, a change in path leads to a changed value for the action, given by

$$I + \delta I = \int_{t'}^{t''} L(\dot{x}(t) + \delta\dot{x}(t), x(t) + \delta x(t), t)\, dt \, .$$

We now seek specific paths for which the variation of the action *vanishes* to first order. Specifically, we insist on

$$0 = \delta I = \int_{t'}^{t''} \left[\frac{\partial L}{\partial \dot{x}(t)} \delta\dot{x}(t) + \frac{\partial L}{\partial x(t)} \delta x(t) \right] dt$$

$$= \int_{t'}^{t''} \left[\frac{\partial L}{\partial x(t)} - \frac{d}{dt}\frac{\partial L}{\partial \dot{x}(t)} \right] \delta x(t)\, dt + \frac{\partial L}{\partial \dot{x}(t)} \delta x(t) \Big|_{t'}^{t''}$$

$$= \int_{t'}^{t''} \left[\frac{\partial L}{\partial x(t)} - \frac{d}{dt}\frac{\partial L}{\partial \dot{x}(t)} \right] \delta x(t)\, dt$$

thanks to the vanishing of the variation at the end points of integration. Stationarity for arbitrary first-order variations then leads to the classical equations of motion,

$$\frac{d}{dt}\frac{\partial L}{\partial \dot{x}(t)} - \frac{\partial L}{\partial x(t)} = 0 \, .$$

It is important to note that these equations of motion are unchanged if a total derivative of a function of $x(t)$ is added to the Lagrangian. With the integration limits implicit, consider the action*

$$I = \int [L(\dot{x}(t), x(t)) + \dot{f}(x(t))]\, dt$$

$$= \int L(\dot{x}(t), x(t))\, dt + f(x'') - f(x') \, ,$$

where $x'' = x(t'')$ and $x' = x(t')$, the values at the end points of the range of integration. If these end point values were not held fixed, then the simple

---

* For notational simplicity in the rest of this chapter we let $\dot{Y} \equiv (d/dt)Y$ for a general function Y.

criterion that the first-order variation vanishes would not lead to the desired equations of motion. Observe that if we replace $\dot{f}(x(t))$ by $\dot{f}(\dot{x}(t), x(t))$ in the integrand, then, in general, the simple criterion $\delta I = 0$ would no longer be appropriate to derive the equations of motion since the term added to the action now reads $f(\dot{x}'', x'') - f(\dot{x}', x')$ and its variation need not vanish. The only way to ensure that such a total derivative does not destroy the simple criterion by which one chooses the equations of motion is to insist that all variations have fixed values of $\dot{x}(t)$ as well as fixed values of $x(t)$ at both of the end points. Such a restriction on the variations is of course entirely possible. For a general path $x(t)$, $t' < t < t''$, the action functionals differ from one another for different functions f. However, all such action functionals form an equivalence class in that they all lead to the same equation of motion.

One of the virtues of the Lagrangian formulation of mechanics is that it is covariant under coordinate transformations. Let us introduce a new coordinate $\overline{x}(t)$ in place of the old coordinate according to the rule $x(t) = x(\overline{x}(t))$. It follows that $\dot{x}(t) = (\partial x / \partial \overline{x}) \dot{\overline{x}}(t) \equiv \dot{x}(\dot{\overline{x}}(t), \overline{x}(t))$ and

$$\overline{L}(\dot{\overline{x}}(t), \overline{x}(t), t) \equiv L(\dot{x}(\dot{\overline{x}}(t), \overline{x}(t)), x(\overline{x}(t)), t) = L(\dot{x}(t), x(t), t) .$$

Here in the last relation we have used the fact that L transforms as a scalar under coordinate transformations. As a consequence the action functional is invariant under a coordinate transformation, namely

$$I = \int L(\dot{x}(t), x(t), t) \, dt = \int \overline{L}(\dot{\overline{x}}(t), \overline{x}(t), t) \, dt ,$$

and thus the equation of motion that follows from a stationary variational principle is form invariant,

$$\frac{d}{dt} \frac{\partial \overline{L}}{\partial \dot{\overline{x}}(t)} - \frac{\partial \overline{L}}{\partial \overline{x}(t)} = 0 .$$

This simple example illustrates an important lesson, namely, that deriving equations of motion from actions that are invariant under suitable coordinate transformations automatically leads to equations of motion that transform covariantly under such coordinate transformations.

Solutions of the equations of motion are determined in principle once appropriate boundary data are given. In a first form, the boundary data consist of giving the initial position and velocity, $x(t')$ and $\dot{x}(t')$, respectively. In this case the future behavior of the solution is uniquely determined. Under favorable conditions the solution exists for all time, and this occurs, for example, when, in suitable units, $L = \dot{x}^2/2$, appropriate to the free particle, $L = \dot{x}^2/2 - x^2/2$, appropriate to the harmonic oscillator, $L = \dot{x}^2/2 + x^2/2$, or $L = \dot{x}^2/2 - x^4/4$, the latter being one of the variants of a quartic anharmonic oscillator. More generally, the solution will be well behaved only for a finite time, after which it may become ill-defined (e.g., divergent). Examples of such behavior arise when the Lagrangian is given by $L = \dot{x}^2/2 + x^4/4$ or $L = \dot{x}^2/2 + x^{-2}/2$; see Exercise

2.1. In a second form, the boundary data consists of giving one condition at each end of the time interval, e.g., the initial and final position, $x(t')$ and $x(t'')$, respectively. These data may lead to no solution, to one solution, or possibly to multiple solutions. When solutions do exist, they may be defined for all time or only for a finite time interval.

An important lesson from all this is an appreciation that not every equation of motion enjoys global solutions for arbitrary initial conditions!

Several degrees of freedom

Before leaving Lagrangian mechanics we wish to generalize our discussion from one degree of freedom (one variable) to finitely many degrees of freedom. For this purpose let*

$$x = (x^1, x^2, \ldots, x^N) \in \mathbb{R}^N$$

and we regard our path as a time-dependent point moving in an N-dimensional configuration space

$$x(t) = (x^1(t), x^2(t), \ldots, x^N(t)) \in \mathbb{R}^N .$$

The velocity is likewise given by†

$$\dot{x}(t) = (\dot{x}^1(t), \dot{x}^2(t), \ldots, \dot{x}^N(t)) \in \mathbb{R}^N .$$

With this notation, the Lagrangian for an N-dimensional system is taken to be $L(\dot{x}(t), x(t), t)$ and the action functional is the time integral of the Lagrangian. Variation of the path now consists of varying each component $x^k(t)$, $1 \le k \le N$, $N < \infty$, separately and independently, subject only to the vanishing of the variation at the end points. This procedure leads to N equations of motion given by

$$\frac{d}{dt} \frac{\partial L}{\partial \dot{x}^k(t)} - \frac{\partial L}{\partial x^k(t)} = 0 , \qquad 1 \le k \le N .$$

Initial conditions for these equations typically consist of specifying the initial position and velocity, $x(t')$ and $\dot{x}(t')$, or the initial and final positions, $x(t')$ and $x(t'')$. As in the one-dimensional case, solutions may exist for all time or for only a finite time before they become ill-defined. Changes of coordinates proceed in a completely analogous fashion to the one-dimensional case, and the Lagrangian equations of motion transform covariantly under coordinate transformations.

---

* Here we have assumed that the configuration space is a Euclidean space, namely a flat space as opposed to a curved space [such as the surface of an N-sphere of unit radius defined by $(y^1)^2 + (y^2)^2 + \cdots + (y^{N+1})^2 = 1$].

† In the case of the velocity the space is always flat – hence $\mathbb{R}^N$ – or at least a portion of a flat space, since one is working in the tangent space (thanks to the dx) to the configuration space at each point of the configuration space.

2.2 Hamiltonian classical mechanics

An alternative but essentially equivalent formulation of classical mechanics is that of Hamilton. For simplicity in this discussion we shall assume that the Lagrangian does not explicitly depend on time. In this formulation we first introduce, for $N = 1$, the canonical momentum

$$p(t) \equiv \frac{\partial L}{\partial \dot{x}(t)},$$

and suppose that we can solve for $\dot{x}(t)$ in terms of $x(t)$ and $p(t)$. The necessary requirement for this to be the case is that $\partial^2 L/\partial \dot{x}(t)^2 \neq 0$. There are many interesting cases where this condition is not true, but for present purposes we shall assume that this condition holds and that we can solve for $\dot{x}(t)$, at least in principle. At the same time we will introduce the convention that

$$q(t) \equiv x(t),$$

so that we may attempt to distinguish when we are dealing with the Lagrangian form of mechanics [i.e., $x(t)$] and when we are dealing with the Hamiltonian form of mechanics [i.e., $q(t)$], at least as much as possible. Now we define the fundamentally important Hamiltonian function $H(p(t), q(t))$ by the equation

$$H(p, q) \equiv p\dot{q} - L(\dot{q}, q), \qquad \dot{q} = \dot{q}(p, q),$$

namely

$$H(p, q) \equiv p\dot{q}(p, q) - L(\dot{q}(p, q), q).$$

We derive the equations of motion again by the same variational principle, this time involving the expression

$$I = \int L(\dot{q}(p(t), q(t)), q(t)) \, dt$$

$$= \int [p(t)\dot{q}(t) - H(p(t), q(t))] \, dt.$$

Treating $p(t)$ and $q(t)$ as independent of each other in the variation leads to the equation

$$\delta I = \int \left[\dot{q}(t)\delta p(t) + p(t)\delta\dot{q}(t) - \frac{\partial H}{\partial p(t)}\delta p(t) - \frac{\partial H}{\partial q(t)}\delta q(t) \right] dt$$

$$= \int \left[\dot{q}(t)\delta p(t) - \dot{p}(t)\delta q(t) - \frac{\partial H}{\partial p(t)}\delta p(t) - \frac{\partial H}{\partial q(t)}\delta q(t) \right] dt.$$

To arrive at this expression we have assumed, as before, that $\delta q(t)$ vanishes at both ends of the range of integration in order that the term which arises from the integration by parts does not contribute. Treating $p(t)$ and $q(t)$ as independent variations leads to two equations of motion to achieve stationarity, namely

$$\dot{q}(t) = \frac{\partial H(p,q)}{\partial p(t)} \, ,$$

$$\dot{p}(t) = -\frac{\partial H(p,q)}{\partial q(t)} \, .$$

These are the celebrated Hamiltonian equations of motion. In comparison with the Lagrangian form of the equation of motion, it follows that the first of Hamilton's equations serves to define $p = p(\dot{q}, q)$, while the second of Hamilton's equations leads to the Lagrangian equation of motion. On the other hand, taking the two equations at face value, there is an evident symmetry between the variables $p(t)$ and $q(t)$, which will be quite important in the sequel. Observe that to derive Hamilton's equations it was not necessary to specify the behavior of $\delta p(t)$ at the boundaries, only that of $\delta q(t)$. This seems to be an asymmetric treatment of the two variables. In point of fact, it is extremely useful to also insist on the vanishing of the variation of $p(t)$ at both ends of the integration range, even though it was not necessary to derive Hamilton's equations of motion above. Let us see why this is the case.

Just as in the Lagrangian case, it is important that the equations of motion derived from an action principle are invariant under changes of the Lagrangian by a total derivative. Again, a total derivative amounts to a change of the action by terms that depend only on the values of the variables at the end points of the integration range. For example, consider the alternative action functional given by

$$I = \int [p(t)\dot{q}(t) - H(p(t), q(t)) + \dot{G}(p(t), q(t))] \, dt$$

$$= \int [p(t)\dot{q}(t) - H(p(t), q(t))] \, dt + G(p'', q'') - G(p', q') \, .$$

Simple examples such as $G(p,q) = -pq$ or $G(p,q) = -pq/2$ illustrate the kinds of situation that may arise. If the variation of these terms did not vanish, then the criterion to choose the equations of motion would depend on G and not be the simple and universal criterion that $\delta I = 0$. The only way to ensure such terms do not affect the manner by which the equations of motion are chosen is to insist that variations are made that hold *both* $p(t)$ and $q(t)$ fixed at *both* end points of the range of integration. [The analogy, in the Lagrangian context, with variations that hold $\dot{x}(t)$ and $x(t)$ fixed at both end points should be evident.] Observe that the action functional is different in actual value for different functions G, but each such action leads to identical equations of motion. Thus, we again deal with an equivalence class of action functionals.

In Hamiltonian mechanics the coordinates $p(t)$ and $q(t)$ we have used are referred to as *canonical coordinates*. Coordinate transformations between one set of canonical coordinates and another set of canonical coordinates – called *canonical coordinate transformations* – play a major role in the classical theory of Hamiltonian mechanics. If $\bar{p}(t) = \bar{p}(p(t), q(t))$ and $\bar{q}(t) = \bar{q}(p(t), q(t))$ denote new canonical coordinates, then they may generically be related to the original canonical coordinates by means of the differential relation

$$\bar{p}(t)d\bar{q}(t) = p(t)dq(t) + dF(\bar{q}(t), q(t)),$$

where F is called the generator of the canonical coordinate transformation. In this equation, the variables $\bar{q}(t)$ and $q(t)$ are regarded as the independent variables while $\bar{p}(t)$ and $p(t)$ are treated as the dependent variables.* Under a coordinate transformation, H transforms as a scalar and consequently $\overline{H}(\bar{p}(t), \bar{q}(t)) = H(p(t), q(t))$. Therefore, after a canonical coordinate transformation the action functional assumes the form

$$I = \int [\bar{p}(t)\dot{\bar{q}}(t) - \overline{H}(\bar{p}(t), \bar{q}(t)) + \dot{\overline{G}}(\bar{p}(t), \bar{q}(t))]\, dt\,,$$

for some function $\overline{G}$ that incorporates the effects of both F and G. Stationary variation of the action in this form, holding both $\bar{p}(t)$ and $\bar{q}(t)$ fixed at both end points, leads to the new version of the equations of motion,

$$\dot{\bar{q}}(t) = \frac{\partial \overline{H}(\bar{p}, \bar{q})}{\partial \bar{p}(t)}\,,$$

$$\dot{\bar{p}}(t) = -\frac{\partial \overline{H}(\bar{p}, \bar{q})}{\partial \bar{q}(t)}\,.$$

Evidently these equations are form invariant under canonical coordinate transformations, and it is this important property that gives to these transformations their major significance. Among such canonical transformations is the elementary example where $F = \bar{q}q$, which leads to $\bar{p} = \partial F/\partial \bar{q} = q$ and $p = -\partial F/\partial q = -\bar{q}$. This transformation, in effect, interchanges p and q, emphasizing the symmetry between the two canonical variables.

When suitable initial conditions are given, then solutions of the Hamiltonian equations of motion may be considered. As two first-order equations, a solution is uniquely defined by specifying $q(t')$ and $p(t')$. Such a solution may be well defined for all time, as illustrated by the Hamiltonians $H = p^2/2$, the free particle, $H = p^2/2 + q^2/2$, the harmonic oscillator, $H = p^2/2 - q^2/2$, or $H = p^2/2 + q^4/4$, an anharmonic oscillator. Alternatively, the solution may become ill-defined (e.g., divergent or even complex) after a finite time, as occurs for the examples $H = p^2/2 - q^4/4$ or $H = pq^3$; again, see Exercise 2.1.

Just as in the case of Lagrangian mechanics, we wish to generalize our discussion from one degree of freedom, i.e., one p and one q, or two variables, to finitely

---

* When $\bar{q}$ and q are *dependent*, i.e., $\bar{q} = \bar{q}(q)$, then this differential relation must be changed. For example, one may then consider

$$-\bar{q}(t)\, d\bar{p}(t) = p(t)\, dq(t) + d\tilde{F}(\bar{p}(t), q(t))\,.$$

many degrees of freedom. For this purpose we assume that the phase space is a Euclidean space and that our system is described by a phase-space point

$$(p, q) = (p_1, p_2, \ldots, p_N, q^1, q^2, \ldots, q^N) \in \mathbb{R}^{2N} ,$$

and we regard our path as a time-dependent point moving in the same $2N$-dimensional phase space

$$(p(t), q(t)) = (p_1(t), \ldots, p_N(t), q^1(t), \ldots, q^N(t)) \in \mathbb{R}^{2N} .$$

With this notation, the Hamiltonian for the N degree-of-freedom system is taken to be $H(p(t), q(t))$, and the action functional is the time integral of the Lagrangian, $L = p \cdot \dot{q} - H$, where $p \cdot \dot{q} = \Sigma p_k \dot{q}^k$, $1 \leq k \leq N$. Variation of the path now consists of varying each component $p_k(t)$ and $q^k(t)$, $1 \leq k \leq N$, separately and independently, subject only to the vanishing of the variation at the end points. This procedure leads to $2N$ Hamiltonian equations of motion, given by

$$\dot{q}^k(t) = \frac{\partial H(p, q)}{\partial p_k(t)}, \qquad 1 \leq k \leq N ,$$

$$\dot{p}_k(t) = -\frac{\partial H(p, q)}{\partial q^k(t)}, \qquad 1 \leq k \leq N .$$

Initial conditions for these equations typically consist of specifying the initial positions and momenta, $q(t')$ and $p(t')$, or the initial and final positions, $q(t')$ and $q(t'')$. As in the one-dimensional case, the solutions may exist for all time or for only a finite time before they become ill-defined. Changes of canonical coordinates proceed in a completely analogous fashion as in the single degree-of-freedom case, and the Hamiltonian equations of motion transform covariantly under canonical coordinate transformations.

Poisson brackets

It is often useful to reexpress Hamiltonian mechanics in the language of Poisson brackets. Let us directly treat the N degree-of-freedom situation. If $A(p, q)$ and $B(p, q)$ denote two smooth phase-space functions, then the Poisson bracket of A and B is defined by

$$\{A, B\} \equiv \sum \left[\frac{\partial A(p, q)}{\partial q^k} \frac{\partial B(p, q)}{\partial p_k} - \frac{\partial A(p, q)}{\partial p_k} \frac{\partial B(p, q)}{\partial q^k} \right] ,$$

an expression that exhibits the important symmetry $\{B, A\} = -\{A, B\}$. For example, it follows that $\{q^j, p_k\} = \delta_k^j [\equiv 1 \ (j = k); \equiv 0 \ (j \neq k)]$, and that

$$\{q^j, H\} = \frac{\partial H(p, q)}{\partial p_j(t)} = \dot{q}^j(t) ,$$

$$\{p_k, H\} = -\frac{\partial H(p, q)}{\partial q^k(t)} = \dot{p}_k(t) .$$

Thus for any $A(p,q)$,

$$\dot{A}(p,q) = \sum \left[\frac{\partial A(p,q)}{\partial p_k(t)} \dot{p}_k(t) + \frac{\partial A(p,q)}{\partial q^k(t)} \dot{q}^k(t) \right] = \{A, H\} \ .$$

In particular, $\dot{H} = \{H, H\} = 0$, which asserts that if the Hamiltonian is not explicitly time dependent (as we have assumed), then the Hamiltonian is a constant of the motion usually identified with the energy. These relations assert that the dynamical equations may be entirely rewritten in terms of Poisson brackets.

It is an important fact that the Poisson bracket may be evaluated in any canonical coordinate system and it leads to the same answer; this property is very useful since it lets one choose the most convenient canonical coordinate system in which to evaluate the result. This equality holds because of the way in which canonical transformations are defined. Initially, note that the existence of the generator F leads to the fact that $\overline{p}_k = \partial F / \partial \overline{q}^k$ and $p_k = -\partial F / \partial q^k$; consequently, $\partial \overline{p}_j / \partial q^l = -\partial p_l / \partial \overline{q}^j$. Similarly, the relation $\overline{p}_j d\overline{q}^j = -q^j dp_j + d\hat{F}(\overline{q}, p)$ implies that $\partial \overline{p}_j / \partial p_l = \partial q^l / \partial \overline{q}^j$. Therefore,

$$\{\overline{q}^k, \overline{p}_j\} = \sum \left[\frac{\partial \overline{q}^k}{\partial q^l} \frac{\partial \overline{p}_j}{\partial p_l} - \frac{\partial \overline{q}^k}{\partial p_l} \frac{\partial \overline{p}_j}{\partial q^l} \right]$$

$$= \sum \left[\frac{\partial \overline{q}^k}{\partial q^l} \frac{\partial q_l}{\partial \overline{q}^j} + \frac{\partial \overline{q}^k}{\partial p_l} \frac{\partial p_l}{\partial \overline{q}^j} \right] = \frac{\partial \overline{q}^k}{\partial \overline{q}^j} = \delta_j^k \ .$$

In like manner, it follows that $\{\overline{q}^k, \overline{q}^j\} = 0 = \{\overline{p}_j, \overline{p}_k\}$, which, together with $\{\overline{q}^k, \overline{p}_j\} = \delta_j^k$, establishes the equality of Poisson brackets evaluated in any canonical coordinate system. Furthermore, it may be shown that the natural volume element on phase space is form invariant in all canonical coordinates, i.e.,

$$\prod d\overline{p}_k \, d\overline{q}^k = \prod dp_k \, dq^k \ .$$

An important class of canonical transformations is given by those of the form

$$\overline{q}^k = \overline{q}^k(q) \ , \qquad p_k = \sum \left(\frac{\partial \overline{q}^l}{\partial q^k} \right) \overline{p}_l \ ,$$

for which the generator $\tilde{F}(\overline{p}, q) = -\overline{p}_l \overline{q}^l(q)$ (note $F = 0$ in this case), each of which corresponds to a multidimensional coordinate transformation (just as in the Lagrangian formulation). For $N = 1$, another example is given (for $F = \overline{q}^3 q^3 / 3$) by

$$\overline{p} = p^{2/3} q^{5/3} \ , \qquad \overline{q} = -p^{1/3} q^{-2/3} \ .$$

These last two examples refer to macroscopic coordinate transformations, and indeed, as the last example shows, they may even contain points of singularity. It is also useful to consider *infinitesimal* canonical coordinate transformations, namely those that change the coordinates by infinitesimal quantities. To do so we may let $\overline{p}_k = p_k + \epsilon R_k(p,q)$ and $\overline{q}^j = q^j + \epsilon S^j(p,q)$, where $0 < \epsilon \ll 1$. To satisfy the conditions that we have new canonical coordinates it is sufficient that to lowest order

$$\delta_k^j = \{\overline{q}^j, \overline{p}_k\} = \sum \left[(\delta_l^j + \epsilon \frac{\partial S^j}{\partial q^l})(\delta_k^l + \epsilon \frac{\partial R_k}{\partial p_l}) - \epsilon^2 \frac{\partial S^j}{\partial p_l} \frac{\partial R_k}{\partial q^l} \right]$$

$$= \delta_k^j + \epsilon (\frac{\partial S^j}{\partial q^k} + \frac{\partial R_k}{\partial p_j}) + O(\epsilon^2) \,,$$

from which it follows that $S^j = \partial T / \partial p_j$ and $R_k = -\partial T / \partial q^k$ for some scalar function T. Hence, the equations to generate infinitesimal canonical coordinate transformations read

$$\delta p_k = -\epsilon \frac{\partial T(p,q)}{\partial q^k} \,,$$

$$\delta q^k = \epsilon \frac{\partial T(p,q)}{\partial p_k} \,.$$

It is important to observe that the Hamiltonian equations of motion themselves may be written in the form of infinitesimal canonical coordinate transformations, i.e.,

$$dp_k = -dt \frac{\partial H(p,q)}{\partial q^k} \,,$$

$$dq^k = dt \frac{\partial H(p,q)}{\partial p_k} \,,$$

and therefore the temporal evolution of the dynamical solution, i.e., $p(t)$ and $q(t)$ for $t' \leq t \leq t''$, is actually nothing but the continuous unfolding of a family of canonical coordinate transformations. This statement holds true in any system of canonical coordinates. Suppose we choose new canonical coordinates such that in the new coordinate system the Hamiltonian reads $\Sigma \omega^k \overline{p}_k = H(p,q)$, specifically that the new Hamiltonian is simply $\overline{H}(\overline{p}, \overline{q}) = \Sigma \omega^k \overline{p}_k$, where the parameters ω^k, $1 \leq k \leq N$, are constants. In such a case Hamilton's equations of motion would read

$$\dot{\overline{q}}^k = \frac{\partial \overline{H}}{\partial \overline{p}_k} = \omega^k \,,$$

$$\dot{\overline{p}}_k = -\frac{\partial \overline{H}}{\partial \overline{q}^k} = 0 \,,$$

and these trivial equations of motion lead to the immediate solution

$$\overline{q}^k(t) = \omega^k(t - t') + \overline{q}^k(t'), \qquad \overline{p}_k(t) = \overline{p}_k(t') \,.$$

This formulation seems to assert that a global solution exists for all times, and yet there are counterexamples to this statement. In such cases, infinitesimal coordinate transformations that are repeatedly iterated so as to generate finite canonical transformations would lead to a resultant macroscopic transformation that is not globally defined; for $N = 1$, such would be the case for $T = pq^3$. In cases that do not have global solutions, then the required macroscopic canonical transformation simply does not exist. The validity of such a transformation assumes the global existence of solutions of the original Hamiltonian equations of motion. A sufficient condition for global solutions to exist is that the Hamiltonian is bounded below. Even if a global solution exists, it is generally difficult to determine the appropriate canonical transformation to new coordinates that makes such a simple solution hold; after all, this transformation is tantamount to solving the equations of motion in the first place. For $N = 1$, one simple example where the appropriate canonical transformation is known is for the harmonic oscillator, which then reads

$$\bar{p} = \tfrac{1}{2}(p^2 + q^2), \qquad \bar{q} = \arctan\left(q/p\right).$$

The simplicity of the harmonic oscillator makes it a very convenient "minilaboratory" for many investigations; in effect, we shall appeal to it often!

2.3 Statistical description

For present purposes let us assume that we deal with a Hamiltonian and its associated Hamiltonian equations of motion that enjoy global solutions. Therefore, given the initial data for the equations of motion, the future evolution exists and is uniquely determined. Let us denote such solutions to the equations of motion by

$$p(t) \equiv p(t; p', q'), \qquad q(t) \equiv q(t; p', q'),$$

which are the solutions conditioned on the fact that the initial data is given by $p(t') = p(t'; p', q') \equiv p'$ and $q(t') = q(t'; p', q') \equiv q'$. However, it may be the case that we do not know the initial data uniquely, but rather only in a statistical sense. That is, we know only a probability distribution for the initial values, say $\rho_0(p', q')$, subject to positivity and the normalization requirement, when integrated over the entire phase space, that $\int \rho_0(p', q')\, dp'dq' = 1$. This initial distribution evolves into a time-dependent distribution given by

$$\rho(p, q, t) \equiv \int \delta(p - p(t; p', q'))\, \delta(q - q(t; p', q'))\, \rho_0(p', q')\, dp'dq',$$

which represents the distribution of system momenta and coordinates at time t.*
Clearly this function also satisfies the initial condition that $\rho(p, q, t') = \rho_0(p, q)$.

---

* In this equation we have introduced the Dirac δ-function, a generalized function distinguished by the fact that $\delta(x) = 0$ for $x \neq 0$ and yet $\int \delta(x)\, dx = 1$, provided the interval of integration contains the origin. More generally,

$$\int \delta(x - y) f(y)\, dy = f(x)$$

for any continuous function f so long as the point x is contained within the domain of integration. Additionally, $\delta(ax) = |a|^{-1}\delta(x)$, as follows from a change of variables within an integral.

An equation of motion for ρ may easily be found knowing that at each point along the trajectory the system follows the Hamiltonian equations of motion. Thus

$$\frac{d\rho(p,q,t)}{dt} = -\frac{\partial\rho(p,q,t)}{\partial q(t)}\dot{q}(t) - \frac{\partial\rho(p,q,t)}{\partial p(t)}\dot{p}(t)$$

$$= -\frac{\partial\rho(p,q,t)}{\partial q(t)}\frac{\partial H(p,q)}{\partial p(t)} + \frac{\partial\rho(p,q,t)}{\partial p(t)}\frac{\partial H(p,q)}{\partial q(t)}$$

$$= -\{\rho(p,q,t), H(p,q)\} \ .$$

Given a solution to this equation, it may be used to compute statistical averages of various phase-space functions. For example, if $f(p,q)$ denotes some function of interest, e.g., $f = p$ or $f = q$, then the average of f at time t is given by

$$\langle f \rangle(t) \equiv \int f(p,q)\,\rho(p,q,t)\,dp\,dq \ .$$

It is useful to reexpress this equation in another fashion. Suppose that we make a canonical transformation at each time t from the variables $p(t), q(t)$ back to the original variables p', q'. Since the integral representing $\langle f \rangle$ is invariant under such transformations, we would find that

$$\langle f \rangle(t) = \int f(p',q',t)\,\rho_0(p',q')\,dp'\,dq' \ .$$

This change would have the effect that the time dependence of the right-hand side now resides in the function f and not in the distribution ρ. The equation of motion for f in this case follows from the equality of the time derivative of the two expressions, i.e.,

$$\int \frac{df(p,q,t)}{dt}\,\rho_0(p,q)\,dp\,dq = \int f(p,q)\,\frac{d\rho(p,q,t)}{dt}\,dp\,dq$$

$$= -\int f(p,q)\left[\frac{\partial\rho(p,q,t)}{\partial q}\dot{q} + \frac{\partial\rho(p,q,t)}{\partial p}\dot{p}\right]dp\,dq$$

$$= \int \left[\frac{\partial f(p,q)}{\partial q}\dot{q} + \frac{\partial f(p,q)}{\partial p}\dot{p}\right]\rho(p,q,t)\,dp\,dq$$

$$= \int \{f(p,q), H(p,q)\}\,\rho(p,q,t)\,dp\,dq$$

$$= \int \{f(p,q,t), H(p,q)\}\,\rho_0(p,q)\,dp\,dq \ .$$

In deriving this equation we have integrated by parts and have assumed that the distribution ρ vanishes sufficiently fast at infinity. Since this equation holds for a general ρ_0, it follows that in this picture f satisfies the equation of motion

$$\frac{df(p,q,t)}{dt} = \{f(p,q,t), H(p,q)\} \ ,$$

which is the time-reversed form of the equation of motion from the perspective of $\rho(t)$. These two complementary ways of description will reappear when we study the quantum equations of motion.

Exercises

2.1 For the following four systems characterized by their Hamiltonians, determine whether or not the single degree-of-freedom system admits global solutions, i.e., whether or not solutions exist for all time for an arbitrary initial condition:

(a) $H(p,q) = p^2 + q^4$,   (b) $H(p,q) = p^2 - q^2$,

(c) $H(p,q) = p^2 - q^4$,   (d) $H(p,q) = pq^3$.

Reformulate as many of these examples as you can in terms of Lagrangian mechanics.

2.2 Let $T(p,q)$ denote the infinitesimal generator of a canonical transformation for a single degree-of-freedom system. Discuss the global character of the finite canonical transformations determined by such infinitesimal generators in the four cases when T is given by the expression denoted by H in the preceding exercise.

2.3 Let $p = \{p_n\}$ and $q = \{q_n\}$, $1 \leq n \leq N$, denote the momenta and coordinates for an N degree-of-freedom system, and let $p^2 = \Sigma_1^N p_n^2$ and $q^2 = \Sigma_1^N q_n^2$. For the system Hamiltonian choose

$$H(p,q) = \tfrac{1}{2}(p^2 + q^2) + \lambda(q^2)^2 , \qquad \lambda \geq 0 .$$

Discuss the symmetry of the underlying system, and determine that the form of the general solution for any $N > 2$ is qualitatively similar to the form of the general solution for $N = 2$.

3

Hilbert Space:
The Arena of Quantum Mechanics

WHAT TO LOOK FOR

Vectors and linear operators that transform them, whether in an abstract form
or in one or another concrete representation, lie at the heart of the mathematical
machinery needed to describe quantum theory. Canonical operators that satisfy
some basic set of commutation rules form the heart of quantum kinematics. An
especially convenient choice for basic operators are the so-called creation and
annihilation operators, and closely associated with them is an important set of
states – the coherent states – which are introduced in this chapter and figure
significantly in essentially all later chapters.

3.1 Hilbert space

The quantum analog of the classical phase space is a complex Hilbert space,
and the quantum analog of a phase-space point is a vector lying in this Hilbert
space.* A complex Hilbert space has many properties in common with a complex
finite-dimensional Euclidean space. In particular, such a space is a linear vector
space with an inner product. In the elegant notation of Dirac [Di 76], Hilbert
space vectors are denoted by "kets", such as $|\psi\rangle$ and $|\phi\rangle$. These vectors lie in a
Hilbert space $\mathfrak{H}$, as well as their sum with arbitrary complex coefficients,

$$a|\psi\rangle + b|\phi\rangle \in \mathfrak{H}, \qquad a, b \in \mathbf{C}.$$

As $\mathfrak{H}$ is a vector space there exists a unique zero vector $0 \in \mathfrak{H}$ which has the
property that $|\phi\rangle + 0 = |\phi\rangle$ and $0|\phi\rangle = 0$; in this last expression, the first
$0 \in \mathbf{C}$ while the second $0 \in \mathfrak{H}$. With each "ket" vector $|\chi\rangle$ there is associated
an adjoint form of the vector denoted by $\langle\chi|$, which is referred to as a "bra"

---

* A recent survey of Hilbert space for quantum mechanics is given by [BlEH 94].

vector. The adjoint vector for the given sum is $a^*\langle\psi| + b^*\langle\phi|$ and involves the complex conjugate of the coefficients a and b. The inner product of two vectors involves a bra, say $\langle\chi|$, and a ket, say $|\phi\rangle$, and is a complex number denoted by $\langle\chi|\phi\rangle \in \mathbf{C}$, which is sometimes referred to as a "bra-ket". The inner product satisfies $\langle\phi|\chi\rangle = \langle\chi|\phi\rangle^*$ involving the complex conjugate number. Furthermore, the inner product is linear in the right-hand argument, specifically

$$\langle\chi|(a|\psi\rangle + b|\phi\rangle) = a\langle\chi|\psi\rangle + b\langle\chi|\phi\rangle \ .$$

It follows that the inner product of any vector with the zero vector vanishes (set $b = 0$ in the preceding expression). Hilbert space has a positive-definite inner product, which means that the inner product satisfies $0 \leq \langle\phi|\phi\rangle < \infty$ for any $|\phi\rangle \in \mathfrak{H}$; furthermore, the condition $\langle\phi|\phi\rangle = 0$ holds if and only if $|\phi\rangle = 0$, the zero vector.

One advantage of the Dirac notation is that the unadorned symbol $|\ \rangle$ already denotes a ket vector, and thus one is free to place any label inside one likes. Thus we may use $|0\rangle, |n\rangle, |\heartsuit\rangle$, etc., all of which denote vectors with one or another label. Normally, the meaning of the label is clear from the context; otherwise it should be explained. The length, or norm, of a vector $|\psi\rangle$ is given by $\||\psi\rangle\| \equiv +\sqrt{\langle\psi|\psi\rangle}$, which is always finite. Schwarz's inequality states that $|\langle\phi|\psi\rangle|^2 \leq \langle\phi|\phi\rangle\langle\psi|\psi\rangle$ with equality holding only if $|\phi\rangle = c|\psi\rangle$ for some complex coefficient c. A vector $|\psi\rangle$ is called normalized if $\langle\psi|\psi\rangle = 1$. Two vectors $|\psi\rangle$ and $|\phi\rangle$ are said to be orthogonal if $\langle\phi|\psi\rangle = 0$. A set of vectors $\{|n\rangle\}_{n=1}^N$ is called an orthonormal set if it satisfies the property that $\langle n|m\rangle = \delta_{nm}$ for all $1 \leq n, m \leq N$, $N \leq \infty$. Linear sums of the vectors in an N-dimensional orthonormal set span an N-dimensional space. If every vector in $\mathfrak{H}$ may be represented in this fashion then the orthonormal set is called complete, or more simply a "basis"; it should be noted that there exist many (if $N > 1$, uncountably many!) distinct bases. The number N then becomes the dimension of the Hilbert space, and if $N = \infty$ then the Hilbert space is infinite dimensional and is called "separable". Many Hilbert spaces in quantum theory are infinite dimensional. In summary, for an infinite-dimensional Hilbert space, for example, there exists a basis (a complete orthonormal set) $\{|n\rangle\}_{n=1}^\infty$ with the property that every vector in $\mathfrak{H}$ admits a unique representation in the form

$$|\psi\rangle = \sum_{n=1}^\infty \psi_n |n\rangle, \qquad \psi_n \in \mathbf{C} \ .$$

Due to the orthonormality, it follows that $\psi_n = \langle n|\psi\rangle$. Of course, if the basis is changed, then the coefficients ψ_n are generally different. However, in any basis, it is necessary that

$$0 \leq \langle\psi|\psi\rangle = \sum_{m,n=1}^\infty \psi_m^*\psi_n\langle m|n\rangle = \sum_{n=1}^\infty |\psi_n|^2 < \infty \ .$$

Thus the coefficients ψ_n, namely the "coordinates" in the present basis, must be square summable, or as one says, $\{\psi_n\}_{n=1}^{\infty} \in l^2$, in order for the vector $|\psi\rangle \in \mathfrak{H}$. If $|\psi\rangle = 0$, then $\psi_n = 0$ for all $n \geq 1$. This example shows that we can "represent" each abstract Hilbert space vector $|\psi\rangle$ by means of a unique set of square summable complex numbers $\{\psi_n\}_{n=1}^{\infty}$, which are nothing but the coordinates of $|\psi\rangle$ with respect to the basis $\{|n\rangle\}$. If we represent $|\phi\rangle$ in the same basis by $\{\phi_n\}_{n=1}^{\infty}$, then it follows that the inner product is given by

$$\langle\phi|\psi\rangle = \sum_{n=1}^{\infty} \phi_n^* \psi_n .$$

Although we have indexed our components from $n = 1$ to ∞, there is nothing very special about such a choice. In other cases it may be more convenient to index components from $n = 0$ to ∞, or even from $-\infty$ to ∞. Such changes would make evident modifications in the expressions already given. We begin with just such an example.

An alternative representation for the vectors in $\mathfrak{H}$ can readily be given. One method to do so is provided by the usual Fourier series. Let us introduce the transform pair given by*

$$f(x) = \frac{1}{\sqrt{2\pi}} \sum_{-\infty}^{\infty} e^{inx} \psi_n ,$$

$$\psi_n = \frac{1}{\sqrt{2\pi}} \int_{-\pi}^{\pi} e^{-inx} f(x) \, dx .$$

It readily follows that

$$\langle\psi|\psi\rangle = \sum_{-\infty}^{\infty} |\psi_n|^2 = \int_{-\pi}^{\pi} |f(x)|^2 \, dx ,$$

which asserts that the functions f are square integrable on the interval $-\pi$ to π, i.e., $f \in L^2([-\pi, \pi])$. For two such functions the inner product reads

$$\langle\phi|\psi\rangle = \sum_{-\infty}^{\infty} \phi_n^* \psi_n = \int_{-\pi}^{\pi} g(x)^* f(x) \, dx ,$$

where g is defined in terms of the coefficients ϕ_n in the same way that f is defined by the coefficients ψ_n.

Another set of examples is readily given as well. Let $\{h_n(x)\}_{n=0}^{\infty}$, $x \in \mathbb{R}$, denote a complete set of orthonormal functions on the real line, $-\infty < x < \infty$. When integrated over the real line such functions have the integral property that

$$\int_{-\infty}^{\infty} h_m(x)^* h_n(x) \, dx = \delta_{mn} ,$$

---

* In general, the function $f(x)$ defined by the following series is only convergent in the mean.

and the completeness property that

$$\sum_{n=0}^{\infty} h_n(x)^* h_n(y) = \delta(x-y) ,$$

where $\delta(x)$ again denotes the Dirac δ-function. The well-known Hermite functions, $h_n(x)$, $n \geq 0$, implicitly defined through the expression

$$\exp(-s^2 + 2sx - \tfrac{1}{2}x^2) = \pi^{\frac{1}{4}} \sum_{n=0}^{\infty} (n!)^{-\frac{1}{2}} (s\sqrt{2})^n h_n(x) ,$$

provide a standard example. Using this, or any other complete orthonormal set of functions, we can introduce a functional representation for the vector $|\psi\rangle$ according to the transform pair

$$\psi(x) = \sum_{n=0}^{\infty} h_n(x) \psi_n ,$$

$$\psi_n = \int_{-\infty}^{\infty} h_n(x)^* \psi(x) \, dx .$$

We note further that these relations imply that

$$\langle\psi|\psi\rangle = \sum |\psi_n|^2 = \int |\psi(x)|^2 \, dx,$$

showing that the functions $\psi(x)$ are square integrable on the real line, i.e., $\psi \in L^2(\mathbb{R})$. The inner product of two distinct vectors is given by a similar expression, namely

$$\langle\phi|\psi\rangle = \sum \phi_n^* \psi_n = \int \phi(x)^* \psi(x) \, dx .$$

Here in the last two expressions we have begun to leave the limits of summation and the limits of integration implicit when they are sufficiently clear from the context.

A completely similar representation holds for square-integrable functions $f(x)$ when $x \in \mathbb{R}^N$. To see this we need only generalize the orthonormal set of functions to a set $\{g_n(x)\}_{n=1}^{\infty}$ that satisfies

$$\int g_m(x)^* g_n(x) \, d^N x = \delta_{mn} ,$$

$$\sum_{n=1}^{\infty} g_n(y)^* g_n(x) = \delta(y-x) .$$

Here $\delta(x)$ denotes an N-dimensional δ-function with the property that it vanishes if $x \neq 0$ and $\int \delta(x) \, d^N x = 1$ when the point $x = 0$ is within the domain of integration.

Let us illustrate one further representation of the Hilbert space $\mathfrak{H}$. Let $z = x + iy \in \mathbf{C}$ denote a complex variable, and introduce the transform pair*

$$F(z) = \sum_{n=0}^{\infty} \frac{z^n \psi_n}{\sqrt{n!}} \,,$$

$$\psi_n = \int \frac{z^{*n}}{\sqrt{n!}} \, F(z) \, e^{-|z|^2} \frac{dx\,dy}{\pi} \,,$$

where the integration runs over the entire two-dimensional plane. It is straightforward to verify that

$$\langle \psi | \psi \rangle = \sum |\psi_n|^2 = \int |F(z)|^2 \, e^{-|z|^2} \frac{dx\,dy}{\pi} \,.$$

If for convenience we introduce the abbreviation

$$d\mu = e^{-|z|^2} \frac{dx\,dy}{\pi} \,,$$

then we see that each of the analytic functions $F \in L^2(\mathbb{R}^2, d\mu)$. It follows that the inner product of two such analytic function representatives is given by

$$\langle \phi | \psi \rangle = \int G(z)^* \, F(z) \, d\mu \,.$$

This kind of example can be readily extended to analytic functions of several complex variables as well, and such representations are often referred to as Segal–Bargmann representations [Se 63, Ba 61].

The preceding examples illustrate several different functional representations of an infinite-dimensional Hilbert space, and it is clear from their construction that they are all isomorphic to one another, i.e., they are all equivalent to one another under the given equations of transformation. There are, of course, still other functional representations, but the examples we have introduced above will serve our present purposes.[†]

3.2 Operators in Hilbert space

Simply put, linear operators are elements of an associative algebra, and thus have the properties of a linear vector space with the extra property of an associative multiplication rule. These properties are just those that are familiar from a set of $N \times N$ matrices. For example, if R, S, and T denote linear operators, then $(aR + bS)T = aRT + bST$ denotes another operator, where $a, b \in \mathbf{C}$ and RT

---

* Unlike the previous functional representations, the present sum converges absolutely for every vector in $\mathfrak{H}$.

† The important notions of direct sum and direct product of Hilbert spaces are taken up in the Exercises.

represents the product of two operators. The overworked symbol 0 now also stands for the zero operator with evident properties, and we let **1** denote the unit operator for which $\mathbf{1}R = R\mathbf{1} = R$ for every operator R. Following standard convention, however, we sometimes omit the symbol for the unit operator when it is clear from the context exactly what is meant. Associativity means that $(RS)T = R(ST) \equiv RST$ and so any parentheses are unnecessary.

The role of operators in a Hilbert space is to map vectors into other vectors. A linear operator, say R, has the property that

$$R(a|\phi\rangle + b|\psi\rangle) = aR|\phi\rangle + bR|\psi\rangle \ .$$

We shall only be concerned with linear operators and will generally refer to them simply as operators hereafter. If an operator R takes every vector in $\mathfrak{H}$ into a vector in $\mathfrak{H}$, then the operator is defined everywhere. This is true, in particular, if the operator is bounded, that is if, for a fixed c, where $0 \le c < \infty$, then $\|R|\psi\rangle\| \le c\,\|\,|\psi\rangle\|$ holds for all $|\psi\rangle \in \mathfrak{H}$. An operator – let us say T – may also be unbounded, in which case there is no finite c for which the previous inequality holds. In that case the operator can act only on a subset of vectors in $\mathfrak{H}$ and still yield vectors in $\mathfrak{H}$. To deal with this fact it is necessary to ascribe to T a domain $\mathfrak{D}(T) \subset \mathfrak{H}$ composed of vectors that are transformed by T into vectors in $\mathfrak{H}$. An example may help clarify the situation. In the sequence space l^2, the transformation R that takes the vector $\{\psi_n\}$, $n \in \mathbb{N} \equiv \{1, 2, 3, \ldots\}$, into the vector $\{[n/(1 + n)]\psi_n\}$ is a bounded operator (for any $c \ge 1$), while the transformation T that takes the vector $\{\psi_n\}$ into the sequence $\{n\psi_n\}$ does not result in a vector for all initial vectors. In this case it is necessary that $\sum n^2 |\psi_n|^2 < \infty$ in order that $\{n\psi_n\}$ actually represent a vector. In the present example, therefore

$$\mathfrak{D}(T) = \{|\psi\rangle : \{n\psi_n\}_{n=1}^{\infty} \in l^2\} \ .$$

Unbounded operators play an important role in quantum mechanics, but the need to deal with domains is very often a nuisance, so much so that they are all too often ignored. Following the custom in most quantum texts we will not make any fuss about domains in this chapter; however, they will begin to figure prominently beginning with Chapter 6.

Several definitions and properties of operators deserve mention. An operator R has an eigenvalue $\lambda_r \in \mathbb{C}$ and an eigenvector $|r\rangle$, which lies in $\mathfrak{H}$, provided that $R|r\rangle = \lambda_r|r\rangle$.* An operator may have a number of distinct eigenvectors and several different eigenvalues. For example, every vector is an eigenvector of the zero operator with eigenvalue zero, and every vector is also an eigenvector of the unit operator with eigenvalue one. If R denotes an operator then the adjoint

---

* It may also happen that this eigenvalue equation holds not for a vector in $\mathfrak{H}$ but for a generalized eigenvector that strictly speaking is not in $\mathfrak{H}$. We deal with this more general situation below.

operator, denoted by $R^\dagger$, is defined by the equation

$$\langle\psi|R^\dagger|\phi\rangle = \langle\phi|R|\psi\rangle^* \ .$$

A Hermitian operator has the property that $R^\dagger = R$. Thus eigenvectors of Hermitian operators satisfy $\langle r|R|r\rangle = \lambda_r\langle r|r\rangle = \lambda_r^*\langle r|r\rangle$, and so their eigenvalues are real. If $|r\rangle$ and $|s\rangle$ denote two distinct eigenvectors for a Hermitian operator, then it follows that

$$\langle s|R|r\rangle = \lambda_r\langle s|r\rangle = \langle r|R|s\rangle^* = \lambda_s\langle s|r\rangle \ ,$$

and thus if $\lambda_r \neq \lambda_s$, $|r\rangle$ and $|s\rangle$ are orthogonal, $\langle s|r\rangle = 0$. Even when $\lambda_r = \lambda_s$ one may choose a suitable linear combination of the two eigenvectors so that they are orthogonal. Consequently, it is generally asserted that any Hermitian operator, say R, has a complete set of orthonormal eigenvectors $\{|r\rangle\}$, $r \in \mathbb{N}$, namely a set that constitutes a basis. Indeed, this is one of the most common ways of defining and choosing a basis set with which to work. Based on the representation generated by that basis,

$$|\psi\rangle = \Sigma\,|r\rangle\psi_r = \Sigma\,|r\rangle\langle r|\psi\rangle \ ,$$
$$\langle\phi|\psi\rangle = \Sigma\,\langle\phi|r\rangle\langle r|\psi\rangle \equiv \langle\phi|\mathbf{1}|\psi\rangle \ ,$$

and it is useful to introduce a representation for the unit operator itself given by

$$\mathbf{1} = \Sigma\,|r\rangle\langle r| \ ,$$

which is called a resolution of unity. Every Hermitian operator leads to a basis with which a resolution of unity may be constructed. Clearly we have

$$R = R\mathbf{1} = R\Sigma\,|r\rangle\langle r| = \Sigma\,R|r\rangle\langle r| = \Sigma\,\lambda_r|r\rangle\langle r| \ ,$$
$$R^2 = \Sigma\,\lambda_r^2|r\rangle\langle r| \ ,$$
$$f(R) = \Sigma\,f(\lambda_r)|r\rangle\langle r| \ ,$$

etc., for any reasonable function f. In such a basis one has diagonalized the operator R inasmuch as it acts by simple multiplication.

If the eigenvectors of the Hermitian operator, say X in this case, are generalized eigenvectors, then the eigenvalues comprise an open set in $\mathbb{R}$. In that case orthonormality is taken to mean $\langle x|y\rangle = \delta(x - y)$, and the generalized eigenvectors $|x\rangle$ are said to be δ-function normalized. Under these circumstances, one has

$$\mathbf{1} = \int|x\rangle\langle x|\,dx \ ,$$

and therefore

$$X = X\mathbf{1} = \int X|x\rangle\langle x|\,dx = \int x|x\rangle\langle x|\,dx \ ,$$
$$X^2 = \int x^2|x\rangle\langle x|\,dx \ ,$$
$$f(X) = \int f(x)|x\rangle\langle x|\,dx \ ,$$

where the integral runs over the possible continuous range of eigenvalues of X, i.e., the continuous spectrum of X. These generalized eigenvectors give rise to L^2 representation spaces by means of functions $\psi(x) = \langle x|\psi\rangle$. In this language the function $\psi(x)$ represents the "coordinates" of the abstract vector $|\psi\rangle$, although some care must be exercised in pursuing this analogy with the discretely labelled "coordinates". A few technical remarks may be in order. In particular, the zero vector may be represented not only by the zero function, but by any other function that is zero almost everywhere so that its square integral vanishes. In point of fact, the representatives in the continuum case are equivalence classes composed of functions that differ one from another only on sets of measure zero. This seemingly "minor technicality" is responsible for reducing the uncountable number of dimensions in the basis $\{|x\rangle\}, x \in \mathbb{R}$, to the countable number of dimensions in the basis $\{|n\rangle\}, n \in \mathbb{N}$, that represents the true dimension of a separable infinite-dimensional Hilbert space. Indeed, there also are Hilbert spaces of uncountably many dimensions, the so-called "nonseparable" Hilbert spaces (see Chapter 6), but they have a distinctly limited physical applicability.

Another important class of operators is the unitary operators. An operator U is unitary provided that $U^\dagger U = UU^\dagger = \mathbf{1}$. A unitary operator has the property that the inner product of $U|\psi\rangle$ and $U|\phi\rangle$ is identical to the inner product of $|\psi\rangle$ and $|\phi\rangle$ for any pair of vectors; the same statement holds with U replaced by $U^\dagger$. It is important to note that every unitary operator can be written in the form $U = e^{-iF}$ for some Hermitian operator F (more precisely, F is a self-adjoint operator; see Chapter 6). The action of a unitary operator may be thought of as a rigid rotation of Hilbert space; indeed, unitary operators are the natural extension to an infinite-dimensional complex space of the family of orthogonal rotations, familiar in three-dimensional space, which have the property that the inner product of any two three-vectors is invariant under such rotations. In three dimensions it is evident that one may rotate any orthonormal frame into any other orthonormal frame, including if necessary an improper rotation (e.g., a reflection through any one plane). Likewise, in Hilbert space, a unitary operator can be found that can map any orthonormal basis set into any other orthonormal basis set. Let $|r\rangle$ and $|s'\rangle$ denote two arbitrary vectors of two such bases. Then the operator U defined by the matrix in the $|r\rangle$-basis with components $U_{sr} = \langle s|U|r\rangle \equiv \langle s'|r\rangle$, $r, s \in \mathbb{N}$, generates the required transformation. In particular, as follows from the resolution of unity, the transformed vector components read

$$\phi'_s = \langle s'|\phi\rangle = \Sigma \langle s'|r\rangle\langle r|\phi\rangle = \Sigma U_{sr}\, \phi_r \; .$$

Just as in three space, physical quantities of interest cannot depend on the artificial choice of coordinates but must be expressed in a coordinate invariant form. In quantum mechanics, therefore, the physical results will be expressed in terms of inner products and not in terms of the vector representatives by themselves. In other words, the physical answers must be independent of the representation of the Hilbert space. This fact has two consequences. The first consequence is that much of quantum mechanics can be discussed in the language of an abstract Hilbert space, i.e., in terms of abstract operators and bra- and ket-vectors. The second consequence is that in order to compute something – and one generally needs to choose a concrete representation to do so – the physical answers will not depend on the choice of representation. Thus the choice of representation can be made for the convenience of the calculation.

The commutator of any two operators A and B is defined by

$$[A, B] \equiv AB - BA$$

and in general is nonvanishing. Any two operators for which $[A, B] = 0$ are said to commute. By definition, the unit operator, $\mathbf{1}$, commutes with all operators, $[\mathbf{1}, R] = 0$. A set of operators $\{R_n\} \equiv \{R_1, R_2, R_3, \ldots\}$ is called irreducible if the only operator B that commutes with every operator in the set is a multiple of the identity operator, i.e., $\{R_n\}$ is irreducible if and only if $[B, R_n] = 0$ for all n implies that $B = b\mathbf{1}$ for some $b \in \mathbf{C}$. A minimal irreducible set consists of a set of operators from which all other operators can be constructed by algebraic operations or perhaps by suitable limits thereof, or as we shall simply say, as a function of those operators. A subset $\{C_n\}$ of a minimal irreducible set is called a complete set of commuting operators provided that $[C_n, C_m] = 0$ for every n and m in the set, and any other operator which commutes with all other operators in the set is a function of the operators in the complete set of commuting operators. With this definition, a complete set of commuting operators is at the same time a minimal set of commuting operators. All operators in a complete set of commuting operators that can be diagonalized can be simultaneously diagonalized, and for these it follows that

$$C_n C_m |c\rangle = C_n c_m |c\rangle = c_m C_n |c\rangle = c_n c_m |c\rangle \, ,$$

where we have put $|c\rangle \equiv |\{c_n\}\rangle \equiv |c_1, c_2, \ldots\rangle$. For systems with a finite number of degrees of freedom, a minimal irreducible set $\{R_n\}$ contains only a finite number of operators. Likewise, for finitely many degrees of freedom, a complete set of commuting operators contains only a finite number of operators. The simultaneous (possibly generalized) eigenvectors of a complete set of commuting operators may be chosen to be orthonormal and they therefore constitute a suitable basis set of vectors with which to define a representation of Hilbert space.

Additionally, we introduce the concept of the trace. For a finite, square matrix the trace is simply the sum of the diagonal elements. This concept is invariant under orthogonal transformations and so it is an intrinsic property of the matrix and does not depend on the particular representation that has been chosen. In an infinite-dimensional Hilbert space, each operator may be represented by a semi-infinite, square matrix. In this case the trace is again simply the sum of the diagonal elements, but this sum is independent of the representation and therefore an intrinsic property of the operator only for a subset of operators.

Every operator A for which the sum will be intrinsic admits a canonical form given by

$$A = \Sigma_n a_n |\alpha_n\rangle \langle \alpha_n'| , \qquad \Sigma_n |a_n| < \infty .$$

Here $\{|\alpha_n\rangle\}$ and $\{|\alpha_n'\rangle\}$ denote two, possibly identical, complete orthonormal bases. Such operators are said to be "trace class", and up to trivial phase factors this decomposition itself is intrinsic to the operator A. To test whether an operator A has such a representation it suffices to require that

$$\mathrm{Tr}((A^\dagger A)^{1/2}) \equiv \Sigma_m \langle m|(A^\dagger A)^{1/2}|m\rangle = \Sigma_n |a_n| < \infty .$$

Here $\{|m\rangle\}$ denotes any complete orthonormal basis. If the operator of interest satisfies this criterion, then the trace of A is defined by

$$\mathrm{Tr}(A) \equiv \Sigma_r \langle r|A|r\rangle = \Sigma_n a_n \langle \alpha_n'|\alpha_n\rangle ,$$

which is a number independent of the particular complete orthonormal basis set with which it is evaluated. If the operator in question is not trace class then the indicated sum may converge but the result will in general depend not only on the operator but also on the particular basis set with which the sum has been evaluated!

Coherent states, and annihilation and creation operators

Let $\{|n\rangle\}$, $0 \leq n < \infty$, denote a complete orthonormal set of vectors, and for each $\alpha \in \mathbf{C}$, we define the nonnormalized state

$$|\alpha\rangle \equiv \sum_{n=0}^{\infty} \frac{\alpha^n}{\sqrt{n!}} |n\rangle .$$

Observe that the set of vectors $\{|\alpha\rangle\}$ for all $\alpha \in \mathbf{C}$ is a *total set*, by which we mean that if $\langle \psi|\alpha\rangle = 0$ for all α, then it follows that $\langle \psi| \equiv 0$; in fact since $\langle \psi|\alpha\rangle$ is an entire analytic function of α, it vanishes everywhere if only it vanishes, for example, on a line segment such as $\alpha - \alpha^* = 0$ and $0 < \alpha + \alpha^* < 1$ – or even more generously if $\langle \psi|\alpha\rangle$ vanishes for all α with $\Im(\alpha) = 0$; here $\Im$ denotes the "imaginary part". One consequence of the set $\{|\alpha\rangle\}$ being total is that the vector $\langle \psi|$ is uniquely determined by the values of $\langle \psi|\alpha\rangle$ for all α; likewise for $\langle \alpha|\psi\rangle \equiv \langle \psi|\alpha\rangle^*$. Here we have introduced $\langle \alpha|$, the adjoint vector (bra) associated with the vector $|\alpha\rangle$ (ket), and – with a confessed abuse of notation – set

$$\langle \alpha| \equiv \sum_{n=0}^{\infty} \frac{\alpha^{*n}}{\sqrt{n!}} \langle n| .$$

The norm squared for $|\alpha\rangle$ is given by

$$\langle \alpha|\alpha\rangle = \sum_{n=0}^{\infty} \frac{|\alpha|^{2n}}{n!} = e^{|\alpha|^2} .$$

If $|\beta\rangle$ denotes another such vector, then the inner product of two such vectors reads

$$\langle\alpha|\beta\rangle = \sum_{n=0}^{\infty} \frac{(\alpha^*\beta)^n}{n!} = e^{\alpha^*\beta} .$$

These vectors are the so-called (unnormalized, canonical) *coherent states* [KlS 68, KlS 85].

Let us introduce an operator a defined by the property that

$$a|\alpha\rangle \equiv \alpha|\alpha\rangle ,$$

i.e., $|\alpha\rangle$ is an eigenvector for the operator a with the *complex eigenvalue* α for all $\alpha \in \mathbf{C}$. As a consequence, $a^2|\alpha\rangle = \alpha^2|\alpha\rangle$, $\langle\alpha|a^\dagger = \alpha^*\langle\alpha|$, etc., as well as the basic relation

$$e^{\lambda a}|\alpha\rangle = e^{\lambda\alpha}|\alpha\rangle .$$

Thus it follows that

$$\begin{aligned}
\langle\beta|e^{\lambda a}|\alpha\rangle^* &= e^{\lambda^*\alpha^*}\langle\alpha|\beta\rangle \\
&= e^{\alpha^*(\lambda^*+\beta)} \\
&= \langle\alpha|\lambda^* + \beta\rangle .
\end{aligned}$$

But this expression is also equal to

$$\langle\beta|e^{\lambda a}|\alpha\rangle^* = \langle\alpha|e^{\lambda^* a^\dagger}|\beta\rangle ,$$

where $a^\dagger$ denotes the Hermitian adjoint of a. Therefore, we learn from these expressions that

$$e^{\gamma a^\dagger}|\alpha\rangle \equiv |\gamma + \alpha\rangle .$$

Furthermore, we see that

$$\begin{aligned}
\langle\alpha|e^{\gamma a}e^{\delta a^\dagger}|\beta\rangle &= e^{(\alpha^*+\gamma)(\delta+\beta)} , \\
\langle\alpha|e^{\delta a^\dagger}e^{\gamma a}|\beta\rangle &= e^{\gamma\beta+\alpha^*\delta+\alpha^*\beta} .
\end{aligned}$$

On comparing these two expressions it follows that

$$e^{\gamma a}e^{\delta a^\dagger} = e^{\gamma\delta}e^{\delta a^\dagger}e^{\gamma a} .$$

This relation is an identity in γ and δ, and on expansion to the term linear in each factor, i.e., $(\gamma\delta)$, implies that $aa^\dagger = a^\dagger a + \mathbf{1}$, or as more commonly stated,

$$[a, a^\dagger] = aa^\dagger - a^\dagger a = 1 .$$

The operators a and $a^\dagger$ are the so-called *annihilation* and *creation* operators, respectively. Let us investigate their properties further and see why they merit these names.

Two evident properties that follow from the foregoing discussion are given by

$$a|0\rangle = 0 , \qquad e^{\alpha a^\dagger}|0\rangle = |\alpha\rangle .$$

Expansion of the second relation in powers of α leads to the equation

$$\sum_{n=0}^{\infty} \frac{\alpha^n}{n!} a^{\dagger n}|0\rangle = \sum_{n=0}^{\infty} \frac{\alpha^n}{\sqrt{n!}}|n\rangle \,,$$

which, by identifying like powers of α, implies that

$$|n\rangle = (n!)^{-1/2} a^{\dagger n}|0\rangle \,.$$

We may either assume that $|0\rangle$ is (up to a factor) the unique eigenvector of a with eigenvalue zero, or, as we have done, that the set of coherent states is a total set. In either case it follows that the operators a and $a^\dagger$ form an irreducible set and all operators in the Hilbert space can be made as functions of this pair, i.e., as polynomials and limits thereof.

A convenient way to analyze polynomials and analytic functions of a and $a^\dagger$ is by means of *normal ordering*. Normal ordering is a linear operation that orders any monomial in $a^\dagger$ and a so that all $a^\dagger$ operators are placed to the left of all a operators without regard for the fact that the basic operators do not commute. If we denote normal ordering by the symbols : :, then, for example,

$$: a^r a^{\dagger s} a^t a^{\dagger u} :\, \equiv a^{\dagger(s+u)} a^{(r+t)} \,,$$

and

$$: e^{\tau a^\dagger a} :\, = \sum_{n=0}^{\infty} : \frac{(\tau a^\dagger a)^n}{n!} :\, = \sum_{n=0}^{\infty} \frac{\tau^n a^{\dagger n} a^n}{n!} \,.$$

It follows therefore that

$$\langle \beta | : B(a^\dagger, a) : |\alpha\rangle = B(\beta^*, \alpha)\,\langle \beta|\alpha\rangle = B(\beta^*, \alpha)\, e^{\beta^* \alpha} \,,$$

which provides an automatic way to determine an operator $\mathcal{B} \equiv\, : B(a^\dagger, a) :$ from its coherent state matrix elements. In fact, the operator can be uniquely determined from just the *diagonal* coherent state matrix elements, namely

$$\langle \alpha | : B(a^\dagger, a) : |\alpha\rangle = B(\alpha^*, \alpha)\, e^{\alpha^* \alpha} \,,$$

since the functional dependence of α^* and α can be uniquely separated out from the function $B(\alpha^*, \alpha)$. Let us give an example.

We introduce a family of operators $M(s)$ defined by the fact that

$$M(s)|\alpha\rangle \equiv |e^s \alpha\rangle$$

for all $\alpha \in \mathbf{C}$ and all $s \in \mathbb{R}$. It follows that $M(0) = 1$ and $M(t)M(s) = M(t+s)$, and thanks to evident continuity, the solution necessarily has the form

$$M(s) \equiv e^{s\mathsf{N}}$$

for some operator N. We may readily determine the operator N as follows. Observe that

$$\langle \alpha | e^{s\mathsf{N}} |\alpha\rangle = \langle \alpha | e^s \alpha\rangle = e^{\alpha^* e^s \alpha} \,,$$

and therefore

$$\frac{d}{ds}\langle\alpha|e^{s\mathsf{N}}|\alpha\rangle\Big|_{s=0} = \langle\alpha|\mathsf{N}|\alpha\rangle$$

$$= \frac{d}{ds}e^{\alpha^*e^s\alpha}\Big|_{s=0}$$

$$= \alpha^*\alpha\, e^{\alpha^*\alpha}\,,$$

which implies that

$$\mathsf{N} \equiv a^\dagger a\,.$$

The operator N is evidently a Hermitian operator. The eigenvectors for N may be determined by

$$\frac{d}{ds}M(s)|\alpha\rangle\Big|_{s=0} = \mathsf{N}|\alpha\rangle$$

$$= \mathsf{N}\sum\frac{\alpha^n}{\sqrt{n!}}|n\rangle$$

$$= \frac{d}{ds}\sum\frac{(e^s\alpha)^n}{\sqrt{n!}}|n\rangle\Big|_{s=0}$$

$$= \sum\frac{\alpha^n}{\sqrt{n!}}n|n\rangle\,,$$

which implies that

$$\mathsf{N}|n\rangle = n|n\rangle\,,$$

for all n, $0 \le n < \infty$. The operator N is referred to as the *number operator*. Lastly we observe that

$$a\sum\frac{\alpha^n}{\sqrt{n!}}|n\rangle = \alpha\sum\frac{\alpha^n}{\sqrt{n!}}|n\rangle\,,$$

which, on identifying like powers of α, implies that

$$a|n\rangle = \sqrt{n}\,|n-1\rangle\,.$$

In a similar manner, since $a^{\dagger n}|0\rangle = \sqrt{n!}\,|n\rangle$, it follows that

$$a^\dagger|n\rangle = \sqrt{n+1}\,|n+1\rangle\,.$$

Observe that the operators a and $a^\dagger$ either lower or raise the eigenvalue of N by one; it is for this reason that they are called the *annihilation* and *creation* operators, respectively. It is an elementary exercise to verify that

$$[a,\mathsf{N}] = a\,, \qquad [a^\dagger,\mathsf{N}] = -a^\dagger\,.$$

The introduction of annihilation and creation operators for several and then for infinitely-many variables will be of utmost significance in subsequent chapters.

Exercises

3.1 Derive the *Schwarz inequality* $|\langle\phi|\psi\rangle| \leq \||\phi\rangle\| \||\psi\rangle\|$ for any two vectors in the Hilbert space. In addition, derive the *triangle inequality* $\||\phi\rangle + |\psi\rangle\| \leq \||\phi\rangle\| + \||\psi\rangle\|$ which also holds for any two vectors in the Hilbert space.

3.2 Let $A = \{A_{jk}\}$ denote a matrix representation of an operator, with the complex number A_{jk} as the entry in the jth row and k^{th} column, which acts on the sequence Hilbert space l^2. In each case determine $\mathfrak{D}(A)$, the domain of the operator A, and determine whether or not $\mathfrak{D}(A)$ is dense; if it is dense determine the adjoint operator $A^\dagger$ and its domain $\mathfrak{D}(A^\dagger)$. [Note that a dense set $\mathfrak{D}$ is one for which for any $|\psi\rangle \in \mathfrak{H}$, there exists a vector $|\phi_\delta\rangle \in \mathfrak{D}$ such that $\||\psi\rangle - |\phi_\delta\rangle\| < \delta$ for any $\delta > 0$.]
  (a) Let $A_{jk} = \delta_{j1}k$.
  (b) Let $A_{jk} = 1$ for all j, k.

3.3 A bounded operator B is called positive, i.e., $0 < B$, if $0 < \langle\psi|B|\psi\rangle$ for all nonzero $|\psi\rangle \in \mathfrak{H}$. For two positive operators B_1 and B_2, we write $B_1 < B_2$ provided $B_2 - B_1$ is a positive operator. Let A_n be an increasing sequence of positive operators such that $0 < A_1 < \cdots < A_n \cdots < \mathbf{1}$. Assume that the $\lim_{n\to\infty} \langle\phi|(A_n - \mathbf{1})|\psi\rangle = 0$ holds for all $|\psi\rangle, |\phi\rangle \in \mathfrak{H}$. In that case, show that

$$\lim_{n\to\infty} \|(A_n - \mathbf{1})|\psi\rangle\| = 0$$

holds for all $|\psi\rangle \in \mathfrak{H}$.

3.4 An operator E which satisfies $E^\dagger = E = E^2$ is called a projection operator. Show that the eigenvalues of E are 0 and 1. The range of E is the subspace where E has the eigenvalue 1. If the range of E is finite dimensional then E is a trace class operator and $\text{Tr}(E)$ is the dimensionality of that subspace.

3.5 The direct sum of two Hilbert spaces $\mathfrak{H}_1$ and $\mathfrak{H}_2$ is denoted by $\mathfrak{H} = \mathfrak{H}_1 \oplus \mathfrak{H}_2$ and is defined as follows. Let $|\psi_j\rangle \in \mathfrak{H}_j$, $j = 1, 2$, and set $|\psi\rangle \equiv |\psi_1\rangle \oplus |\psi_2\rangle \in \mathfrak{H}$. Let $|\phi\rangle \equiv |\phi_1\rangle \oplus |\phi_2\rangle \in \mathfrak{H}$ be another such vector. We define the inner product of these two vectors to be

$$\langle\phi|\psi\rangle = \langle\phi_1|\psi_1\rangle + \langle\phi_2|\psi_2\rangle \,.$$

Note well that all vectors $|\psi\rangle \in \mathfrak{H}$ have the form $|\psi_1\rangle \oplus |\psi_2\rangle$ for some vectors $|\psi_j\rangle \in \mathfrak{H}_j$.

Show that this process may be generalized to a direct sum of finitely many Hilbert spaces, in which case

$$|\psi\rangle \equiv \oplus_{n=1}^N |\psi_n\rangle \in \mathfrak{H} \equiv \oplus_{n=1}^N \mathfrak{H}_n \,,$$
$$\langle\phi|\psi\rangle = \Sigma_{n=1}^N \langle\phi_n|\psi_n\rangle \,.$$

Show that this process may be extended to a direct sum of *infinitely* many Hilbert spaces provided that proper convergence criteria are maintained. In particular, show that

$$|\psi\rangle \equiv \oplus_{n=1}^{\infty}|\psi_n\rangle \in \mathfrak{H} \equiv \oplus_{n=1}^{\infty}\mathfrak{H}_n \ ,$$

provided that attention is restricted to those contributions for which

$$\langle\psi|\psi\rangle = \Sigma_{n=1}^{\infty}\langle\psi_n|\psi_n\rangle < \infty \ .$$

Given any two such vectors, say $|\psi\rangle$ and $|\phi\rangle$, show that the inner product

$$\langle\phi|\psi\rangle = \Sigma_{n=1}^{\infty}\langle\phi_n|\psi_n\rangle$$

is well defined for all elements in $\mathfrak{H}$. As an example of such a space, show that the space l^2 of square-summable sequences is nothing but the infinite direct sum of one-dimensional complex Hilbert spaces, each of which may be identified with **C**.

3.6 The direct integral with respect to a measure σ of a set of Hilbert spaces indexed by a parameter $u \in \mathbb{R}$ is patterned after the direct sum and can be obtained from the direct sum of an infinite number of Hilbert spaces in an appropriate limit. More directly, these notions are defined as follows

$$|\psi\rangle \equiv \int^{\oplus}|\psi_u\rangle\, d\sigma(u) \in \mathfrak{H} \equiv \int^{\oplus}\mathfrak{H}_u\, d\sigma(u) \ ,$$

for those ingredients for which

$$\langle\psi|\psi\rangle \equiv \int\langle\psi_u|\psi_u\rangle\, d\sigma(u) < \infty \ .$$

Here, $|\psi_u\rangle \in \mathfrak{H}_u$ for almost all u as determined by the measure σ. Every vector in $\mathfrak{H}$ is given in this form. The inner product of two such vectors, say $|\phi\rangle$ and $|\psi\rangle$, is given by

$$\langle\phi|\psi\rangle \equiv \int\langle\phi_u|\psi_u\rangle\, d\sigma(u) \ .$$

Note that unlike the case of the direct sum, vectors that differ on sets of measure zero are equivalent to one another and are therefore identified.

Show that the complex Hilbert space $L^2(\mathbb{R})$ is equivalent to the direct integral of one-dimensional complex Hilbert spaces (i.e., **C**) over the real line, i.e., for $u \in \mathbb{R}$ and $d\sigma(u) = du$.

3.7 The direct product of two Hilbert spaces $\mathfrak{H}_1$ and $\mathfrak{H}_2$ is denoted by $\mathfrak{H} = \mathfrak{H}_1 \otimes \mathfrak{H}_2$ and is defined as follows. Let $|\psi_j\rangle \in \mathfrak{H}_j$, $j = 1, 2$, and set $|\psi\rangle \equiv |\psi_1\rangle \otimes |\psi_2\rangle \in \mathfrak{H}$. Let $|\phi\rangle \equiv |\phi_1\rangle \otimes |\phi_2\rangle \in \mathfrak{H}$ be another such vector. Such vectors are called product vectors, and the inner product of two product vectors is defined to be

$$\langle\phi|\psi\rangle = \langle\phi_1|\psi_1\rangle\langle\phi_2|\psi_2\rangle \ .$$

Note well that *not* all vectors in $\mathfrak{H}$ are defined as product vectors. However, product vectors do form a total set of vectors for $\mathfrak{H}$, i.e., they span the space $\mathfrak{H}$. Thus, an arbitrary vector, say $|\chi\rangle \in \mathfrak{H}$, is necessarily given by

$$|\chi\rangle = \Sigma_l|\psi_{1,l}\rangle \otimes |\psi_{2,l}\rangle \ ,$$

where the sum in question must satisfy

$$\langle \chi | \chi \rangle = \Sigma_{l,m} \langle \psi_{1,l} | \psi_{1,m} \rangle \langle \psi_{2,l} | \psi_{2,m} \rangle < \infty \, .$$

Show that this process may be generalized to a direct product of finitely many Hilbert spaces, in which case

$$| \psi \rangle \equiv \otimes_{n=1}^{N} | \psi_n \rangle \in \mathfrak{H} \equiv \otimes_{n=1}^{N} \mathfrak{H}_n \, ,$$
$$\langle \phi | \psi \rangle = \Pi_{n=1}^{N} \langle \phi_n | \psi_n \rangle \, .$$

Given that product vectors form a total set in $\mathfrak{H}$, show that an arbitrary vector $| \chi \rangle \in \mathfrak{H}$ is necessarily given by

$$| \chi \rangle = \Sigma_l \otimes_{n=1}^{N} | \psi_{n,l} \rangle \, ,$$

provided that

$$\langle \chi | \chi \rangle = \Sigma_{l,m} \Pi_{n=1}^{N} \langle \psi_{n,l} | \psi_{n,m} \rangle < \infty \, .$$

(The generalization to the direct product of an infinite number of Hilbert spaces is rather technical; to the extend that this topic will be needed, it will be developed in Chapter 6.)

*   *   *

Direct product spaces play an important role in quantum mechanics since the Hilbert space for N degrees of freedom is just the direct product of the Hilbert spaces for each separate degree of freedom. Moreover, product vectors are often useful in discussing systems with dynamically independent degrees of freedom.

4

Quantum Mechanics

WHAT TO LOOK FOR

Although the classical theory developed in Chapter 2 can be used to describe a great many phenomena in the real world, there is little doubt that the proper description of the world is quantum. To a large measure we perceive the world classically and we must find ways to uncover the underlying quantum theory. The process of "quantization" normally consists of turning a classical theory into a corresponding quantum theory. In turn, the "classical limit" is the process by which a quantum theory is brought to its associated classical theory. The parameter $\hbar = 1.0545 \times 10^{-27}$ ($\simeq 10^{-27}$) erg-seconds (cgs units), referred to as Planck's constant, sets the scale of quantum phenomena. In many applications it will prove useful to use "natural" units in which $\hbar = 1$ just so that formulas are less cumbersome. For most typical formulations of quantum mechanics the classical limit means the limit that $\hbar \to 0$, and in this text we shall have many occasions to use this definition as well. In addition, we shall also learn to speak about the classical and quantum theories simultaneously coexisting – as they do in the real world – without the need to take the limit $\hbar \to 0$.

There are three generally accepted "royal" routes to quantization, due principally to Heisenberg, Schrödinger, and Feynman, respectively, and we shall exploit them all. In each case we begin with a single degree of freedom ($N = 1$). For the most part the generalization to several degrees of freedom ($1 < N < \infty$) is relatively straightforward, just as was the case for the classical theory treated in Chapter 2. However, before we begin with any of the quantization schemes it is pedagogically useful to present an overview of the principles involved, and that is most efficiently accomplished in the framework of an abstract formulation of Hilbert space and therefore also of quantum mechanics.

4.1 Abstract quantum mechanics

Just as dynamics in classical mechanics corresponds to the continuous time evolution of a point in phase space, $p(t)$, $q(t)$, $t' \leq t \leq t''$, dynamics in quantum mechanics is described by the continuous time evolution of a unit vector in Hilbert space, i.e.,

$$|\psi(t)\rangle , \qquad \langle\psi(t)|\psi(t)\rangle = 1 , \qquad t' \leq t \leq t'' .$$

In turn, this evolution can always be represented with the help of a time-dependent family of unitary operators. For a wide class of problems, this evolution can be imaged by means of a family of time-dependent unitary operators of the form

$$|\psi(t)\rangle = U(t - t')|\psi(t')\rangle = e^{-i(t-t')\mathcal{H}/\hbar} |\psi(t')\rangle .$$

Here $\mathcal{H}$ denotes the Hamiltonian operator which must be a Hermitian, indeed, a self-adjoint operator. We may derive a differential equation for this temporal evolution as follows. Since

$$|\psi(t + \delta t)\rangle - |\psi(t)\rangle = [e^{-i\delta t\mathcal{H}/\hbar} - 1] |\psi(t)\rangle ,$$

we find, by dividing by δt, taking the limit $\delta t \to 0$, and rearranging constants, that

$$i\hbar \frac{d}{dt}|\psi(t)\rangle = \mathcal{H} |\psi(t)\rangle .$$

This is the fundamental Schrödinger equation, and its solution determines the time evolution of a vector in Hilbert space. This differential equation will play a central role in the Schrödinger formulation of quantum mechanics.* In turn, in the Feynman formulation the emphasis will be on the solution to this differential equation.

The interpretation of the quantum mechanical formalism is statistical, and the generally accepted interpretation of the unit vector $|\psi(t)\rangle$ is that of a probability amplitude. In particular, if X denotes a Hermitian operator then the expected value for the measured output of the physical quantity to which this operator corresponds is represented by

$$\langle X\rangle(t) \equiv \langle\psi(t)|X|\psi(t)\rangle .$$

Suppose that $X = \Sigma x_n|n\rangle\langle n|$, with $\langle n|m\rangle = \delta_{nm}$, then it follows that

$$\langle X\rangle(t) = \Sigma x_n|\langle n|\psi(t)\rangle|^2,$$

in which one may recognize $|\langle n|\psi(t)\rangle|^2$ as the probability that the system is found in the state $|n\rangle$. If, on the other hand, $X = \int x|x\rangle\langle x| \, dx$, with $\langle x|y\rangle = \delta(x - y)$,

---

* A good treatment of traditional quantum mechanics is offered by [Da 68].

then $\langle X \rangle(t) = \int x |\langle x|\psi(t)\rangle|^2 \, dx$, in which case we may interpret $|\langle x|\psi(t)\rangle|^2$ as the probability density for the system to be found in the generalized state $|x\rangle$. After a measurement, the state of the system reflects the outcome of the measurement. If, for example, the discrete, nondegenerate value x_n is found, then the state of the system is immediately transformed to the state $|n\rangle$. This change of the state is the so-called "collapse of the wavepacket". There is nothing mysterious in this sudden change of state when one appreciates the fact that the state represents *probability*, a quantity that corresponds to a lack of information, and which changes instantly for every person dozens of times each day. In like manner one has

$$\langle X^2 \rangle(t) = \Sigma \, x_n^2 \, |\langle n|\psi(t)\rangle|^2 \, ,$$
$$\langle f(X) \rangle(t) = \Sigma \, f(x_n)|\langle n|\psi(t)\rangle|^2 \, ,$$
$$\langle e^{isX} \rangle(t) = \Sigma \, e^{isx_n}|\langle n|\psi(t)\rangle|^2 \, ,$$

with a corresponding set of relations in the case of a continuous spectrum.

It is useful to give another expression for such expectation values. For this purpose we define a (pure state) *density matrix*, which in the present case is the projection operator

$$\rho(t) \equiv |\psi(t)\rangle\langle\psi(t)| \, .$$

It follows that $\langle X \rangle(t) = \text{Tr}\,(X\rho(t))$ for a general operator X. Observe further that

$$\rho(t) = e^{-i(t-t')\mathcal{H}/\hbar}\rho(t')e^{i(t-t')\mathcal{H}/\hbar} \, ,$$

and therefore this quantity obeys the equation of motion

$$i\hbar\frac{d}{dt}\rho(t) = -[\rho(t), \mathcal{H}]$$

subject to the initial condition that $\rho(t') \equiv |\psi(t')\rangle\langle\psi(t')|$. If we are uncertain about the initial state of the system this formalism permits us to introduce a more general (impure state) initial condition for the differential equation satisfied by the density matrix. This more general initial condition is defined by

$$\rho(t') \equiv \Sigma_n \lambda_n|n\rangle\langle n|, \qquad \lambda_n \geq 0, \qquad \Sigma_n \lambda_n = 1 \, ,$$

for some orthonormal basis set $\{|n\rangle\}_{n=0}^{\infty}$. Observe that $\rho(t')$ satisfies the criteria to be a trace-class operator and that $\text{Tr}\,(\rho(t')) = 1$. For a general operator X the relation $\langle X \rangle(t) = \text{Tr}\,(X\rho(t))$ still holds true.

The definition of the temporal behavior of these expectation values allows for an alternative interpretation. If for simplicity we set $t' = 0$ in the temporal evolution of the state vector, then we may simply write

$$\langle X \rangle(t) = \langle\psi(t)|X|\psi(t)\rangle$$
$$= \langle\psi|e^{it\mathcal{H}/\hbar}Xe^{-it\mathcal{H}/\hbar}|\psi\rangle$$
$$\equiv \langle\psi|X(t)|\psi\rangle \, ,$$

where in the last line we have introduced $X(t) \equiv e^{it\mathcal{H}/\hbar} X e^{-it\mathcal{H}/\hbar}$. Observe that this time dependence is captured equally well by the differential equation

$$i\hbar \frac{d}{dt} X(t) = [X(t), \mathcal{H}]$$

satisfied by the time-dependent operator. This formula is central to the Heisenberg formulation of quantum mechanics. Observe that the sign in this equation of motion for the operator X is opposite to the one in the equation for the density matrix ρ. These two signs parallel the two signs already seen in the classical equations of motion as expressed in terms of Poisson brackets.

The relations just developed have given us two ways to calculate the same quantity. In the initial picture – the so-called Schrödinger picture – the operators are time independent and the state vectors are time dependent. In the latter picture – the so-called Heisenberg picture – the operators are time dependent and the state vectors are time independent. This distinction is rather like that of active and passive transformations in various disciplines. For example, in evaluating the inner product of two three-dimensional vectors, the result is the same if either both vectors are rotated a common amount, or if instead the coordinate axes to which they are referenced are rotated a corresponding amount. Both pictures are valid and they both lead to the same physical predictions. One point of view may be more useful for some problems, while the other point of view may be more useful for different problems. We shall find both pictures to be of value.

In the foregoing paragraphs we have sketched the basic ingredients that go to make up the formalism of quantum mechanics. Now it is time to illustrate how this formalism is applied in practice.

4.2 Canonical quantization

The program of canonical quantization attempts to take a classical theory described by the phase-space variables p and q and a Hamiltonian $H(p,q)$ and to quantize this classical theory. This is achieved by "promoting" the classical variables (c-numbers) p and q to the quantum operators (q-numbers) P and Q, i.e. two operators that satisfy the fundamental canonical commutation relation

$$[Q, P] = QP - PQ = i\hbar \mathbb{1} = i\hbar .$$

Furthermore, these two canonical operators are assumed to be irreducible. Recall that irreducible means that every operator in the space may be given as a function of the pair (P, Q). It is entirely reasonable to assume that the canonical operators are irreducible. For instance, in the classical theory of a single degree-of-freedom system, every function on phase space is given by a function of the classical canonical variables p and q. Therefore, it is natural to assume that in the quantum theory every operator can be constructed from the basic quantum canonical operators P and Q. In this chapter we shall accept this logic and adopt an irreducible representation for the canonical operators. In particular, the quantum Hamiltonian is given by the classical Hamiltonian with p replaced by P and with q replaced by Q. Thus the classical Hamiltonian $H(p, q)$ is promoted to the operator

$$\mathcal{H} \equiv H(P, Q) .$$

Two problems arise in this procedure. The first problem has to do with the choice of classical canonical coordinates in which this promotion takes place, and the second has to do with an operator ordering ambiguity. As stated clearly by Dirac: this procedure of generating the quantum Hamiltonian from the classical Hamiltonian works only if the classical variables refer to a system of Cartesian coordinates [Di 76]. We accept this requirement on the choice of coordinates. The second problem deals with the operator ordering ambiguity and refers to the fact that $pq - qp = 0$ for the classical variables while $PQ - QP = -i\hbar \neq 0$ for the quantum variables. Hence, if the classical Hamiltonian had a term of the form $p^2 q^2$, among others, then the quantum Hamiltonian might contain the Hermitian combination $(P^2 Q^2 + Q^2 P^2)/2$ or $PQ^2 P$, which are in fact not equal to each other since

$$PQ^2 P = \tfrac{1}{2}(P^2 Q^2 + Q^2 P^2) + \hbar^2 .$$

In any case, in the limit that $\hbar \to 0$, generally referred to as the "classical limit", the difference in these quantum expressions disappear. In the limit $\hbar \to 0$ it is clear that $PQ - QP \to 0$ and so any difference among quantum expressions due to ordering would disappear. On the other hand, if there is an ordering ambiguity and therefore there are several choices for the quantum Hamiltonian, then it is generally accepted that there is no classical criterion to choose between the different possibilities, and, as one says, "it is necessary to appeal to experiment" to decide which is the correct one for the case at hand. We shall have much more to say about the classical limit later.

Although operator ordering ambiguity is the only uncertainty generally acknowledged in the quantization procedure, it is by no means the only possible one. In principle, one could add to the quantum Hamiltonian essentially *any* other operator with a coefficient that is $O(\hbar)$, for in the classical limit $\hbar \to 0$ this additional term will simply disappear. That is, we may consider the quantum Hamiltonian given by $H(P, Q) + \hbar Y(P, Q)$, which for any reasonable choice of Y, will have the same classical limit as the quantum Hamiltonian $H(P, Q)$ when $\hbar \to 0$. What is to prevent such a wide family of uncertainties from arising? In principle, nothing prevents the appearance of such a term, and indeed when it comes to the quantum theory of an infinite number of degrees of freedom, namely a field theory, we shall actually exploit this level of ambiguity to avoid undesirable divergences. However, for one, or for a finite number of degrees of freedom, no such divergences arise, so this argument for the inclusion or noninclusion of such a term is not available. Instead, one must resort to the statement that the theory described by the quantum Hamiltonian $H(P, Q) + \hbar Y(P, Q)$ does not, in general, "accurately represent" the original theory under consideration. How are we to make such a judgment? We can do so by isolating the desired functional dependence of the original theory. For example, suppose that we start with the classical harmonic oscillator Hamiltonian in Cartesian coordinates given by $H(p, q) = \frac{1}{2}(p^2 + q^2)$. Then suppose we choose as our quantum operator $\mathcal{H} = \frac{1}{2}(P^2 + Q^2) + c\hbar|Q|^\beta$, where β is an $\hbar$-independent constant while c is a constant with a possible (positive power) $\hbar$ dependence. Since the classical Hamiltonian was nonnegative it is reasonable to insist that $c \geq 0$. Now under a scaling transformation where $p \to s^{-1}p$, $q \to sq$, $0 < s < \infty$, and correspondingly, $P \to s^{-1}P$, $Q \to sQ$ – which, incidentally, has the property that the basic Poisson bracket $\{q, p\}$ and the basic commutation relation $[Q, P]$ remain invariant – it follows that the classical Hamiltonian has been rescaled to $H \to \frac{1}{2}(s^{-2}p^2 + s^2q^2)$ while the quantum Hamiltonian has been rescaled to

$$\mathcal{H} \to \tfrac{1}{2}(s^{-2}P^2 + s^2Q^2) + c\hbar s^\beta|Q|^\beta .$$

We can use the dependence of these two expressions on the scaling variable s to limit the ambiguity in the quantization process. In particular, when $s \to \infty$, then $H \propto s^2$. This should be an invariant feature of the quantum theory as well as of the classical theory. Consequently, we are led to exclude any behavior where $\beta > 2$, and so we conclude that $\beta \leq 2$. On the other hand, if $s \to 0$, then $H \propto s^{-2}$. Thus we exclude $\beta < -2$. If we further assert that only s^2 and s^{-2} must be the characteristic behavior of the "true" theory, then we conclude that only two possibilities remain, namely $\beta = 2$ and $\beta = -2$ (although $\beta = 0$ would be pretty harmless). The case $\beta = 2$ represents an $O(\hbar)$ correction to the coefficient of the term Q^2 in the quantum Hamiltonian; this kind of uncertainty is generally accepted and constitutes a potential modification – a finite renormalization – of the coefficient of this term. On the other hand, the remaining possibility, that of $\beta = -2$, is rather different in kind. In the same language used for s^2, we would say that the term Q^{-2} constitutes an $O(\hbar)$ renormalization of the kinetic energy term. To rule out such a term in the quantum theory of one or a finite number of degrees of freedom, is a very definite assumption, the validity of which needs to be tested by experiment. As the reader may imagine, such experiments show that this kind of modification is not necessary for finitely many degrees of freedom and therefore we may safely assume that $c = 0$ when $\beta = -2$. On the other hand, when it comes to the quantum theory of infinitely many degrees of freedom, we shall find that just this kind of renormalization of the kinetic energy term is in fact required on certain occasions!

The quantum Hamiltonian for the harmonic oscillator is therefore taken to be $\mathcal{H} = \frac{1}{2}(P^2 + Q^2)$, that for a quartic anharmonic oscillator is taken as $\mathcal{H} = \frac{1}{2}(P^2 + Q^2) + Q^4$, etc. These operators each have a discrete spectrum and a complete set of normalizable eigenvectors with nondegenerate eigenvalues. They are examples of self-adjoint operators in that when they are defined on a natural domain of vectors, e.g., $\overline{\mathfrak{D}} = \mathfrak{D}(P^2) \cap \mathfrak{D}(Q^4)$, they uniquely define unitary transformations in the Hilbert space according to the definition $U(t) = e^{-it\mathcal{H}/\hbar}$. On the other hand, the quantum Hamiltonian given by $\mathcal{H} = \frac{1}{2}(P^2 + Q^2) - Q^4$ defined on a natural domain of vectors (say $\overline{\mathfrak{D}}$) does *not* uniquely define a unitary operator acting on the Hilbert space; one must specify additional data – four real parameters – for the generator of this unitary operator to be determined. The specific value of these additional parameters actually affects the spectrum of the resultant generator and thus they are entirely physical. Furthermore, the quantum Hamiltonian given by $\mathcal{H} = (PQ^3 + Q^3P)/2$ can *never* serve as the generator of a unitary transformation; this is a quantum reflection of the fact that the classical solutions for the corresponding classical Hamiltonian are never entirely real but must – whenever the classical energy is nonzero – become *complex.* See Exercises 4.1 and 4.2.

Thus not every quantum Hamiltonian serves to generate unitary transformations, but it is safe to say that the cases of major physical interest are limited to those Hamiltonian operators that do generate unitary transformations. As stressed in Chapter 6, it is just those Hamiltonian operators which are self-adjoint operators that constitute the class in which one is interested.

Uncertainty relation

Let the expression $\langle F \rangle \equiv \langle \psi | F | \psi \rangle$ represent the mean of the operator F in the normalized state $|\psi\rangle$. Suppose that P and Q satisfy the canonical commutation relation $[Q, P] = i\hbar$. We introduce $\Delta Q \equiv Q - \langle Q \rangle$ and $\Delta P \equiv P - \langle P \rangle$. Then it follows from the triangle and from the Schwarz inequalities that

$$\hbar = |\langle (QP - PQ) \rangle|$$
$$= |\langle [(Q - \langle Q \rangle)(P - \langle P \rangle) - (P - \langle P \rangle)(Q - \langle Q \rangle)] \rangle|$$
$$\leq |\langle \Delta Q \, \Delta P \rangle| + |\langle \Delta P \, \Delta Q \rangle|$$
$$\leq 2 \, \|\Delta Q | \psi \rangle\| \, \|\Delta P | \psi \rangle\| \, .$$

Stated otherwise, this expression implies that

$$\langle (\Delta Q)^2 \rangle \langle (\Delta P)^2 \rangle \geq \tfrac{1}{4} \hbar^2 \, ,$$

which is the usual form of the uncertainty relation between two canonical variables. Since $\langle Q \rangle$ and $\langle Q^2 \rangle$ represent the first and second moment, it follows that $\langle (\Delta Q)^2 \rangle$ denotes the *variance* of the distribution with respect to the variable Q. A small value for the variance means that the distribution is rather tightly distributed about its mean value. The meaning of the uncertainty relation is that it is impossible to make the variance of any state arbitrarily small with respect to Q and with respect to P simultaneously since the product of these two separate variances must always be greater than or equal to $\hbar^2/4$.

Equality in the uncertainty relationship requires that $\langle \Delta Q \, \Delta P \rangle = -\langle \Delta P \, \Delta Q \rangle$ and $\Delta P |\psi\rangle = i\omega \Delta Q |\psi\rangle$ for some real parameter ω. In turn, these relations imply that

$$\| \Delta P |\psi\rangle \|^2 + \omega^2 \| \Delta Q |\psi\rangle \|^2 = \omega \hbar \, ,$$

and thus it is necessary that $\omega > 0$. Consequently, equality holds if and only if

$$(P - i\omega Q)|\psi\rangle = (\langle P \rangle - i\omega \langle Q \rangle)|\psi\rangle \, , \qquad \omega > 0 \, .$$

Therefore, these states are eigenstates of the annihilation operator $a = (\omega Q + iP)/\sqrt{2\omega\hbar}$, and such states are exactly the coherent states which have already been introduced in Chapter 3. Of course, in the present case we require the normalized form of the coherent states.

4.3 The Schrödinger formulation

We have now agreed to accept the operator $\frac{1}{2}(P^2 + \omega^2 Q^2) + gQ^4$ to be the quantum Hamiltonian for an (an)harmonic oscillator, where we have introduced two additional parameters, an angular frequency $\omega > 0$ and a coupling constant $g \geq 0$. How does one proceed to solve the associated quantum dynamics, or to put it otherwise, how does one determine the eigenvalues and eigenvectors associated with this Hamiltonian. The way proposed by Schrödinger – and a very good way it is – is to introduce a natural and useful *irreducible representation* for the basic operators in the form

$$P \to -i\hbar \frac{\partial}{\partial x} \, , \qquad Q \to x \, ,$$

with $x \in \mathbb{R}$, acting on elements of the space $L^2(\mathbb{R}, dx)$ composed of complex, square-integrable functions $\psi(x)$, $\int |\psi(x)|^2 \, dx < \infty$. All operators acting on this space are expressible as functions of the basic operator pair: *differentiation* by $-i\hbar(\partial/\partial x)$ and *multiplication* by x. Stated otherwise, any operator B that commutes with both $-i\hbar(\partial/\partial x)$ and x is necessarily proportional to the identity. If we adopt this representation for P and Q, then the Hamiltonian operator becomes

$$\mathcal{H} = -\tfrac{1}{2}\hbar^2 \frac{\partial^2}{\partial x^2} + \tfrac{1}{2}\omega^2 x^2 + gx^4 \, .$$

Consequently, the *time-dependent Schrödinger equation* becomes

$$i\hbar \frac{\partial}{\partial t} \psi(x, t) = [-\tfrac{1}{2}\hbar^2 \frac{\partial^2}{\partial x^2} + \tfrac{1}{2}\omega^2 x^2 + gx^4] \, \psi(x, t) \, ,$$

and we seek solutions to this differential equation subject to the condition that $\int |\psi(x, t)|^2 \, dx < \infty$, which will permit us to normalize the solution initially, say at $t = 0$. The fact that the Hamiltonian is self adjoint will guarantee that the initial normalization is preserved for all time thereafter.

If we assume that $\psi(x,t) = \phi_n(x)e^{-iE_nt/\hbar}$, then we are led to the *time-independent Schrödinger equation* given by

$$E_n\phi_n(x) = [-\tfrac{1}{2}\hbar^2\frac{d^2}{dx^2} + \tfrac{1}{2}\omega^2x^2 + gx^4]\,\phi_n(x)\,,$$

$n \in \{0, 1, 2, \ldots\}$, an equation which, for each n, is subject to the requirement that

$$\int |\phi_n(x)|^2\,dx < \infty\,,$$

which is the central criterion that singles out acceptable energy eigenvalues, E_n. As discussed previously, the eigenfunctions belonging to unequal eigenvalues are always orthogonal – and one may always choose those belonging to equal eigenvalues to be orthogonal as well – so, without loss of generality, we may assume that the solutions to the eigenvalue equation are chosen to satisfy the condition of orthonormality embodied in the relations

$$\int \phi_n(x)^*\phi_m(x)\,dx = \delta_{nm}\,,$$

$$\sum_n \phi_n(x)^*\phi_n(y) = \delta(x-y)\,,$$

i.e., that the set of functions $\{\phi_n\}_{n=0}^{\infty}$ is a complete orthonormal set of functions.

As a homogeneous linear equation the sum of any number of solutions of the time-dependent Schrödinger equation is still a solution. As a consequence the general solution of the time-dependent Schrödinger equation follows from the solutions of the time-independent Schrödinger equation. In particular, the most general normalized solution to the time-dependent equation is given by

$$\psi(x,t) = \sum_{n=0}^{\infty} b_n\,\phi_n(x)e^{-itE_n/\hbar}\,,$$

subject only to the normalization criterion that

$$\int |\psi(x,t)|^2\,dx = \sum_{n=0}^{\infty} |b_n|^2 = 1\,.$$

From the point of view of differential equations, the time-independent Schrödinger equation is a second-order differential equation with two linearly independent solutions for any value of the energy E. In general, however, both of these solutions fail to be square integrable; that very special requirement occurs for selected and isolated values of the energy eigenvalue. It is the condition of square integrability – specifically the condition of being in the Hilbert space – that effectively singles out the physically important energy spectral values. If the energy levels are nondegenerate, as is the case for the one-dimensional (an)harmonic oscillator, and an energy eigenvalue occurs, then only one particular combination of the linearly independent solutions is square integrable and any other linearly independent combination is not square integrable. This feature persists in any representation of the Schrödinger equation that one chooses. Suppose, for example, that one chooses to work in another representation of the basic canonical operators given by

$$P \to p \, , \qquad Q \to i\hbar \frac{\partial}{\partial p} \, ,$$

acting on functions in the space $L^2(\mathbb{R}, dp)$. In that case the time-independent Schrödinger equation for the anharmonic oscillator would assume the form

$$E_n \tilde{\phi}_n(p) = [\frac{1}{2}p^2 - \frac{\hbar^2}{2}\omega^2 \frac{d^2}{dp^2} + g\hbar^4 \frac{d^4}{dp^4}] \, \tilde{\phi}_n(p) \, ,$$

where these two representations are simply related by a Fourier transformation, i.e.,

$$\tilde{\phi}_n(p) = \frac{1}{\sqrt{2\pi\hbar}} \int e^{-ipx/\hbar} \phi_n(x) \, dx \, .$$

In the p-representation the differential equation is *fourth* order and there are *four* linearly independent solutions for each energy value. For a general energy value none of these four solutions is square integrable. Only for special energy values – *the very same energy eigenvalues that arose in the x-representation* – do we have any solutions that are square integrable, and again for a nondegenerate system only one of the four linearly independent solutions can be square integrable at any one time, while the other three solutions are not square integrable. Since the Fourier transformation takes normalized square-integrable functions into normalized square-integrable functions, it is evident that the proper eigenfunction in the p-representation and the proper eigenfunction in the x-representation are genuine Fourier transforms of each other. But how can it be that in the p-representation the Schrödinger equation has four linearly independent solutions while in the x-representation it has only two linearly independent solutions; what happens to two of the solutions in the process of taking the Fourier transformation?

To understand the answer to this question it is necessary to appreciate that when the solutions to the differential equation are not square integrable it is more natural to regard them as generalized functions, i.e., as distributions. Distributions generally need test functions in order to be evaluated, and the class of suitable test functions very much depends on the character of the distributions themselves. Let us first start in x space. For large values of $\pm x$, $|x| \gg 0$, the form of the solution is entirely governed by the second derivative term and the quartic term, which leads to an asymptotic behavior given by [here $c \equiv (2g/9\hbar^2)^{1/2}$]

$$e^{-c|x|^3} \, , \qquad\qquad e^{c|x|^3} \, .$$

At an energy value for which there are no square-integrable solutions, then the asymptotic solution diverges exponentially either as $x \to \infty$ or as $x \to -\infty$, or both, but nevertheless it gives rise to a functional form that we can readily appreciate. Now let us repeat a similar analysis in p space. At large values of $\pm p$, $|p| \gg 0$, the form of the solution is entirely governed by the fourth derivative term and the quadratic term, which lead to an asymptotic behavior given by [here $C \equiv (2/3\hbar)(-1/2g)^{1/4}$]

$$e^{iC|p|^{3/2}} \; , \quad e^{-C|p|^{3/2}} \; , \quad e^{-iC|p|^{3/2}} \; , \quad e^{C|p|^{3/2}} \; ,$$

which involves the four fourth roots of unity. Just as before, at an energy value for which there are no square integrable solutions, then the asymptotic solution diverges exponentially either as $p \to \infty$ or as $p \to -\infty$, or both, but nevertheless it gives rise to a functional form that we can readily appreciate. On the other hand, the Fourier transformation of any such distributional solution is nearly impossible to visualize in functional terms. As a consequence, the Fourier transform of the four functional, but nonsquare integrable solutions in p space have *nothing to do* with either of the two functional, but nonsquare integrable solutions that exist in x-space! Thus the paradox about what happened to the additional solutions is resolved by the statement that the *non*square integrable solutions to the Schrödinger equation in the p-representation are completely independent of the solutions in the x-representation and vice versa. The presence or absence of additional solutions is strictly *an artifact of the form of* **representation** *of the operators* and it is *not* a property of the operators themselves. This remark adds further significance to the assertion that it is only the square-integrable functions – those that constitute vectors in Hilbert space – that are the ones of physical interest.

The previous discussion dealt with Hamiltonians with discrete spectra and one may well wonder what is the situation regarding Hamiltonians with continuous spectra. In particular, consider the operator P itself. In the x-representation the eigenvalue equation reads

$$P\lambda(x) = -i\hbar \frac{d}{dx}\lambda(x) = p'\lambda(x) \; ,$$

$p' \in \mathbb{R}$, with the evident solution that

$$\lambda(x) = \frac{1}{\sqrt{2\pi\hbar}} \, e^{ip'x/\hbar} \; ,$$

which is *not* square integrable. On the other hand, in Fourier space the eigenvalue equation reads

$$P\tilde{\lambda}(p) = p\tilde{\lambda}(p) = p'\tilde{\lambda}(p)$$

with solution given by

$$\tilde{\lambda}(p) = \delta(p - p') \; .$$

In this case, even though neither solution is square integrable, the nature of the distributional solutions are sufficiently mild so that the two solutions are in fact Fourier transforms of one another. Thus in this case the behavior is completely different from the nonsquare-integrable solutions in the case of discrete spectra and rather more like the behavior of the square-integrable solutions for discrete spectra [GeS 64].

We stress these points because, occasionally, a casual application of the usual formalism can lead to an apparent paradox. As an illustration, consider the following example. Let P and Q denote canonical self-adjoint operators, and let $D \equiv (PQ + QP)/2$, which is also self adjoint on a natural domain. It follows that $[Q, D] = i\hbar Q$ holds as well. Let $|d\rangle$ denote an eigenvector for D, namely, $D|d\rangle = d|d\rangle$, for some real d value; after all, as a self-adjoint operator D must have only real eigenvalues. However, consider the relation

$$DQ|d\rangle = QD|d\rangle - i\hbar Q|d\rangle$$
$$= (d - i\hbar)Q|d\rangle \,,$$

which asserts that if $|d\rangle$ is an eigenvector with the real eigenvalue d, then $Q|d\rangle$ is also an eigenvector with the manifestly *complex* eigenvalue $d - i\hbar$. How can a self-adjoint operator such as D have an eigenvector with a *complex* eigenvalue? Of course, the proper conclusion must be that it *cannot* have an eigenvector with a complex eigenvalue.

In the language of the preceding discussion, the resolution of this paradox lies in the observation that the states $|d\rangle$ are analogs of the square-integrable eigenstates of an anharmonic oscillator, while the states $Q|d\rangle$ are analogs of the non square-integrable solutions, as recourse to a representation demonstrates.

Time-reversal transformation

Time-reversed motion is well known in classical mechanics and is obtained by changing t to $-t$, reversing the signs of all the momenta (or velocities), and letting the system develop in time. In addition, for some dynamical systems (but not all) the time-reversed motion is an actual evolution of the original system. This state of affairs holds for systems with a so-called *time-reversal invariant* Hamiltonian. There are corresponding notions in quantum mechanics as well.

For a system with a general canonical Hamiltonian, Schrödinger's equation reads

$$i\hbar \frac{\partial}{\partial t}\psi(x, t) = \mathcal{H}(-i\hbar \frac{\partial}{\partial x}, x)\psi(x, t) \,.$$

For a real Hamiltonian function – one for which the only $i = \sqrt{-1}$ terms are those arising from the chosen representation of P and Q – the complex conjugate wave function evidently satisfies the differential equation

$$-i\hbar \frac{\partial}{\partial t}\psi(x, t)^* = \mathcal{H}(i\hbar \frac{\partial}{\partial x}, x)\psi(x, t)^* \,.$$

If we introduce $\psi'(x,t) \equiv \psi(x,-t)^*$, then it follows that this transformed function satisfies the differential equation

$$i\hbar\frac{\partial}{\partial t}\psi'(x,t) = \mathcal{H}(i\hbar\frac{\partial}{\partial x}, x)\psi'(x,t) .$$

For Hamiltonians such that $\mathcal{H}(-P,Q) = \mathcal{H}(P,Q)$ – referred to as time-reversal invariant Hamiltonians – then it follows that $\psi'(x,t)$ and $\psi(x,t)$ satisfy *identical* Schrödinger equations. As a consequence, the energy eigenfunctions for time-reversal invariant Hamiltonians may, without loss of generality, be assumed to be real. This will have useful consequences, as we can see immediately.

The time-reversal operation for a time-independent state in the Schrödinger representation is complex conjugation. Let $\phi(x)$ be such a time-reversal invariant state, namely one that is *real*, i.e., $\phi(x)^* = \phi(x)$, and which is nonzero save perhaps on a set of measure zero. Consider the expectation function given by

$$E(p) \equiv \langle 0|p\rangle \equiv \langle 0|e^{ipQ/\hbar}|0\rangle = \int\phi(x)e^{ipx/\hbar}\phi(x)\,dx ,$$

where we assume that ϕ is normalized and thus $E(0) = 1$. In addition we consider

$$\langle p'|P|p\rangle \equiv -i\hbar\int\phi(x)e^{-ixp'/\hbar}\frac{\partial}{\partial x}e^{ixp/\hbar}\phi(x)\,dx .$$

Evidently, this expression is also given by

$$\langle p'|P|p\rangle \equiv +i\hbar\int\phi(x)e^{ixp/\hbar}\frac{\partial}{\partial x}e^{-ixp'/\hbar}\phi(x)\,dx$$

which arises from complex conjugation and an interchange of p and p'. The average of these two expressions readily leads to

$$\langle p'|P|p\rangle = \tfrac{1}{2}(p'+p)\langle p'|p\rangle = \tfrac{1}{2}(p'+p)E(p-p') ,$$

which expresses the matrix elements of the momentum operator in terms of the expectation function E.

We can also go one step further. Let us choose the state ϕ to be a stationary state, specifically, an eigenstate of the time-reversal invariant Hamiltonian for which $\mathcal{H}|0\rangle = E_0|0\rangle$, and, moreover, we assume that the Hamiltonian is of the form $\mathcal{H} = P^2/2 + V(Q)$. Then it follows that $(i/\hbar)[\mathcal{H},Q] = P$, and hence

$$(\frac{\partial}{\partial p'} + \frac{\partial}{\partial p})\langle p'|\mathcal{H}|p\rangle = \langle p'|P|p\rangle = \tfrac{1}{2}(p'+p)\,E(p-p') ,$$

which has a unique solution given by

$$\langle p'|\mathcal{H}|p\rangle = [\tfrac{1}{2}p'p + E_0]\langle p'|p\rangle = [\tfrac{1}{2}p'p + E_0]E(p-p') .$$

In applications, we shall generally suppose that $|0\rangle$ is the ground state and that the ground state energy vanishes, $E_0 = 0$. In that case

$$\langle p'|\mathcal{H}|p\rangle = \tfrac{1}{2}p'p\,\langle p'|p\rangle = \tfrac{1}{2}p'p\,E(p-p') .$$

For most Hamiltonians of interest these matrix elements are sufficient to uniquely determine the Hamiltonian operator itself. This fact should not be too surprising since knowledge of the function $E(p)$ is equivalent (by Fourier transformation) to knowledge of the ground state $\phi(x)$, which in turn determines the form of the Hamiltonian uniquely since $V(x) = \hbar^2 \phi''(x)/2\phi(x)$.*

The essence of time reversal may also be given entirely in the abstract language of quantum theory as well. To that end we introduce a linear, but anti-unitary, time-reversal operator T with the following properties:

$$ T^{-1} = T , \quad T^2 = 1 , \quad T^\dagger bQT = b^*Q , \quad T^\dagger cPT = -c^*P , $$

where b and c are complex numbers. An anti-unitary operator is one for which for any vectors $|\lambda'\rangle \equiv T|\lambda\rangle$ and $|\chi'\rangle \equiv T|\chi\rangle$, it follows that

$$ \langle \lambda'|\chi'\rangle \equiv \langle \lambda|\chi\rangle^* = \langle \chi|\lambda\rangle . $$

In application, we often focus on time-reversal invariant states, which are the analog of the real functions of the Schrödinger representation. If $|\psi\rangle = T|\psi\rangle$ denotes such a state, then for a real operator $A(P,Q)$, i.e., one with no extra i factors,

$$ \langle \psi|A(P,Q)|\psi\rangle = \langle \psi|A(-P,Q)|\psi\rangle . $$

From the point of view of the Schrödinger representation this identity is basically self evident, as we have seen from the previous discussion.

The harmonic oscillator, and the power of symmetry

Although it is impossible to solve the quartic anharmonic oscillator in closed form, it is entirely possible to do so for the harmonic oscillator, i.e., when $g = 0$, and it is instructive to do so. First we eliminate inessential complications by choosing natural units in which both $\omega = 1$ and $\hbar = 1$. Let us not approach this problem head on but rather in an indirect manner exploiting *symmetry* as much as possible. Symmetry is a powerful tool and it should be used to simplify, if not outright solve, a problem whenever possible; we shall use it often throughout this text. To this end let us not consider a single oscillator but *two* uncoupled, identical oscillators. Thus the Hamiltonian we will consider is given by

$$ \mathcal{H} = \tfrac{1}{2}(P_1^2 + Q_1^2) + \tfrac{1}{2}(P_2^2 + Q_2^2) = \tfrac{1}{2}(P_1^2 + P_2^2 + Q_1^2 + Q_2^2) . $$

---

* In the context of canonical quantum field theory, the important relations connecting matrix elements of the canonical momentum and the Hamiltonian to the expectation functional for the field operator have been initially identified and exploited by Araki [Ar 60a, Ar 60b]; see also Chapter 6.

Here we have two independent sets of self-adjoint, canonical operators that commute among themselves save for

$$[Q_1, P_1] = [Q_2, P_2] = i .$$

Let $|0\rangle$ denote the normalized ground state for this problem which we shall assume is nondegenerate. Let us consider the expression

$$F(p_1, p_2) \equiv \langle 0|e^{i(p_1 Q_1 + p_2 Q_2)}|0\rangle ,$$

where p_1 and p_2 are two real parameters at our disposal. It follows from Schwarz's inequality that $|F| \leq 1 = F(0,0)$. On the one hand, as two identical but *independent* oscillators it follows that the ground state is a product state and therefore

$$F(p_1, p_2) = F(p_1, 0)F(0, p_2) .$$

Moreover, the original Hamiltonian is invariant under either $Q_1 \to -Q_1$ or $Q_2 \to -Q_2$, and thus $F(-p_1, 0) = F(p_1, 0) = F(p_1, 0)^*$ as well as $F(0, -p_2) = F(0, p_2) = F(0, p_2)^*$. On the other hand, the original Hamiltonian is *invariant under rotations* in the (1-2)-plane, i.e., invariant under the rotation given by

$$P_1' = \cos(\theta)P_1 - \sin(\theta)P_2 ,$$
$$P_2' = \sin(\theta)P_1 + \cos(\theta)P_2 ,$$

coupled with

$$Q_1' = \cos(\theta)Q_1 - \sin(\theta)Q_2 ,$$
$$Q_2' = \sin(\theta)Q_1 + \cos(\theta)Q_2 ,$$

where θ is an arbitrary c-number angle. Since the ground state is unique it is invariant under such rotations, and as a consequence

$$F(p_1, p_2) = F(\cos(\theta)p_1 - \sin(\theta)p_2, \sin(\theta)p_1 + \cos(\theta)p_2)$$

which must hold for all θ. It follows therefore that

$$F(p_1, p_2) \equiv G(p_1^2 + p_2^2) = G(p_1^2)G(p_2^2)$$

for some function G. Since both Q_1 and Q_2 are self adjoint it follows that the function G is necessarily continuous, and the only continuous function G that fulfills the condition $G(0) = 1$ and satisfies this functional equation has the form $G(y) = \exp(Cy)$ for some $C \in \mathbf{C}$; see Exercise 4.3. In our particular case the function of interest is given by

$$F(p_1, p_2) = e^{-(1/4B)(p_1^2 + p_2^2)}$$

for some constant $B > 0$. Next we invoke the Schrödinger representation when we observe that

$$e^{-(1/4B)(p_1^2 + p_2^2)} = \frac{B}{\pi} \int e^{i(p_1 x_1 + p_2 x_2) - B(x_1^2 + x_2^2)} \, dx_1 dx_2 ,$$

which tells us, apart from the unknown factor B, that the ground state wave function for a single oscillator is proportional to $\exp(-Bx^2/2)$. To find B we argue as follows. The original Hamiltonian is also invariant under the transformation $P_l \to Q_l$ coupled with $Q_l \to -P_l$, $l = 1, 2$. Since

$$\int e^{-ikx} e^{-Bx^2/2} \, dx \propto e^{-k^2/2B} ,$$

and since the result must be invariant under this transformation, it follows that in the chosen units $B = 1$. Hence we conclude that the ground state wave function is given by

$$h_0(x) = \pi^{-1/4} e^{-x^2/2} ,$$

where the appropriate normalization factor has been appended. In physical units – rather than natural units – the ground state reads

$$\phi_0(x) = (\omega/\pi\hbar)^{1/4} e^{-(\omega/2\hbar)x^2} .$$

We take this opportunity to obtain another quantity of interest regarding the ground state for a single harmonic oscillator, specifically, again using natural units, the expression

$$E(p,q) \equiv \langle 0|e^{i(pQ-qP)}|0\rangle ,$$

where p and q are two real parameters. On the one hand, when $q = 0$ this expression reduces to $F(p,0) = \exp(-p^2/4)$. On the other hand, the harmonic oscillator Hamiltonian expressed in natural units is invariant under the rotation given by

$$P' = \cos(\gamma)P - \sin(\gamma)Q ,$$
$$Q' = \sin(\gamma)P + \cos(\gamma)Q ,$$

where γ is an arbitrary c-number angle. Thus, rather like the case previously considered, $E(p,q) = D(p^2 + q^2)$ for some function D, and it follows directly, therefore, that

$$E(p,q) = e^{-(p^2+q^2)/4} .$$

In physical units we have

$$E(p,q) \equiv \langle 0|e^{i(pQ-qP)/\hbar}|0\rangle = e^{-(1/4\hbar)(\omega^{-1}p^2+\omega q^2)} ,$$

an expression that will play an important role subsequently.

Besides obtaining the functional form of the ground state for the harmonic oscillator, we can say even more in the present case. First, for Hamiltonians given in the form

$$\mathcal{H} = \tfrac{1}{2}P^2 + \mathcal{V}(Q) ,$$

we make the obvious remark that the energy eigenfunctions and eigenvalues are completely determined in principle once the potential $\mathcal{V}$ is specified. In addition, if we know the ground state to this problem – let us say $\phi_0(x)$ in the Schrödinger representation – then the potential is in principle determined, in natural units and apart from an additive constant E_0, by the relation

$$\mathcal{V}(x) \equiv \frac{1}{2}\frac{1}{\phi_0(x)}\frac{d^2\phi_0(x)}{dx^2} + E_0 .$$

This definition is unambiguous only for the ground state since it never vanishes and does not cause difficulties with a division by zero. As a consequence, the ground state for such Hamiltonians determines the functional form of *all* the other eigenfunctions and also determines *all* their eigenvalues apart from the one additive constant E_0.

Recall that we have previously given the generating function for the orthonormal Hermite functions as

$$\exp(-s^2 + 2sx - \tfrac{1}{2}x^2) = \pi^{\frac{1}{4}} \sum_{n=0}^{\infty} (n!)^{-\frac{1}{2}} (s\sqrt{2})^n \, h_n(x) .$$

When $s = 0$ we see that $h_0(x) \propto \exp(-\tfrac{1}{2}x^2)$ which, as we have seen, is equal to the ground state of the harmonic oscillator. Thus it is natural to conjecture that the eigenfunctions for the harmonic oscillator are given by the set of functions $\{h_n(x)\}_{n=0}^{\infty}$. To verify this conjecture and at the same time determine the energy spectra in natural units, we may observe that

$$[-\tfrac{1}{2}\frac{\partial^2}{\partial x^2} + \tfrac{1}{2}x^2] \, e^{(-s^2+2sx-\frac{1}{2}x^2)}$$
$$= [2sx - 2s^2 + \tfrac{1}{2}] \, e^{(-s^2+2sx-\frac{1}{2}x^2)}$$
$$= [s\frac{\partial}{\partial s} + \tfrac{1}{2}] \, e^{(-s^2+2sx-\frac{1}{2}x^2)}$$
$$= \pi^{\frac{1}{4}} \sum_{n=0}^{\infty} (n!)^{-\frac{1}{2}} [n + \tfrac{1}{2}] \, (s\sqrt{2})^n \, h_n(x) ,$$

from which we learn not only that each function $h_n(x)$ is indeed an eigenfunction of the harmonic oscillator in natural units but that the energy eigenvalue for $h_n(x)$ is given by $n + 1/2$. Finally, in physical units, the eigenfunctions and eigenvalues are given by

$$\phi_n(x) = (\omega/\hbar)^{1/4} h_n((\omega/\hbar)^{1/2}x) ,$$
$$E_n = \hbar\omega(n + \tfrac{1}{2}) .$$

4.4 The Heisenberg formulation

In contrast to the Schrödinger formulation in which the operators are time independent and the state vectors evolve in time according to the Schrödinger equation, the Heisenberg formulation of quantum mechanics freezes the state vectors in time and places the time evolution on the operators. In particular, if $\mathcal{H}$ denotes a time-independent Hamiltonian operator then it follows that

$$P(t) = e^{it\mathcal{H}/\hbar} P e^{-it\mathcal{H}/\hbar} , \qquad Q(t) = e^{it\mathcal{H}/\hbar} Q e^{-it\mathcal{H}/\hbar} ,$$

which automatically implies the invariance in time of the basic commutation relation,

$$[Q(t), P(t)] = [Q, P] = i\hbar .$$

Furthermore, the Hamiltonian is a constant of the motion, and we find that

$$\mathcal{H} = \mathcal{H}(P, Q) = \mathcal{H}(P(t), Q(t))$$

holds for all t. For a simple canonical system with kinematical variables P and Q and a Hamiltonian given by $\mathcal{H} = P^2/2 + \mathcal{V}(Q)$ it follows that

$$\dot{Q}(t) = -\frac{i}{\hbar}[Q(t), \mathcal{H}]$$
$$= -\frac{i}{\hbar}[Q(t), P^2(t)/2 + \mathcal{V}(Q(t))]$$
$$= P(t) ,$$
$$\dot{P}(t) = -\frac{i}{\hbar}[P(t), P^2(t)/2 + \mathcal{V}(Q(t))]$$
$$= -\mathcal{V}'(Q(t)) .$$

Observe that these equations of motion are just the classical Hamiltonian equations of motion promoted to operator form! This feature of the Heisenberg formalism is quite universal. Just like their classical counterpart, these differential equations may be solved subject to suitable initial conditions, say at $t = 0$, which in the quantum case are given by $P(0) = P$ and $Q(0) = Q$, where the two operators in question must obey the basic commutation relation $[Q, P] = i\hbar$. For example, if $\mathcal{V}(Q) = \omega^2 Q^2/2 - \hbar\omega/2$ appropriate to a harmonic oscillator with the ground state energy subtracted – a useful and common renormalization of the energy levels – then the Heisenberg equations of motion read

$$\dot{Q}(t) = P(t) , \qquad \dot{P}(t) = -\omega^2 Q(t) .$$

Notice how these equations of motion are totally unaffected by the energy renormalization as that only added a multiple of the identity operator to the Hamiltonian which disappeared when taking the commutators. These equations have an immediate solution consistent with the required boundary conditions given by

$$Q(t) = \cos(\omega t)Q + \omega^{-1}\sin(\omega t)P ,$$
$$P(t) = -\omega\sin(\omega t)Q + \cos(\omega t)P .$$

Armed with the solution to the Heisenberg equations one may study a variety of expectation values. In contrast to the Schrödinger picture, the Heisenberg picture is ideally suited to the discussion of expectations involving operators with several different time values. Although one may ask similar questions for essentially any dynamical system, it is especially convenient to do so for simple systems like the harmonic oscillator. Therefore, let us focus on the harmonic oscillator, and denote by $|0\rangle$ the ground state of the oscillator for which (with the zero point energy subtracted) $\exp(-it\mathcal{H}/\hbar)|0\rangle = |0\rangle$ for any t. In that case it follows that

$$
\begin{aligned}
\langle 0|Q(t)Q(s)|0\rangle &= \langle 0|Qe^{-i(t-s)\mathcal{H}/\hbar}Q|0\rangle \\
&= \langle 0|[\cos(\omega(t-s))Q + \omega^{-1}\sin(\omega(t-s))P]Q|0\rangle \\
&= \cos(\omega(t-s))\langle 0|Q^2|0\rangle + \omega^{-1}\sin(\omega(t-s))\langle 0|PQ|0\rangle \\
&= \frac{\hbar}{2\omega}e^{-i\omega(t-s)} \ ,
\end{aligned}
$$

where in the last line we have used the two relations $\langle 0|PQ|0\rangle = -i\hbar/2$ and $\langle 0|Q^2|0\rangle = \hbar/(2\omega)$. For the first relation we use the fact that $PQ = (PQ - QP)/2 + (PQ + QP)/2$, while time-reversal invariance of $|0\rangle$ ensures us that

$$
\langle 0|(PQ + QP)|0\rangle = -\langle 0|(PQ + QP)|0\rangle \ ,
$$

hence equal to zero. The second relation follows from expansion to order p^2 of the expression for

$$
\langle 0|e^{ipQ/\hbar}|0\rangle = e^{-p^2/(4\hbar\omega)}
$$

derived in the previous section.

It is also possible to investigate higher-order correlations. A convenient way to do so is by means of a *generating function* such as

$$
\langle 0|\exp[(i/\hbar)\int s(t)Q(t)\,dt]|0\rangle = \exp[-\tfrac{1}{2\hbar^2}\int s(t)\langle 0|Q(t)Q(u)|0\rangle s(u)\,dt\,du] \ .
$$

The verification of this expression is quite simple when it is recognized that for the harmonic oscillator

$$
\int s(t)Q(t)\,dt = \int s(t)[\cos(t)Q + \omega^{-1}\sin(t)P]\,dt \equiv p(s)Q - q(s)P \ .
$$

This relation means that we can appeal to the evaluation of $E(p, q)$ derived earlier, and the essential fact we need to use is that, in the exponent, the answer is *quadratic* in the variables p and q, or in our case the answer is quadratic in s. The answer given for the generating function has been chosen so that the quadratic terms on each side agree. Only the symmetric part of the two-point correlation function survives due to the symmetry between the integration variables t and u, and as a consequence

$$
\langle 0|\exp[(i/\hbar)\int s(t)Q(t)\,dt]|0\rangle = \exp[-\tfrac{1}{4\hbar\omega}\int s(t)\cos(\omega(t-u))s(u)\,dt\,du] \ .
$$

It is convenient to reexpress this quantity in frequency space by means of a Fourier transformation. If we set

$$
\tilde{s}(v) = \frac{1}{\sqrt{2\pi}}\int e^{ivt}s(t)\,dt \ ,
$$

then it follows that

$$\langle 0| \exp[(i/\hbar)\int s(t)Q(t)\,dt]|0\rangle = \exp[-(\pi/2\hbar\omega)|\tilde{s}(\omega)|^2]$$
$$= \exp[-(\pi/\hbar)\int|\tilde{s}(v)|^2\delta(v^2-\omega^2)\,dv]\ .$$

In this last expression $0 < v < \infty$, and in presenting this form we have anticipated a similar expression that arises in the study of a free relativistic scalar field. Observe, that upon expansion of the left-hand side of these relations we find symmetrized multitime correlation functions, and most importantly, these are all fully determined by special combinations of the two-point correlation function. Essentially, this significant simplification holds only for the harmonic oscillator.

Another generating function of considerable interest is the *time-ordered generating function* given by

$$\langle 0|\mathsf{T}\exp[(i/\hbar)\int s(t)Q(t)\,dt]|0\rangle$$
$$= \exp[-(1/2\hbar^2)\int s(t)\langle 0|\mathsf{T}Q(t)Q(u)|0\rangle s(u)\,dt\,du]$$
$$= \exp[-(1/4\hbar\omega)\int s(t)e^{-i\omega|t-u|}s(u)\,dt\,du]\ ,$$

where the important *time-ordering operator* T is defined by

$$\mathsf{T}[Q(t_1)Q(t_2)\cdots Q(t_p)] \equiv Q(t_{\pi(1)})Q(t_{\pi(2)})\cdots Q(t_{\pi(p)})\ ,$$

where $\pi(\cdot)$ denotes a permutation of the times such that

$$t_{\pi(1)} \geq t_{\pi(2)} \geq \cdots \geq t_{\pi(p)}\ .$$

That is, the operators are ordered from left to right by decreasing times. For example,

$$\langle 0|\mathsf{T}Q(t)Q(u)|0\rangle = \langle 0|Q(t)Q(u)|0\rangle = \frac{\hbar}{2\omega}e^{-i\omega(t-u)}\ , \qquad t \geq u\ ,$$
$$= \langle 0|Q(u)Q(t)|0\rangle = \frac{\hbar}{2\omega}e^{-i\omega(u-t)}\ , \qquad u \geq t\ .$$

Stated otherwise,

$$\langle 0|\mathsf{T}Q(t)Q(u)|0\rangle \equiv \frac{\hbar}{2\omega}e^{-i\omega|t-u|}\ ,$$

which explains the final step in the evaluation of the generating function.

We observe that the functional form for the time-ordered generating function is established if we can show, like the previous case, that the answer must be quadratic in s in the exponent. Let us assume the support of s is confined between $t = 0$ and $t = T < \infty$, an interval we divide into N equal steps each of length $\epsilon = T/N$. Then, if one approximates the integral involved by a Riemann sum as an intermediate step, we are led to consider

$$\langle 0|\mathsf{T}e^{i\Sigma s_k Q_k\epsilon/\hbar}|0\rangle = \langle 0|e^{is_N Q_N\epsilon/\hbar}\cdots e^{is_2 Q_2\epsilon/\hbar}e^{is_1 Q_1\epsilon/\hbar}|0\rangle\ .$$

Note that it is the time ordering that has separated the various terms into individual, and time-ordered, exponents. In each factor

$$Q_k = \cos(\omega k \epsilon) Q + \omega^{-1} \sin(\omega k \epsilon) P \, ,$$

and we determine that the result depends on how one combines a number of similar factors each of the form $\exp[i(aQ - bP)/\hbar]$; these operators appear frequently and are known as *Weyl operators*.

To determine the combination of Weyl operators, let us first consider two operators A and B and the expression

$$B(t) \equiv e^{-tA} B e^{tA} \, .$$

for which we have

$$\dot{B}(t) = e^{-tA}[B, A]e^{tA} \, ,$$
$$\ddot{B}(t) = e^{-tA}[[B, A]A]e^{tA} \, ,$$

etc. Thus a Taylor series leads to

$$\begin{aligned} B(t) &= B(0) + t\dot{B}(0) + t^2\ddot{B}(0)/2! + \cdots \\ &= B + t[B, A] + t^2[[B, A]A]/2! + \cdots \, . \end{aligned}$$

Next specialize to the case that $C \equiv [A, B]$ commutes with A and B, i.e., $[C, A] = [C, B] = 0$, in which case $B(t) = B - tC$. With these relations as background, let us introduce

$$V(t) \equiv e^{-tA} e^{t(A+B)}$$

for which

$$\dot{V}(t) = e^{-tA} B e^{tA} V(t) = (B - tC) V(t) \, ,$$

and the solution of this differential equation, subject to the initial condition $V(0) = 1$, is given by $V(t) = \exp(tB - t^2C/2)$, from which we conclude the important relation (after setting $t = 1$) that

$$e^A e^B = e^{[A,B]/2} e^{A+B} \, ,$$

which holds whenever $[[A, B], A] = [[A, B], B] = 0$. This relation is one of the elementary Baker–Campbell–Hausdorff relations [Gi 74] that figure so importantly in the study of groups, etc.*

---

* Observe that a two-fold application of this relation leads to

$$e^A e^B = e^{[A,B]/2} e^{A+B} = e^{[A,B]} e^B e^A \, .$$

We are interested in applying these formulas to combine two Weyl operators. Given the fact that $[aQ - bP, cQ - dP] = -i\hbar(ad - bc)$, it follows that

$$\exp\left[i(aQ - bP)/\hbar\right]\exp[i(cQ - dP)/\hbar]$$
$$= \exp[i(ad - bc)/2\hbar]\exp[i((a + c)Q - (b + d)P)/\hbar] .$$

This relation further shows that many such factors can be combined with the only modification being a factor that is quadratic in the exponent. This result leads to the conclusion that the time-ordered generating function is indeed quadratic in the exponent – first for the Riemann approximation, and then also in the limit as $\epsilon \to 0$ – and justifies our earlier assumption regarding the form of the time-ordered generating function.

Incidentally, the rule for multiplying two Weyl operators given above serves as a useful substitute for the Heisenberg commutation relation, because in the Weyl form one deals with unitary hence bounded operators, while in the Heisenberg form one should properly take into account the domains of the operators involved. Most careful studies tend to work with the Weyl form of the commutation relations, and we shall see them again in our study of model field theories.

4.5 The Feynman formulation

The Schrödinger and Heisenberg formulations initially deal with differential formulations of the basic dynamical equations of quantum theory. In contrast, the Feynman approach focuses directly on the solution of the differential equation, especially the solution of the Schrödinger equation. We recall the Schrödinger equation for a particle in the presence of an external potential given by

$$i\hbar\frac{\partial}{\partial t}\psi(x, t) = -\frac{\hbar^2}{2m}\frac{\partial^2}{\partial x^2}\psi(x, t) + V(x)\psi(x, t) ,$$

where we have introduced the mass m of the particle as well. As a first-order differential equation in time, the solution is determined once the initial wave function, say at time t', is given. As a linear equation, the solution depends linearly on that initial wave function. In particular, there exists an integral kernel K, called the *propagator*, such that

$$\psi(x, t) = \int K(x, t; x', t')\psi(x', t') \, dx' ,$$

and furthermore this kernel K is itself a solution of Schrödinger's equation subject to the initial condition that

$$\lim_{t \to t'} K(x, t; x', t') = \delta(x - x') .$$

In dealing with the propagator, we shall always assume that the first time argument is later, or at least not earlier, in time than the second time argument; for $K(x, t; y, s)$ this means that $t \geq s$.

Feynman addressed himself to the propagator K, and he proposed a *path-integral expression* to determine it, formally given by

$$K(x'', t''; x', t') = \mathcal{N} \int e^{\frac{i}{\hbar} \int_{t'}^{t''} [\frac{1}{2}m\dot{x}(t)^2 - V(x(t))] \, dt} \, \mathcal{D}x \, .$$

This expression is formal because several quantities that appear therein need definition; this we shall certainly do. But first, let us focus on the implicit structure of the overall formula itself. In the exponent is a factor that is nothing but the classical action for the system under investigation. Planck's constant $\hbar$ enters as a divisor of the classical action. The factor $\mathcal{N}$ denotes a formal normalization constant. The integral symbolized by $\int \cdots \mathcal{D}x$ is designed to uniformly integrate over all paths $x(t)$ such that $x(t') = x'$ and $x(t'') = x''$. Clearly, this is a transcendental concept that needs clarification. No matter how this definition is resolved, it is the spirit of this formulation that the propagator may be represented as a kind of sum over all paths of a phase factor involving the classical action. In this approach there are no operators, no differential equations, just an "integral" representation for the *solution* itself. It is small wonder that this formulation of quantum mechanics has intrigued several generations of physicists and mathematicians alike!

In order to give better meaning to this integral representation for the propagator, we appeal to the operator formulation of quantum mechanics and observe that the propagator may also be written, with $T \equiv t'' - t'$, as

$$\langle x'' | e^{-(iT/\hbar)[(1/2m)P^2 + V(Q)]} | x' \rangle \, .$$

We next observe that the relation

$$e^{i(A+B)} = \lim_{N \to \infty} [e^{iA/(N+1)} e^{iB/(N+1)}]^{(N+1)}$$

holds whenever both A and B are self adjoint and $A + B$ is essentially self adjoint on the domain $\mathfrak{D}(A) \cap \mathfrak{D}(B)$. In brief, an operator is essentially self adjoint if the unitary operator it generates is uniquely determined by the action of the generator on the indicated domain. This product formula is due principally to Trotter [Tr 58]. Following Nelson [Ne 64] we apply the Trotter product formula to obtain the intermediate expression

$$\langle x'' | e^{-\frac{i\epsilon}{\hbar} \frac{1}{2m} P^2} e^{-\frac{i\epsilon}{\hbar} V(Q)} \cdots e^{-\frac{i\epsilon}{\hbar} \frac{1}{2m} P^2} e^{-\frac{i\epsilon}{\hbar} V(Q)} | x' \rangle \, ,$$

where there are $N + 1$ pairs of factors between the bra and ket, and $\epsilon \equiv T/(N+1)$. Between each pair of factors insert a resolution of unity given by

$$\int |x\rangle \langle x| \, dx = \mathbf{1} \, ,$$

in which case we are led to the multiple-integral representation (although we use only one integration sign)

$$\int \prod_{l=0}^{N} \langle x_{l+1} | e^{-\frac{i\epsilon}{\hbar}\frac{1}{2m}P^2} e^{-\frac{i\epsilon}{\hbar}V(Q)} | x_l \rangle \prod_{l=1}^{N} dx_l \,,$$

where we have introduced $x_{N+1} \equiv x''$ and $x_0 \equiv x'$. Now each factor may be given the functional representation

$$\langle x_{l+1} | e^{-\frac{i\epsilon}{\hbar}\frac{1}{2m}P^2} e^{-\frac{i\epsilon}{\hbar}V(Q)} | x_l \rangle$$

$$= \langle x_{l+1} | e^{-\frac{i\epsilon}{\hbar}\frac{1}{2m}P^2} | x_l \rangle \, e^{-\frac{i\epsilon}{\hbar}V(x_l)}$$

$$= \int \langle x_{l+1} | p \rangle \langle p | x_l \rangle \, e^{-\frac{i\epsilon}{\hbar}\frac{1}{2m}p^2} e^{-\frac{i\epsilon}{\hbar}V(x_l)} \, dp$$

$$= \sqrt{\frac{m}{2\pi i \hbar \epsilon}} \, e^{\frac{im}{2\epsilon\hbar}(x_{l+1}-x_l)^2} e^{-\frac{i\epsilon}{\hbar}V(x_l)} \,.$$

When this relation is introduced we learn that the propagator is given by

$$\lim_{N\to\infty} N_\epsilon \int e^{\frac{i}{\hbar}\Sigma_{l=0}^{N}[\frac{1}{2\epsilon}m(x_{l+1}-x_l)^2 - \epsilon V(x_l)]} \prod_{l=1}^{N} dx_l \,,$$

where $N_\epsilon \equiv [m/(2\pi i\hbar\epsilon)]^{(N+1)/2}$. At this point one recognizes a Riemann sum approximation to the classical action appearing in the exponent. If one were allowed to interchange the order of integration and the limit on N – *which in point of fact does* **not** *lead to a well-defined mathematical entity* – then one would arrive at the symbolic expression

$$\int \lim_{N\to\infty} N_\epsilon \, e^{\frac{i}{\hbar}\Sigma_{l=0}^{N}[\frac{1}{2\epsilon}m(x_{l+1}-x_l)^2 - \epsilon V(x_l)]} \prod_{l=1}^{N} dx_l$$

$$= \mathcal{N} \int e^{\frac{i}{\hbar}\int_{t'}^{t''}[\frac{1}{2}m\dot{x}(t)^2 - V(x(t))]\, dt} \, \mathcal{D}x \,,$$

where $\mathcal{N} = \lim N_\epsilon$, $\mathcal{D}x = \lim \Pi \, dx_l$, and we have written for the exponent the form it assumes for *continuous and differentiable paths*, namely the classical action. The proper definition for the path integral representation of the propagator is the one in which the limit is taken as the final step after integration. This proper formulation is often called a *lattice formulation* inasmuch as one has replaced the temporal continuum by a discrete lattice of time values in dealing with the classical action. In so doing we have *regularized* the formal path integral and given it meaning.

Regularization is not in the least confined to path-integral expressions, but is implicitly involved even when one considers a conditionally convergent one-dimensional integral, for real $a \neq 0$, of the form

$$\int_{-\infty}^{\infty} e^{iax^2} \, dx \, ,$$

which is usually understood as the limit of integrals taken between finite upper and lower limits as those values tend to infinity. Another useful way to regularize such a one-dimensional integral, and which gives the same result, is by considering, for $\epsilon > 0$, the regularized expression

$$\int_{-\infty}^{\infty} e^{(ia-\epsilon)x^2} \, dx$$

followed by the limit $\epsilon \to 0$ outside the integral. Either of these methods is a suitable way to define each of the infinite-range integrals that remain in the regularized form of the path integral. Although we shall properly understand the formal path integral by means of its lattice regularization, we shall nevertheless continue to write such expressions in the shorthand of the formal notation.

The propagator contains all the information about the eigenfunctions and the eigenvalues. If, for a system with discrete energy levels, we denote by $\{|n\rangle\}$ the set of eigenvectors and by $\{E_n\}$ the set of eigenvalues for the Hamiltonian, then it follows that

$$\begin{aligned}
K(x'', t''; x', t') &= \sum_n \langle x''| \, e^{-(i/\hbar)(t''-t')\mathcal{H}} |n\rangle \langle n|x'\rangle \\
&= \sum_n \langle x''|n\rangle \, e^{-(i/\hbar)(t''-t')E_n} \, \langle n|x'\rangle \\
&= \sum_n \phi_n(x'') \, e^{-(i/\hbar)(t''-t')E_n} \, \phi_n(x')^* \, .
\end{aligned}$$

The path integral may be readily evaluated for a few cases, notably those Hamiltonians which are quadratic such as the harmonic oscillator. To help in the case of the harmonic oscillator, let us recall the evaluation of a general N-dimensional Gaussian integral given by

$$\int e^{-\frac{1}{2}\Sigma x_k A_{kl} x_l + \Sigma b_k x_k} \prod dx_k = \frac{(2\pi)^{N/2}}{\sqrt{\det(A)}} \, e^{\frac{1}{2}\Sigma b_k A_{kl}^{-1} b_l} \, ,$$

where the sums and product run from 1 to N. The matrix A is symmetric and positive definite so that its inverse A^{-1} exists. The proof of this relation is straightforward if one first makes an orthogonal transformation to new variables, i.e., $x_k = \Sigma O_{kl} y_l$, where $O = \{O_{kl}\}$ denotes an orthogonal matrix with unit determinant, and chooses the transformation so as to diagonalize the matrix A, i.e., $O^T A O = D$, where the superscript T denotes transpose, and D is a diagonal matrix with diagonal elements $d_1, \ldots, d_N$. With $c_k \equiv \Sigma O_{kl} b_l$ the integral in question becomes

$$\int e^{-\frac{1}{2}\Sigma d_k y_k^2 + \Sigma c_k y_k} \prod dy_k$$

$$= \prod \int e^{-\frac{1}{2}d_k y^2 + c_k y} \, dy$$

$$= \frac{(2\pi)^{N/2}}{\sqrt{\prod d_k}} e^{\frac{1}{2}\Sigma d_k^{-1} c_k^2}$$

$$= \frac{(2\pi)^{N/2}}{\sqrt{\det(A)}} e^{\frac{1}{2}\Sigma b_k A_{kl}^{-1} b_l} .$$

This result has two parts, a prefactor and an exponent. It is useful to observe that the exponent is equivalent to the extremal of the original exponent. In particular, let us focus on the term in the exponent of the integrand given by

$$-\tfrac{1}{2}\Sigma x_k A_{kl} x_l + \Sigma b_k x_k .$$

Observe that the extremal value of this expression is located at the point $\Sigma A_{kl} x_l = b_k$. And at that point the exponent reads

$$\tfrac{1}{2}\Sigma b_k A_{kl}^{-1} b_l ,$$

which is just the exponent found upon integration.

Let us apply the formula for multiple Gaussian integrations to the evaluation of the path integral for the harmonic oscillator. The result of the integration will be a prefactor and an exponent. The exponent is properly determined from a lattice path integral, but the continuum limit will be the same if we had already sought the extremal in the continuum to begin with. Thus, we may evaluate the exponent by evaluating the classical action for the extremal path. The extremal path in this case satisfies the classical equation of motion (with $m = 1$), i.e.,

$$\ddot{x}(t) = -\omega^2 x(t),$$

subject to the appropriate boundary conditions. The desired solution, valid for generic times, is given by

$$x(t) = \frac{\sin(\omega(t - t'))x'' + \sin(\omega(t'' - t))x'}{\sin(\omega(t'' - t'))} ,$$

and evaluation of the classical action is made easier by the observation that

$$\tfrac{1}{2}\int [\dot{x}^2(t) - \omega^2 x^2(t)] \, dt$$

$$= \tfrac{1}{2}\{\int [-\ddot{x}(t) - \omega^2 x(t)]x(t) \, dt + x(t)\dot{x}(t)|\}$$

$$= \tfrac{1}{2}\omega[\,(x''^2 + x'^2)\cot(\omega(t'' - t')) - 2x''x' \csc(\omega(t'' - t'))\,] .$$

For the prefactor to the propagator, we note that for quadratic problems the prefactor is always given by the following recipe: if we denote the exponent for the evaluated path integral by $iS(x'', x')/\hbar$, then the prefactor is given by

$$\sqrt{\frac{i}{2\pi\hbar}\frac{\partial^2 S(x'', x')}{\partial x''\partial x'}} .$$

For the harmonic oscillator, the prefactor becomes

$$\sqrt{\frac{\omega}{2\pi i\hbar \sin(\omega(t'' - t'))}} .$$

Finally, the propagator for the harmonic oscillator is given by

$$
\begin{aligned}
K(x'', t''; x', t') \\
= \langle x''|e^{-i(t'' - t')[\frac{1}{2\hbar}(P^2 + \omega^2 Q^2)]}|x'\rangle \\
= \sqrt{\frac{\omega}{2\pi i\hbar \sin(\omega(t'' - t'))}} \\
\times \exp\left(\frac{i\omega}{2\hbar}\left\{\frac{(x''^2 + x'^2)\cos(\omega(t'' - t')) - 2x''x'}{\sin(\omega(t'' - t'))}\right\}\right).
\end{aligned}
$$

The validity of the prefactor in this expression for the propagator may be verified after the fact by the composition rule, which requires that any propagator fulfill

$$K(x'', t''; x', t') = \int K(x'', t''; x, t)\, K(x, t; x', t')\, dx ,$$

where $t'' > t > t'$, which is a nonlinear restriction on the propagator that largely fixes the normalization factor. Observe that as $\omega \to 0$ the propagator for the harmonic oscillator becomes the propagator for a free particle of unit mass, specifically

$$K(x''t''; x', t') = \frac{1}{\sqrt{2\pi i\hbar(t'' - t')}}\, e^{(i/2\hbar)[(x'' - x')^2/(t'' - t')]} .$$

Time-ordered expectation values

The propagator is of fundamental utility in computing matrix elements involving several time-ordered operators. To this end let us first introduce the notation

$$|x, t\rangle \equiv e^{it\mathcal{H}/\hbar}|x\rangle .$$

In that case, for example, we observe that

$$\int K(x'', t''; x, t)A(x)K(x, t; x', t')\, dx = \langle x'', t''|A(Q(t))|x', t'\rangle ,$$

$$\int K(x'', t''; \bar{x}, \bar{t})A(\bar{x})K(\bar{x}, \bar{t}; x, t)B(x)K(x, t; x', t')\, d\bar{x}\, dx$$

$$= \langle x'', t''|A(Q(\bar{t}))B(Q(t))|x', t'\rangle ,$$

etc., based on the ordering $t'' \geq \bar{t} \geq t \geq t'$. More generally, one has the important relation (with $m = 1$) that

$$\langle x'', t''| T e^{(i/\hbar) \int s(t)Q(t)\,dt} |x', t'\rangle$$
$$= \mathcal{N} \int e^{(i/\hbar) \int [s(t)x(t) + \frac{1}{2}\dot{x}(t)^2 - V(x(t))]\,dt} \mathcal{D}x \ .$$

Let us assume that the system possesses a normalizable ground state. Without loss of generality we may assume that the potential V has been adjusted so that the ground state energy is zero. If $\phi_0(x)$ denotes the appropriate ground state, then

$$\langle 0| T e^{(i/\hbar) \int s(t)Q(t)\,dt} |0\rangle = \mathcal{N} \int \phi_0(x'')^*$$
$$\times e^{(i/\hbar) \int [s(t)x(t) + \frac{1}{2}\dot{x}(t)^2 - V(x(t))]\,dt} \phi_0(x') \, \mathcal{D}x \ ,$$

where now $\mathcal{D}x$ also includes an integration over the end variables, x'' and x'. This is a fundamentally important relation.

Let us refine slightly the last equation by extracting and making explicit a quadratic part of the potential so that

$$\langle 0| T e^{(i/\hbar) \int s(t)Q(t)\,dt} |0\rangle = \mathcal{N} \int \phi_0(x'')^*$$
$$\times e^{(i/\hbar) \int [s(t)x(t) + \frac{1}{2}\dot{x}(t)^2 - \frac{1}{2}m^2 x(t)^2 - W(x(t))]\,dt} \phi_0(x') \, \mathcal{D}x \ .$$

Here the coefficient of the quadratic term is called m^2 in anticipation of a similar term in the context of field theory, and we assume that $W(x) \geq C$ where C is adjusted so that the ground state energy is zero. Observe if $s(t) = 0$ that the right side equals unity because it represents

$$\langle 0| e^{-iT\mathcal{H}/\hbar} |0\rangle = 1$$

which holds because the ground state energy equals zero. The path integral on the right-hand side is an oscillatory integral and it can be given a convergence factor – much like the simple one-dimensional example treated above – by the simple expedient of giving the term m^2 a small negative imaginary part, i.e., by replacing m^2 by $m^2 - i\epsilon$, $\epsilon > 0$.* With $\epsilon > 0$ even the ground state energy is complex and C is adjusted to remove even that complex value. In such a case the path integral reads

$$\langle 0| T e^{(i/\hbar) \int s(t)Q(t)\,dt} |0\rangle = \lim_{\epsilon \to 0} \mathcal{N} \int \phi_0(x'')^*$$
$$\times e^{(i/\hbar) \int [s(t)x(t) + \frac{1}{2}\dot{x}(t)^2 - \frac{1}{2}(m^2 - i\epsilon)x(t)^2 - W(x(t))]\,dt} \phi_0(x') \, \mathcal{D}x \ .$$

---

* Note that this use of ϵ is entirely different from the earlier use when $\epsilon = T/(N+1)$.

The consequence of this modification is to introduce a small damping factor in the integrand rendering it absolutely integrable. Observe that the limit $\epsilon \to 0$ is reserved to the end of the computation. If we understand by the state $|0\rangle$ the ground state in the presence of this small imaginary factor, then it follows that all higher energy levels acquire a nonvanishing imaginary part which gives to all the higher-order terms a damping coefficient. In short, if one extended the time integration from $t = -\infty$ to $t = \infty$, then the contributions that were not proportional to the ground state would have damped themselves away, leaving only a term proportional to the ground state expectation value remaining. Stated otherwise, we use the fact that, with $\epsilon > 0$, all energy eigenvalues, except the subtracted ground state energy, have a nonzero negative imaginary part and therefore

$$\lim_{t \to \infty} e^{-i\mathcal{H}_\epsilon t} = |0\rangle\langle 0| \, ,$$

namely, a projection operator onto the ground state. Therefore, with the $\epsilon \to 0$ limit as well as the final infinite time limits both implicit and understood, it follows that

$$\langle 0|Te^{(i/\hbar) \int s(t)Q(t)\,dt}|0\rangle$$
$$= \mathcal{N}' \int e^{(i/\hbar) \int [s(t)x(t) + \frac{1}{2}\dot{x}(t)^2 - \frac{1}{2}(m^2 - i\epsilon)x(t)^2 - W(x(t))]\,dt} \, \mathcal{D}x \, ,$$

without the need to incorporate or even know the ground state wave function at the initial or final times. Indeed, the boundary conditions generally do not matter, and it is traditional to choose $\lim_{|t| \to \infty} x(t) = 0$. Here we have introduced a different normalization factor, $\mathcal{N}'$, defined by the fact that the right-hand side must equal unity when $s \equiv 0$. This formulation is clearly of great significance for it does not require prior knowledge of any properties of the ground state (other than it is an isolated eigenstate) in order to evaluate its matrix elements!

Imaginary-time formulation

In the previous paragraph we have introduced a regularizing factor within the integral by giving the parameter m^2 a small imaginary part. There is another way to regularize the path integral, and that is by *analytic continuation in the time variable*. This approach is of value when the Hamiltonian operator is bounded below, as we have tacitly assumed when we asserted that the ground state energy was zero. The analytic continuation we have in mind is such as to take $t \to -it$, and hence, $T \to -iT$, so that

$$e^{-iT\mathcal{H}/\hbar} = \sum_n |n\rangle\langle n|e^{-iTE_n/\hbar}$$

becomes

$$e^{-T\mathcal{H}/\hbar} = \sum_n |n\rangle\langle n|e^{-TE_n/\hbar} \, ,$$

which is well defined since $E_n \geq 0$ by assumption. The resultant expression is closely related to the Boltzmann distribution for a system in thermal equilibrium at a temperature Θ proportional to $\hbar/T$. Indeed it is common to express such an equation as

$$e^{-\beta \mathcal{H}}$$

in terms of the so-called inverse temperature $\beta = 1/k_B\Theta$, where k_B is Boltzmann's constant. The relevant path integral expression now reads

$$\mathcal{N} \int e^{(1/\hbar) \int [s(t)x(t) - \frac{1}{2}\dot{x}(t)^2 - \frac{1}{2}m^2 x(t)^2 - W(x(t))] \, dt} \, \mathcal{D}x \;.$$

Again normalization is fixed so that this expression equals unity when $s \equiv 0$. To avoid multiple primes, we have reverted to the normalization symbol $\mathcal{N}$ even though it does not have the same meaning as before; indeed, we shall often use the normalization symbol $\mathcal{N}$ for different purposes. Once again the time integration in the exponent is intended to run from $-\infty$ to ∞.

The quantity calculated by this so-called Euclidean, or imaginary time path integral is clearly a generating functional for a stochastic process, that is, for an ensemble of *random functions*. Thus we can set this integral formally equal to the relevant ensemble average,

$$\left\langle e^{\int s(t)x(t) \, dt} \right\rangle \equiv \mathcal{N} \int e^{\int [s(t)x(t) - \frac{1}{2}\dot{x}(t)^2 - \frac{1}{2}m^2 x(t)^2 - W(x(t))] \, dt} \, \mathcal{D}x \;.$$

For convenience, we have also set $\hbar = 1$ in this formula. This choice of units makes the expression seem more like what it really appears to be, namely the integral representation of the partition function for a certain classical statistical problem. This analogy is a powerful one and it permits, for example, the calculation of quantum expressions by techniques that have proven successful in the calculation of classical statistical expressions. Indeed, the formal expression we have obtained may be given a well defined mathematical meaning in terms of an Ornstein–Uhlenbeck (O-U) measure [GlJ 87], μ_{OU}, concentrated on continuous but nowhere differentiable paths. If for the moment we set $W = 0$, then the O-U measure μ_{OU} is defined by the fact that

$$\left\langle e^{\int s(t)x(t) \, dt} \right\rangle_{\mathrm{OU}} \equiv \int e^{\int s(t)x(t) \, dt} \, d\mu_{\mathrm{OU}}(x)$$

$$= e^{\frac{1}{2} \int s(t) \, C(t-u) \, s(u) \, dt \, du} \;,$$

where

$$C(t) \equiv \frac{1}{2m} e^{-m|t|} = \frac{1}{2\pi} \int \frac{e^{-i\omega t}}{(\omega^2 + m^2)} \, d\omega \;.$$

With W present, the formula of interest becomes

$$\left\langle e^{\int s(t)x(t) \, dt} \right\rangle = \mathcal{N} \int e^{\int [s(t)x(t) - W(x(t))] \, dt} \, d\mu_{\mathrm{OU}}(x) \;,$$

or more precisely

$$\langle e^{\int s(t)x(t)\,dt}\rangle = \lim_{T\to\infty} N_T \int e^{\int_{-T}^{T}[s(t)x(t)-W(x(t))]\,dt}\,d\mu_{\text{OU}}(x),$$

due to the fact that in the former case the integrand is ill-defined because $\int W(x(t))\,dt = \infty$ when integrated over an infinite range for almost every O-U path. However, we may indeed introduce yet *another* measure μ which incorporates the interaction term W, so that these equations may also be written as

$$\langle e^{\int s(t)x(t)\,dt}\rangle \equiv \int e^{\int s(t)x(t)\,dt}\,d\mu(x).$$

As is well known, and readily established [Hi 70], the "sets" of path space that carry the support of the measure μ are distinct for different W functions and indeed even for different m values for the O-U measure alone. This property of disjoint support for the measures is due in the present case to the infinite range of integration in (imaginary) time.

Imaginary-time lattice formulation

Although the imaginary-time path integral admits a well-defined interpretation in terms of a path-space measure, it is pedagogically convenient to give an alternative formulation in terms of a lattice much as was the case for the real-time path integral. To obtain such a lattice formulation we need only make a natural Riemann sum approximation to the integral in the exponent – or, equivalently, let the temporal $i\epsilon \to a$ in the real-time lattice path integral – to arrive at the formula

$$\langle e^{\int s(t)x(t)\,dt}\rangle = \lim_{a\to0} N_a \int \exp\{\Sigma[s_k x_k a - \tfrac{1}{2}(x_{k+1} - x_k)^2 a^{-1}$$
$$- \tfrac{1}{2}m^2 x_k^2 a - W(x_k)a]\}\,\Pi\,dx_k.$$

Here we have specialized to a finite time interval $-T \le t \le T$, divided this interval into a number L of equal steps of length $a \equiv 2T/L$, introduced field values at the lattice times, i.e., $s_k = s(ka)$ and $x_k = x(ka)$, and written a Riemann approximation for the exponent. The resultant lattice path integral involves only a finite number of integration variables and is well defined. The limit to be taken after integration is called the *continuum limit*, namely, the limit as the lattice spacing a goes to zero. Of course, as $a \to 0$ it is necessary that $L \to \infty$ so that the resultant time interval remains positive. This can be done in such a way as to keep fixed the upper and lower limit, T and $-T$, respectively, or it can be done so that as part of the same limit $T \to \infty$ occurs as well. The limit $T \to \infty$ may be called the infinite time-interval limit, or the infinite volume limit as it is called when dealing with a field theory. We may assume that the indicated limit includes both the continuum and the infinite volume limit either taken sequentially or together as appropriate. We shall have occasion in our field theory studies to appeal to lattice formulations for imaginary-time path integrals as well.

A lattice representation of the imaginary-time path integral is also well suited to numerical studies on a computer. Of course, on a computer one is limited to a finite number of integrations and so the indicated limit cannot actually be taken. However, for large enough T and small enough a the integral should yield an approximate value for the limiting result. Consequently, one is generally left with an integral with a huge number of integrations to perform, and statistical methods, notably Monte Carlo and Langevin equation techniques, are available to estimate the value of such integrals [Ca 88]. In brief, such approximate statistical methods work by homing in on the region where the integrand is large, estimating the contribution from that region, and ignoring as much as possible other regions where the integrand is small.

4.6 Multiple degrees of freedom

In this section we wish to summarize the basic formulations presented in this chapter and at the same time extend them to finitely many degrees of freedom. Let $P = \{P_j\}$ and $Q = \{Q^k\}$ denote a finite collection of irreducible, self-adjoint, canonical operators that satisfy, for all $1 \leq j, k \leq N < \infty$, the canonical commutation relations

$$[Q^j, Q^k] = [P_j, P_k] = 0 \,,$$
$$[Q^j, P_k] = i\hbar \delta^j_k \mathbb{1} = i\hbar \delta^j_k \,.$$

The Hamiltonian operator

$$\mathcal{H} = H(P, Q) = H(P_1, \ldots, P_N, Q^1, \ldots, Q^N)$$

is a self-adjoint operator, which then defines the abstract Schrödinger equation

$$i\hbar \frac{\partial |\psi(t)\rangle}{\partial t} = \mathcal{H}|\psi(t)\rangle \,,$$

with an abstract solution given by

$$|\psi(t)\rangle = e^{-i(t-t')\mathcal{H}/\hbar}|\psi(t')\rangle \,,$$

which determines a continuous path of unit vectors in Hilbert space. The Schrödinger representation for the canonical operators is given by the realization

$$P_k = -i\hbar \frac{\partial}{\partial x^k} \,, \qquad Q^k = x^k \,,$$

acting on the space of square-integrable functions $L^2(\mathbb{R}^N)$, in which case Schrödinger's equation becomes

$$i\hbar \frac{\partial \psi(x, t)}{\partial t} = H(-i\hbar \frac{\partial}{\partial x}, x)\, \psi(x, t) \,.$$

Attention is focused on normalizable solutions to this differential equation. As a homogeneous linear equation the solution to Schrödinger's equation may be given in the form

$$\psi(x'', t'') = \int K(x'', t''; x', t')\, \psi(x', t')\, d^N x'\,,$$

where K denotes the propagator. By repeated folding together of the propagator for short times, an integral representation for the propagator itself may be given by the equation

$$K(x'', t''; x', t') = \lim_{\epsilon \to 0} N_\epsilon \int e^{(i/\hbar)\Sigma[\frac{1}{2}(x_{l+1}-x_l)^2\epsilon^{-1} - V(x_l)\epsilon]} \Pi_l\, d^N x_l$$

$$= \mathcal{N} \int e^{(i/\hbar)\int[\frac{1}{2}\dot{x}(t)^2 - V(x(t))]\, dt} \Pi_t\, d^N x(t)\,,$$

where the latter, formal expression is a convenient shorthand for the former, well-defined expression. Here $\dot{x}(t)^2 \equiv \Sigma\, \dot{x}^k(t)^2$. Additionally, the generating functional for the ground state expectation of time-ordered operators may be given for smooth functions s by

$$\langle 0|Te^{(i/\hbar)\int s(t)\cdot Q(t)\, dt}|0\rangle$$

$$= \mathcal{N} \int e^{(i/\hbar)\int[s(t)\cdot x(t) + \frac{1}{2}\dot{x}(t)^2 - \frac{1}{2}(m^2 - i\epsilon)x(t)^2 - W(x(t))]\, dt} \Pi_t\, d^N x(t)\,,$$

with the understanding that $\epsilon \to 0$ is reserved until the final step. In this last expression $\mathcal{N}$ is chosen so that the right-hand side equals unity when $s = 0$. The limits of the time integration are assumed to approach infinity in both directions, which has the effect of projecting out any other state but the ground state. An analytic continuation of the previous formula from t to $-it$ leads to the Euclidean, or imaginary-time, path-integral representation given, for $\hbar = 1$, by

$$\langle e^{\int s(t)\cdot x(t)\, dt}\rangle = \mathcal{N} \int \exp\{\int[s(t)\cdot x(t) - \tfrac{1}{2}\dot{x}(t)^2$$

$$-\tfrac{1}{2}m^2 x(t)^2 - W(x(t))]\, dt\}\Pi_t\, d^N x(t)\,.$$

This formal expression may be understood as the continuum limit of the lattice-regularized path integral given by

$$\langle e^{\int s(t)\cdot x(t)\, dt}\rangle = \lim_{a \to 0} N_a \int \exp\{\Sigma_l[s_l \cdot x_l a - \tfrac{1}{2}(x_{l+1} - x_l)^2 a^{-1}$$

$$-\tfrac{1}{2}m^2 x_l^2 a - W(x_l)a]\}\Pi_l\, d^N x_l\,,$$

and this continuum limit implicitly includes a limit as the lattice spacing a goes to zero as well as a limit that the time interval of integration goes to infinity.

If, for example, $W(x(t)) = g\, x^4(t) \equiv g\Sigma_{k=1}^N x^k(t)^4$, where $g \geq 0$ is a coupling constant, then one may introduce a *perturbation expansion*, say of the two-point, imaginary-time correlation function, in the form

$$\langle x^k(t_2)x^l(t_1)\rangle = \sum_{p=0}^{\infty} \frac{(-g)^p}{(p!)} \int x^k(t_2)x^l(t_1)[\int x^4(u)\,du]^p$$
$$\times e^{-\frac{1}{2}\int [\dot{x}(t)^2 + m^2 x(t)^2]\,dt}\Pi_t\,d^N x(t)\,.$$

In turn, this expression may be evaluated order by order in the coupling constant g by means of the information contained in the correlation functions for the O-U distribution as determined by

$$\langle e^{\int s(t)\cdot x(t)\,dt}\rangle = e^{\frac{1}{2}\int s(t)\cdot s(u)\,C(t-u)\,dt\,du}\,,$$

where, as before,

$$C(t-u) = \frac{1}{2m}e^{-m|t-u|} = \frac{1}{2\pi}\int \frac{e^{i\omega(t-u)}}{(\omega^2 + m^2)}\,d\omega\,.$$

Equating powers of s on both sides in a power-series expansion leads to formulas, for $N = 1$, such as

$$\langle x(t)x(u)\rangle = C(t-u)\,,$$
$$\langle x(r)x(s)x(t)x(u)\rangle = C(r-s)C(t-u) + C(r-t)C(s-u)$$
$$+ C(r-u)C(s-t)\,,$$

and more generally,

$$\langle \Pi_{l=1}^{2q} x(t_l)\rangle = 2^{-q}\Sigma_{\pi,\sigma}\Pi_{l=1}^{q}C(t_{\pi(l)} - t_{\sigma(l)})\,,$$

where π is a permutation that chooses any q elements out of $2q$, and thus has $(2q)!/(q!)^2$ distinct choices, and σ is an arbitrary perturbation on q items with $q!$ elements. The prefactor of 2^{-q} eliminates the double counting for each factor due to the symmetry $C(t-u) = C(u-t)$. For generic times, the total number of distinct terms on the right-hand side of the last relation is $(2q)!/(2^q q!) = 1\cdot 3\cdot 5\cdots (2q-1)$.

The generalization of these latter formulas to space-time dimension $n > 1$ is rather straightforward. As we shall observe in the next chapter, it is often convenient to associate with each term in such a perturbation expansion a graph – a so-called Feynman diagram – which graphically depicts just exactly which term is under consideration.

Heisenberg equations of motion

The Heisenberg equations of motion generalize to several degrees of freedom in a straightforward fashion. In particular, for Hamiltonians of the form

$$\mathcal{H} = \tfrac{1}{2}P^2 + V(Q)\,,$$

the operator equations of motion read

$$\dot{Q}^j = \frac{i}{\hbar}[\mathcal{H}, Q^j] = P_j \,,$$

$$\dot{P}_j = \frac{i}{\hbar}[\mathcal{H}, P_j] = -\frac{\partial V(Q)}{\partial Q^j} = -V_j(Q) \,,$$

where in the last line we have introduced the shorthand

$$\frac{\partial V(Q)}{\partial Q^j} \equiv \left.\frac{\partial V(x)}{\partial x^j}\right|_{\{x^l = Q^l\}} \equiv \left. V_j(x)\right|_{\{x^l = Q^l\}} \,.$$

The solution to the Heisenberg equations of motion, subject to the initial conditions that

$$Q^l(0) \equiv Q^l \,, \qquad P_l(0) \equiv P_l \,,$$

which must fulfill the basic nonvanishing commutation relations

$$[Q^k, P_l] = i\hbar \delta_l^k \,,$$

determines the temporal behavior of the operators of interest. Multitime expectations are defined for several degrees of freedom in essentially the same way as was the case for a single degree of freedom.

Exercises

4.1 Let P and Q be self-adjoint Heisenberg operators which satisfy $[Q, P] = i\hbar$. Discuss the eigenproperties of each of the candidate Hamiltonian operators (compare with Exercise 2.1):
  (a) $\mathcal{H} = P^2 + Q^4$,
  (b) $\mathcal{H} = P^2 - Q^2$,
  (c) $\mathcal{H} = P^2 - Q^4$.

4.2 Let $X = (PQ^3 + Q^3P)/2$, where $Q^\dagger = Q$ and $P^\dagger = P$, denote a symmetric (Hermitian) operator on a natural domain. By using the Schrödinger representation, show that $X^\dagger w_\pm = \pm i w_\pm$ admits a square integrable solution for one sign in this pair of equations, but not for both signs. What is the significance of this fact regarding whether or not X admits self-adjoint extensions (compare with Exercise 2.1)?

4.3 Let $W(t)$ be a complex-valued function that satisfies $W(0) = 1$ and $W(s)W(t) = W(s + t)$ for all $s, t \in \mathbb{R}$.
  (a) Assume that $W(t)$ is *continuous*. Show that there exists a complex parameter A such that $W(t) = \exp(At)$ for all t.
  (b) Dropping the assumption of continuity, show that more general solutions are possible, such as, for example, $W(t) = \exp(at)$ for t rational, $W(t) = \exp(bt)$

for t a rational multiple of $\sqrt{2}$, $W(t) = \exp(ct)$ for t a rational multiple of $\sqrt{3}$, and (say) $W(t) = \exp(ft)$ otherwise, where a, b, c and f are complex parameters.

4.4 Consider the Hamiltonian operator in the Schrödinger representation given in natural units by

$$\mathcal{H} = -\frac{1}{2}\frac{\partial^2}{\partial x^2} + \frac{\gamma(\gamma+1)}{2x^2} + \frac{1}{2}x^2 + \kappa x^4 \,, \qquad \kappa \geq 0 \,.$$

Determine the qualitative properties of the eigenfunctions and eigenvalues, especially the functional form of the eigenfunctions for very small and for very large $|x|$, for the eigenvalue equation $\mathcal{H}u_m(x) = E_m u_m(x)$ in the two cases:

(a) $0 \leq \gamma < 1/2$,

(b) $1/2 \leq \gamma$.

As a starting point, consider the case $\kappa = 0$.

4.5 Except for the last line, repeat the preceding problem in the case that

$$\mathcal{H} = -\frac{1}{2}\frac{\partial^2}{\partial x^2} + \frac{\gamma(\gamma+1)}{2x^2} + \frac{1}{2}x^2 + \sigma x^3 + \kappa x^4 \,, \qquad \sigma \neq 0 \,, \ \kappa > \sigma^2/2 \,.$$

5

Scalar Quantum Field Theory

In its simplest characterization, the quantum theory of scalar fields is nothing but the quantum mechanics of N canonical degrees of freedom in the limit that $N \to \infty$. Of course, things are not quite that simple since sequences do not always have the virtue of converging, and even if they do converge they do not always converge to acceptable limits. However, in this chapter we take a simple, pragmatic point of view and assume that any needed limits converge and, in addition, that the resultant answers are acceptable. Partway through this chapter we introduce units in which $\hbar = 1$.

5.1 Classical scalar fields

For purposes of the present section we let $g(x)$ denote a real scalar field defined for $x \in \mathbb{R}^n$, where $n \geq 1$ denotes the dimension of space-time. If $n = 1$ then that single dimension is the time dimension and so there is no space dimension – hence no space – and thus one is really back in the case of classical mechanics. We shall have essentially no occasion to study the case $n = 1$ in this chapter. For $n \geq 2$ the number of space dimensions is $s \equiv n - 1 \geq 1$. (We note that the symbol n has several uses in this chapter, often as a dummy index of summation, but the context is generally clear in each case.) When we need to distinguish the space and time components, we shall set $x = (t, \mathbf{x})$, where, as clear from the notation, $t = x^0$ refers to the time and $\mathbf{x} = (x^1, x^2, \ldots, x^s)$ refers to the spatial components.* We denote the components of the space-time vector by $x = \{x^\mu\} = (x^0, x^1, \ldots, x^s)$, and the spatial vector alone by $\mathbf{x} = \{x^m\} = (x^1, \ldots, x^s)$; i.e.,

---

* When relevant, we shall always use units in which the speed of light $c = 1$.

Greek super/subscripts are used for space-time, while Roman super/subscripts are used for space alone.

Under a coordinate transformation that takes $x \to \bar{x} \equiv \bar{x}(x)$, a scalar field is one which transforms so that $\bar{g}(\bar{x}) = g(x)$. This equation means that the value of g at a point is the same whether it is expressed in the original or in the transformed coordinate system, and hence that value is intrinsic to the point in question. We assume that g is a scalar field under a general set of coordinate transformations, that for the moment remains unspecified as it will in general change from problem to problem. For example, these coordinate transformations may simply be spatial rotations and translations in space, or a subset of spatial rotations and translations such as would leave a lattice invariant, or they may be the Poincaré transformations, i.e., the Lorentz transformations plus space-time translations, of special relativity. Each of these are *linear* transformations so that

$$\bar{x}^\lambda = \sum_{\mu=0}^{s} A^\lambda{}_\mu x^\mu + b^\lambda \equiv A^\lambda{}_\mu x^\mu + b^\lambda \ ,$$

for some vector b and some nonsingular, $n \times n$-matrix $A = \left\{ A^\lambda{}_\mu \right\}$, the set of which form the *transformation group* under study. We have also introduced the *summation convention* whereby repeated indices are assumed summed (unless explicitly stated otherwise). We shall content ourselves with such linear transformations and not consider more general coordinate transformations. The derivatives of the field, denoted sometimes for brevity by

$$\partial_\mu g(x) \equiv \frac{\partial g(x)}{\partial x^\mu} \ ,$$

transform as components of a vector, i.e., as

$$\frac{\partial g(x)}{\partial x^\lambda} = \frac{\partial \bar{x}^\mu}{\partial x^\lambda} \frac{\partial \bar{g}(\bar{x})}{\partial \bar{x}^\mu} = A^\mu{}_\lambda \frac{\partial \bar{g}(\bar{x})}{\partial \bar{x}^\mu} \ ,$$

$$\partial_\lambda g(x) = \frac{\partial \bar{x}^\mu}{\partial x^\lambda} \partial_\mu \bar{g}(\bar{x}) = A^\mu{}_\lambda \partial_\mu \bar{g}(\bar{x}) \ .$$

Each transformation A is chosen so as to preserve the component structure of a metric. For relativistic covariance, the metric components are $\eta_{\mu\lambda}$, which we take to be diagonal with values

$$\eta_{00} = -1 \ , \qquad \eta_{11} = \cdots = \eta_{ss} = 1 \ .$$

In this case, the matrix elements of A are chosen so that

$$\eta_{\alpha\beta} = A^\mu{}_\alpha A^\lambda{}_\beta \eta_{\mu\lambda} \ .$$

The set of transformations A that preserve the space-time metric are the Lorentz transformations which include rotations in space, boosts (or relativity transformations) in a spatial direction, plus reflections and inversions in space and/or time. For a nonrelativistic theory for which transformations among spatial components are the only ones involved it often suffices to assume the metric to be

$$\eta_{ab} = \delta_{ab} \ ,$$

which is just the unit $s \times s$-matrix. The set of transformations in this case are spatial rotations plus reflections and inversions. The theories that may be so described are defined in either a Minkowski space-time or a Euclidean space, respectively.

The classical field equations may be derived from an action principle which in turn is based on the space-time integral of a Lagrangian density. A suitable Lagrangian density for a scalar field is generally given by $\mathcal{L} = \mathcal{L}(\partial_\mu g(x), g(x))$, which, although not indicated here, may also be an explicit function of x. The action functional of interest is

$$I = \int \mathcal{L}(\partial_\mu g(x), g(x)) \, d^n x \ ,$$

and stationarity of this functional under variations that vanish at the boundaries leads to the Euler–Lagrange equations of motion given by

$$\frac{\partial}{\partial x^\mu} \frac{\partial \mathcal{L}}{\partial \, \partial_\mu g(x)} - \frac{\partial \mathcal{L}}{\partial g(x)} = 0 \ .$$

As an example let us focus on the Lagrangian density for the quartic self-coupled relativistic scalar field given by

$$\mathcal{L}(x) = -\tfrac{1}{2}\eta^{\mu\tau}[\partial_\mu g(x)][\partial_\tau g(x)] - \tfrac{1}{2}m^2 g(x)^2 - \lambda g(x)^4 \ ,$$

which leads to the equation of motion

$$\Box g(x) - m^2 g(x) = 4\lambda \, g(x)^3 \ ,$$
$$\Box \equiv \eta^{\mu\tau} \partial_\mu \partial_\tau \ .$$

It is traditional to solve this equation subject to suitable boundary conditions, say at $t = 0$, such as

$$g(\mathbf{x}) \equiv g(0, \mathbf{x}) \ , \qquad f(\mathbf{x}) \equiv \dot{g}(0, \mathbf{x}) \equiv \partial_0 \, g(0, \mathbf{x}) \ .$$

In the equation of motion, λ denotes the coupling constant of the quartic nonlinear coupling. Just as is the case for the quartic anharmonic oscillator for a single degree of freedom, the solution to this field equation depends strongly on the sign of λ. If $\lambda > 0$, then given that the initial data, f and g, are sufficiently well behaved, the solution exists globally and is nontrivial for all $n \geq 2$ [Re 76]. On the other hand, if $\lambda < 0$, then the solution will diverge in a finite time whenever the initial data are not identically zero.

Hamiltonian formulation

As was the case for particle mechanics, we can trade a Lagrangian formulation for that of Hamilton. To that end let us first introduce the canonical momentum field f according to the definition

$$f(x) \equiv \frac{\partial \mathcal{L}}{\partial \dot{g}(x)} \ .$$

We assume that we deal with Lagrangians with the property that we can solve this equation so as to express $\dot{g}(x) = \dot{g}(f(x), g(x))$; indeed for a great many examples of interest – including the quartic self-coupled relativistic scalar field discussed above – it follows that $f(x) = \dot{g}(x)$. Next we introduce the Hamiltonian density through a *Legendre transformation* given by*

$$\mathcal{H}_x(f(x), g(x)) = f(x)\dot{g}(x) - \mathcal{L}(\dot{g}(x), g(x)), \qquad \dot{g}(x) = \dot{g}(f(x), g(x)) \ .$$

In this equation one is instructed to eliminate $\dot{g}$ in favor of f and g. In turn, the (full) Hamiltonian is defined by

$$H(f, g) = \int \mathcal{H}_x(f(x), g(x)) \, d^s x \ ,$$

and the action functional may be reexpressed in these new variables as

$$I = \int [f(x)\dot{g}(x) - \mathcal{H}_x(f(x), g(x))] \, d^n x \ .$$

The Hamiltonian equations of motion are obtained by variation of this expression, holding both f and g fixed at the initial and the final time surface as discussed in Chapter 2 in the case of classical mechanics. As in that simpler case we arrive at two sets of equations given by

$$\dot{g}(x) = \frac{\partial \mathcal{H}_x}{\partial f(x)} \ ,$$

$$\dot{f}(x) = -\frac{\partial \mathcal{H}_x}{\partial g(x)} \ .$$

The first equation effectively (re)defines the momentum f in terms of $\dot{g}$ and g, while the second equation yields the equation of motion. Although one may introduce the entire machinery of canonical transformations into classical field theory just as was the case in classical mechanics, it will not prove so important in our studies. Consequently, we will not pursue the subject of classical canonical transformations further. Rather, we shall take the classical Hamiltonian and the given classical canonical coordinates as the starting point for our transition from a classical to a quantum field theory.

---

* We use the cumbersome notation $\mathcal{H}_x$ for the Hamiltonian density so as not to confuse it with our traditional use of $\mathcal{H}$ for the quantum Hamiltonian operator.

Besides the Hamiltonian there are other important generators in the classical theory. For example, let us consider the classical space-translation generator $\mathbf{P}$ which may be characterized by the Poisson brackets

$$\{g(\mathbf{x}), \mathbf{P}\} = -\boldsymbol{\nabla} g(\mathbf{x}) \,, \qquad \{f(\mathbf{x}), \mathbf{P}\} = -\boldsymbol{\nabla} f(\mathbf{x}) \,.$$

In fact these relations completely define $\mathbf{P}$ – up to an additive constant vector – so that

$$\mathbf{P} = -\int f(\mathbf{x}) \, \boldsymbol{\nabla} g(\mathbf{x}) \, d^s x \,.$$

It is natural to choose the additive constant vector to vanish so that no direction in space is favored. In component form this equation reads

$$P_l = -\int f(\mathbf{x}) \, \partial_l \, g(\mathbf{x}) \, d^s x \,.$$

In a similar fashion we may introduce the classical rotation generators as the components of an antisymmetric tensor J_{lm} characterized by the Poisson brackets

$$\{g(\mathbf{x}), J_{lm}\} = \mathsf{J}_{lm} g(\mathbf{x}) \,, \qquad \{f(\mathbf{x}), J_{lm}\} = \mathsf{J}_{lm} f(\mathbf{x}) \,,$$

where

$$\mathsf{J}_{lm} \equiv \left(x_m \frac{\partial}{\partial x_l} - x_l \frac{\partial}{\partial x_m} \right) \,,$$

denotes the infinitesimal rotation operators acting on $\mathbb{R}^s$. Observe that the operators $\{\mathsf{J}_{ab}\}$ satisfy the commutation relations (not Poisson brackets) given by

$$[\mathsf{J}_{ab}, \mathsf{J}_{lm}] = \delta_{bm} \mathsf{J}_{al} - \delta_{bl} \mathsf{J}_{am} + \delta_{al} \mathsf{J}_{bm} - \delta_{am} \mathsf{J}_{bl},$$

characteristic of the Lie algebra of the rotation group SO(s) acting on $\mathbb{R}^s$. Lastly we observe that these relations imply that

$$J_{ab} = \int f(\mathbf{x}) \, \mathsf{J}_{ab} \, g(\mathbf{x}) \, d^s x \,,$$

up to an additive constant antisymmetric tensor. This constant tensor should be chosen to vanish or otherwise some spatial rotation becomes favored. In addition, with the choice of the constant tensor as given, it follows that the Poisson bracket algebra of the rotation generators $\{J_{ab}\}$ is the same as the commutator algebra of the infinitesimal rotation operators $\{\mathsf{J}_{ab}\}$. Lastly we observe that

$$\{J_{ab}, P_l\} = \delta_{al} P_b - \delta_{bl} P_a \,,$$

which just reflects the fact that $\mathbf{P}$ behaves as a vector under spatial rotation.

Coordinate version of scalar fields

Fields may be traded for an infinite set of coordinates by the very kinds of transformation that were introduced in Chapter 3 between function spaces and sequence spaces. In particular, we can expand the field and momentum as

$$g(x) \equiv g(t, \mathbf{x}) = \sum_{n=1}^{\infty} q_n(t)\, h_n(\mathbf{x}) \;,$$

$$f(x) \equiv f(t, \mathbf{x}) = \sum_{n=1}^{\infty} p_n(t)\, h_n(\mathbf{x}) \;,$$

where the set $\{h_n\}_{n=1}^{\infty}$ is composed of a complete set of real orthonormal functions for the space in question (e.g., $\mathbb{R}^s$). For the sake of discussion, we can imagine that at any fixed t the functions g and f are square integrable and so we may restrict attention to square summable sequences $\{q_n\}$ and $\{p_n\}$. When such an expansion is used, the action functional for a large class of interesting systems is transformed to

$$I = \int [\tfrac{1}{2}\Sigma \dot{q}_n^2(t) - V(q(t))]\, dt \;,$$

where $V(q) = V(q_1, q_2, \ldots)$. In turn, this expression may be given in phase-space form as

$$I = \int [\Sigma p_n(t)\dot{q}_n(t) - \tfrac{1}{2}\Sigma p_n^2(t) - V(q(t))]\, dt \;.$$

The equations of motion that follow from this action principle are simply

$$\dot{q}_n(t) = \frac{\partial H(p, q)}{\partial p_n(t)} = p_n(t) \;,$$

$$\dot{p}_n(t) = -\frac{\partial H(p, q)}{\partial q_n(t)} = -\frac{\partial V}{\partial q_n(t)} \;,$$

where

$$H(p, q) \equiv \tfrac{1}{2}\Sigma p_n^2 + V(q) \;.$$

It should be appreciated that the potential $V(q)$ may generally be a rather complicated function of the many coordinates q_n, $n \in \mathbb{N} \equiv \{1, 2, 3, \ldots\}$; however, for a model that is a polynomial in the field g, as is frequently the case, then V is also a polynomial of the same order.

It is natural to imagine a modified set of classical models in which *all coefficients q_n and p_n for $n > N$ are set equal to zero for all time*. The result then is a *sequence of models*, indexed by $N \in \mathbb{N}$, where N signifies the number of degrees of freedom for each particular model. In seems reasonable to suppose that this sequence of models tends to approximate the real system of interest the larger the value of N, and that in the limit $N \to \infty$ we should be able to recover all solutions of the original classical field theory with an infinite number of coordinates and momenta.

If approaching a system with an infinite number of degrees of freedom through a sequence of models each having finitely many degrees of freedom is a suitable way to proceed in the classical theory, then it stands to reason that it may also prove satisfactory in the quantum theory as well. In the present chapter we shall take the standard view that this sequential limiting procedure – augmented when necessary with renormalization of certain parameters – is an acceptable way to proceed. In later chapters we will show that such a natural and simple procedure is in fact *not* always acceptable, at least in any straightforward manner.

5.2 Quantum scalar fields: preliminary comments

We begin our discussion by adopting the coordinate point of view introduced just above. In quantizing, therefore, we promote each element of the real coordinate and momentum sets $\{q_n\}$ and $\{p_n\}$ to Hermitian operators with the properties that

$$[Q_n, Q_m] = 0 \,, \quad [P_n, P_m] = 0 \,,$$
$$[Q_n, P_m] = i\hbar\delta_{nm} \,,$$

where $n, m \in \mathbb{N}$. As we shall discuss further in the next chapter, we may still realize an infinite set of operators of the kind described in a separable Hilbert space, such as a sequence space. Having the kinematical operators involved, we can next discuss dynamics through the introduction of a Hamiltonian operator

$$\mathcal{H} = H(P, Q) = H(P_1, P_2, \ldots, Q_1, Q_2, \ldots) \,.$$

Although the classical Hamiltonian was a valid function of the classical arguments, it is not at all obvious that the quantum Hamiltonian as proposed above is a valid operator when an infinite number of variables are involved. A very simple example of this situation refers to the model

$$\mathcal{H} = \tfrac{1}{2} \sum_1^\infty (P_n^2 + Q_n^2)$$

because each oscillator contributes a ground state or zero point energy of $\hbar/2$, and although for finitely many degrees of freedom such a term is generally negligible, this is not the case when an infinite number of terms are involved. This problem is resolved by declaring that the correct Hamiltonian operator is instead given by

$$\mathcal{H} = \tfrac{1}{2} \sum_1^\infty (P_n^2 + Q_n^2 - \hbar) \equiv \tfrac{1}{2} \sum_1^\infty : (P_n^2 + Q_n^2) : \,,$$

where, just as before, $:\ :$ denotes normal ordering, which in the present case simply amounts to subtracting off the otherwise infinite zero point energy. Observe, as discussed in Chapter 4, that the quantization procedure is uncertain up to terms of order $\hbar$, and that an additional constant added to the Hamiltonian will not affect the operator equations of motion in any way and is therefore an acceptable change of the Hamiltonian. Hence, such a modification of the classical to quantum Hamiltonian, besides being necessary, is harmless and easily accepted. Observe that the modification involved is proportional to $\hbar$ and thus it cannot be determined from the classical theory alone. This modification of the naive quantum Hamiltonian is just the first – and simplest – of the renormalizations that we will encounter.

If the field form was suitable in the classical theory it stands to reason that it should also be reasonable in the quantum theory. To introduce the quantized kinematical fields we appeal once again to an expansion in a real – to preserve the formal Hermitian nature of the operators – orthonormal basis, specifically

$$\varphi(\mathbf{x}) \equiv \sum_{n=1}^{\infty} Q_n\, h_n(\mathbf{x})\,, \qquad \pi(\mathbf{x}) \equiv \sum_{n=1}^{\infty} P_n\, h_n(\mathbf{x})\,.$$

Although a useful formal expression, it follows that the sums in question cannot generally converge since, for fixed $\mathbf{x}$, neither the functions h_n nor the coefficients in the expansion – here the operators Q_n or P_n – fall to zero due, in the case of the latter, to the commutation relations and the Heisenberg uncertainty relations to which they give rise. This failure of the coefficients to fall to zero as $n \to \infty$ is in marked contrast to the classical situation and is the source of much of the difficulty encountered in quantum field theory.

The indicated expressions for φ and π will be called *local operators*, which is only intended to mean that when integrated over appropriate test functions they do in fact become *bona fide* operators. Let f and g denote two real, c-number test functions, which for the sake of illustration we may take as arbitrary polynomials multiplied by fixed Gaussian factors. Then we define

$$\varphi(f) \equiv \int \varphi(\mathbf{x})f(\mathbf{x})\, d^s\!x \equiv \sum_{n=1}^{\infty} Q_n \int h_n(\mathbf{x})f(\mathbf{x})\, d^s\!x\,,$$

$$\pi(g) \equiv \int \pi(\mathbf{x})g(\mathbf{x})\, d^s\!x \equiv \sum_{n=1}^{\infty} P_n \int h_n(\mathbf{x})g(\mathbf{x})\, d^s\!x\,.$$

Such expressions define operators because the coefficients in the sum, for example, $h_n(f) \equiv \int h_n(\mathbf{x})f(\mathbf{x})\, d^s\!x$, fall rapidly to zero with a proper choice of test function space.

The commutation relation between local field operators is given by

$$[\varphi(\mathbf{x}), \pi(\mathbf{y})] = \sum_{n,m=1}^{\infty} [Q_n, P_m]\, h_n(\mathbf{x})\, h_m(\mathbf{y})$$

$$= i\hbar \sum_{n=1}^{\infty} h_n(\mathbf{x})\, h_n(\mathbf{y})\mathbf{1} = i\hbar\, \delta(\mathbf{x} - \mathbf{y})\mathbf{1}\,,$$

which if $\mathbf{x} = \mathbf{y}$ involves an infinite multiple of the unit operator. This consequence is avoided if we replace this commutator for local fields by one for smeared fields, i.e., by the well-defined expression

$$[\varphi(f), \pi(g)] = i\hbar \int f(\mathbf{x})\, g(\mathbf{x})\, d^s x \,\mathbb{1} \equiv i\hbar\,(f, g)\,\mathbb{1}$$

for which the multiplier of the unit operator is always finite. Although the local operators are generally only operators after smearing, it is nonetheless convenient to continue to work with them, at least formally.

5.3 Fock space and quantum kinematics

Let us resolve our formally Hermitian field and momentum operators into their Fourier components according to the following transformations:

$$\varphi(\mathbf{x}) = \frac{1}{(2\pi)^{s/2}} \int [a(\mathbf{k})^\dagger e^{-i\mathbf{k}\cdot\mathbf{x}} + a(\mathbf{k})e^{i\mathbf{k}\cdot\mathbf{x}}]\, \sqrt{\hbar/2}\, d^s k \,,$$

$$\pi(\mathbf{x}) = \frac{1}{(2\pi)^{s/2}} \int [ia(\mathbf{k})^\dagger e^{-i\mathbf{k}\cdot\mathbf{x}} - ia(\mathbf{k})e^{i\mathbf{k}\cdot\mathbf{x}}]\, \sqrt{\hbar/2}\, d^s k \,.$$

Here the operators $a(\mathbf{k})$ and $a(\mathbf{k})^\dagger$ are analogs of the annihilation and creation operators introduced in Chapter 3, and it is natural to assume that they satisfy the commutation relations

$$[a(\mathbf{k}), a(\mathbf{l})] = 0\,, \qquad [a(\mathbf{k})^\dagger, a(\mathbf{l})^\dagger] = 0\,,$$
$$[a(\mathbf{k}), a(\mathbf{l})^\dagger] = \delta(\mathbf{k} - \mathbf{l})\,,$$

for all $\mathbf{k}, \mathbf{l} \in \mathbb{R}^s$. Conversely, if the operators a and $a^\dagger$ fulfill the indicated commutation relations, then it follows that the operators φ and π will satisfy the commutation relations claimed for them. The Fourier amplitudes $a(\mathbf{k})$ and $a(\mathbf{k})^\dagger$ are also local operators, and the commutator is properly defined if we again introduce test functions. Let us continue to work with the local operators a and $a^\dagger$ for the present.

Following the procedure for finitely many degrees of freedom, we next introduce a *no-particle state*, $|0\rangle$, with the property that

$$a(\mathbf{k})\, |0\rangle \equiv 0$$

which holds for all $\mathbf{k} \in \mathbb{R}^s$. [We note that the adjoint relation $\langle 0|\, a(\mathbf{k})^\dagger = 0$ holds for all $\mathbf{k}$ as well.] We suppose further that the state $|0\rangle$ is the *unique* no-particle state up to a factor, or stated otherwise, if $a(\mathbf{k})\, |\%\rangle = 0$ for all $\mathbf{k} \in \mathbb{R}^s$, then it follows that $|\%\rangle = c|0\rangle$ for some $c \in \mathbf{C}$. We define the smeared creation operator by

$$a^\dagger(\alpha) \equiv \int \alpha(\mathbf{k})\, a(\mathbf{k})^\dagger\, d^s k \,,$$

for α in a suitable class of complex test functions. Next we define states of the form

$$|\alpha\rangle = a^{\dagger}(\alpha)|0\rangle \,,$$
$$|\beta_1, \beta_2\rangle = (1/\sqrt{2!})\, a^{\dagger}(\beta_1)\, a^{\dagger}(\beta_2)|0\rangle \,,$$
$$|\gamma_1, \gamma_2, \gamma_3\rangle = (1/\sqrt{3!})\, a^{\dagger}(\gamma_1)\, a^{\dagger}(\gamma_2)\, a^{\dagger}(\gamma_3)|0\rangle \,,$$

etc., for general test functions and an increasing number of applications of the creation operator to the no-particle state. If we adopt the nomenclature that $a^{\dagger}$ creates "particles", we may refer to the indicated states as a one-particle, two-particle, three-particle state, etc. Because the creation operators commute among themselves it follows that the multiparticle states are symmetric in their arguments. In particular,

$$|\beta_2, \beta_1\rangle = |\beta_1, \beta_2\rangle \,,$$
$$|\gamma_3, \gamma_2, \gamma_1\rangle = |\gamma_2, \gamma_1, \gamma_3\rangle = |\gamma_1, \gamma_2, \gamma_3\rangle \,,$$

etc. A general multiparticle state is composed of a superposition of the indicated states. Thus, for example, an arbitrary three-particle state is given by

$$|\psi_3\rangle \equiv (1/\sqrt{3!})\int \psi_3(\mathbf{k}_1, \mathbf{k}_2, \mathbf{k}_3)$$
$$\times a(\mathbf{k}_1)^{\dagger} a(\mathbf{k}_2)^{\dagger} a(\mathbf{k}_3)^{\dagger}\, |0\rangle\, d^s k_1\, d^s k_2\, d^s k_3 \,,$$

where ψ_3 is symmetric in its arguments and

$$\langle\psi_3|\psi_3\rangle = \int |\psi_3(\mathbf{k}_1, \mathbf{k}_2, \mathbf{k}_3)|^2\, d^s k_1\, d^s k_2\, d^s k_3 < \infty \,.$$

An arbitrary state may be given as a linear superposition of such multi-particle states, including of course the no-particle state as well. Thus an arbitrary state in the Hilbert space of interest has the form

$$|\Psi\rangle \equiv \sum_{p=0}^{\infty}(1/\sqrt{p!})\int \psi_p(\mathbf{k}_1, \ldots, \mathbf{k}_p)\, a(\mathbf{k}_1)^{\dagger}\cdots a(\mathbf{k}_p)^{\dagger}|0\rangle$$
$$\times d^s k_1 \cdots d^s k_p$$
$$\equiv \sum_{p=0}^{\infty}\int \psi_p(\mathbf{k}_1, \ldots, \mathbf{k}_p)\, |\mathbf{k}_1, \ldots, \mathbf{k}_p\rangle\, d^s k_1 \cdots d^s k_p \,,$$

where $\psi_0 \in \mathbf{C}$, ψ_p, $p \geq 2$, are all symmetric functions, and such that

$$\langle\Psi|\Psi\rangle \equiv \sum_{p=0}^{\infty}\int |\psi_p(\mathbf{k}_1, \ldots, \mathbf{k}_p)|^2\, d^s k_1 \cdots d^s k_p < \infty \,.$$

Note that in defining $|\Psi\rangle$ we have introduced the formal states

$$|\mathbf{k}_1, \ldots, \mathbf{k}_p\rangle \equiv (1/\sqrt{p!})a(\mathbf{k}_1)^{\dagger}\cdots a(\mathbf{k}_p)^{\dagger}\, |0\rangle, \qquad p \geq 1 \,.$$

From these equations it is clear that we can represent the abstract vector $|\Psi\rangle$ by the sequence of probability amplitudes $\{\psi_p\}_{p=0}^{\infty}$.

The Hilbert space introduced in this manner is the celebrated *Fock space* [Sc 62] which is a Hilbert space of fundamental importance in quantum field theory.

Total sets

A total set of vectors $\mathcal{T}$ is a set with the property that it spans the space of interest. Alternatively, a set $\mathcal{T}$ is total if and only if $\langle\tau|?\rangle = 0$ for all $|\tau\rangle \in \mathcal{T}$ implies that $|?\rangle \equiv 0$. Let us first examine a specific multiparticle state, e.g., a three-particle state. The whole space in this case is composed of vectors of the form $|\psi_3\rangle$ for an arbitrary symmetric $\psi_3 \in L^2(\mathbb{R}^{3s})$. A total set of vectors is composed of vectors of the form $|\gamma_1, \gamma_2, \gamma_3\rangle$ for arbitrary square-integrable functions, $\gamma_j \in L^2(\mathbb{R}^s)$, $j = 1, 2, 3$. Next we observe that for special states of the form

$$|\gamma, \gamma, \gamma\rangle = (1/\sqrt{3!})\, a^\dagger(\gamma)\, a^\dagger(\gamma)\, a^\dagger(\gamma)|0\rangle = (1/\sqrt{3!})\, a^\dagger(\gamma)^3 |0\rangle\,,$$

for which all three test functions are the *same*, it is quite straightforward to recover the state where all three test functions are *different*. For that purpose all that is needed is to use special test functions that have the form

$$\gamma(\mathbf{k}) = (1/3!)[b_1\, \gamma_1(\mathbf{k}) + b_2\, \gamma_2(\mathbf{k}) + b_3\, \gamma_3(\mathbf{k})]$$

and seek out the coefficient of the factor $b_1 b_2 b_3$, which can be done, for example, by

$$|\gamma_1, \gamma_2, \gamma_3\rangle = \frac{\partial^3}{\partial b_1 \partial b_2 \partial b_3}\, |\gamma, \gamma, \gamma\rangle\,.$$

Thus the set of states of the form $|\gamma, \gamma, \gamma\rangle$, for a general complex test function γ, also constitutes a total set for three-particle states. An analogous construction extends to all multiparticle states. Thus we see that a total set of vectors for the entire set of multiparticle states, i.e., the Fock space itself, is composed of the vectors

$$|0\rangle\,, \qquad a^\dagger(\psi)|0\rangle\,, \qquad a^\dagger(\psi)^2|0\rangle\,, \qquad a^\dagger(\psi)^3|0\rangle\,, \qquad \ldots\,,$$

for a dense set of complex test functions $\psi \in L^2(\mathbb{R}^s)$. Lastly, we may consider the space spanned by the special states

$$|\psi\rangle \equiv e^{\int \psi(\mathbf{k}) a(\mathbf{k})^\dagger\, d^s k}\, |0\rangle = e^{a^\dagger(\psi)}|0\rangle\,,$$

for all complex test functions ψ, which define the unnormalized *coherent states* for the problem at hand. Clearly, an arbitrary state in the Fock space may be obtained by linear combinations and limits involving the coherent states; in other words, *the set of coherent states is a total set for Fock space*. The norm of a coherent state may be obtained as follows

$$\langle\psi|\psi\rangle = \langle 0| e^{\int \psi^*(\mathbf{k})a(\mathbf{k})\,d^s k}\, e^{\int \psi(\mathbf{k})a(\mathbf{k})^\dagger\,d^s k}\,|0\rangle$$
$$= \langle 0| e^{\int \psi^*(\mathbf{k})\psi(\mathbf{k})\,d^s k}\, e^{\int \psi(\mathbf{k})a(\mathbf{k})^\dagger\,d^s k}\, e^{\int \psi^*(\mathbf{k})a(\mathbf{k})\,d^s k}\,|0\rangle$$
$$= e^{\int \psi^*(\mathbf{k})\psi(\mathbf{k})\,d^s k}\,,$$

a result that involves nothing more than the simple relation – derived in Chapter 4 – that if A and B are two operators both of which commute with $[A, B]$, then

$$e^A\, e^B = e^{[A,B]}\, e^B\, e^A\,.$$

Moreover, this calculation of the norm of the coherent state assures us that such states are in fact well defined for all complex $\psi \in L^2(\mathbb{R}^s)$, and thus we can choose any complex test-function space dense in L^2, even including L^2 itself.

5.4 Basic kinematical operators

The operators $a(\mathbf{k})$ and $a(\mathbf{k})^\dagger$, for all $\mathbf{k} \in \mathbb{R}^s$ – in fact, more properly, those operators smeared by all test functions from a suitable class (dense in L^2) – constitute the fundamental *Fock annihilation and creation operators*. Thanks to the uniqueness of the no-particle state $|0\rangle$, the Fock operators form an *irreducible set*. This property may be stated in two ways. First, irreducibility means that any operator $\mathcal{W}$ for which

$$[\mathcal{W}, a(\mathbf{k})] = 0\,, \qquad [\mathcal{W}, a(\mathbf{k})^\dagger] = 0\,,$$

holds for all $\mathbf{k} \in \mathbb{R}^s$, necessarily is a multiple of the unit operator, i.e., $\mathcal{W} = c\mathbb{1}$ for some $c \in \mathbf{C}$. Second, irreducibility means that *any* operator acting in the Hilbert space is necessarily given as a function of the basic Fock operators, where the meaning of function involves algebraic combinations possibly followed by suitable limits.

One operator of importance is the self-adjoint *number operator* N formally represented by

$$\mathsf{N} \equiv \int a(\mathbf{k})^\dagger a(\mathbf{k})\,d^s k\,.$$

It readily follows that

$$[a(\mathbf{k}), \mathsf{N}] = a(\mathbf{k})\,, \qquad [a(\mathbf{k})^\dagger, \mathsf{N}] = -a(\mathbf{k})^\dagger\,,$$
$$\mathsf{N}\,|0\rangle = 0\,, \qquad \mathsf{N}\,|\psi_n\rangle = n|\psi_n\rangle\,.$$

Evidently the no-particle state $|0\rangle$ is the unique state up to a factor that is annihilated by the number operator. An additional consequence of the commutation relations is the fact that

$$e^{i\theta\mathsf{N}}a(\mathbf{k})e^{-i\theta\mathsf{N}} = e^{-i\theta}a(\mathbf{k})\,,$$
$$e^{i\theta\mathsf{N}}a(\mathbf{k})^\dagger e^{-i\theta\mathsf{N}} = e^{i\theta}a(\mathbf{k})^\dagger\,.$$

Another important operator is the self-adjoint *space translation generator* $\mathcal{P}$, an s-dimensional vector operator, defined by the fact that

$$\mathcal{P} \equiv \hbar \int a(\mathbf{k})^\dagger \, \mathbf{k} \, a(\mathbf{k}) \, d^s k \; .$$

This is the same operator density as in the case of the number operator except that we have multiplied by a scale factor, $\hbar \mathbf{k} \equiv \mathbf{p}$, at each point of the integration. From this fact and the properties of the number operator, we can easily deduce the consequences of the operator $\mathcal{P}$. In particular, it follows that

$$[a(\mathbf{k}), \mathcal{P}] = \mathbf{p} \, a(\mathbf{k}) \; , \qquad [a(\mathbf{k})^\dagger, \mathcal{P}] = -\mathbf{p} \, a(\mathbf{k})^\dagger \; .$$

Furthermore, we see that

$$e^{i\mathbf{a}\cdot\mathcal{P}/\hbar} a(\mathbf{k}) e^{-i\mathbf{a}\cdot\mathcal{P}/\hbar} = e^{-i\mathbf{a}\cdot\mathbf{k}} a(\mathbf{k}) \; ,$$
$$e^{i\mathbf{a}\cdot\mathcal{P}/\hbar} a(\mathbf{k})^\dagger e^{-i\mathbf{a}\cdot\mathcal{P}/\hbar} = e^{i\mathbf{a}\cdot\mathbf{k}} a(\mathbf{k})^\dagger \; .$$

We also have the very important relations

$$\mathcal{P} \, |0\rangle = 0 \; , \qquad e^{i\mathbf{a}\cdot\mathcal{P}/\hbar} |0\rangle = |0\rangle \; ,$$

asserting that $|0\rangle$ is invariant under arbitrary spatial translations. Moreover, the no-particle state $|0\rangle$ is the only state up to a factor that is invariant under arbitrary translations; i.e., if $e^{i\mathbf{a}\cdot\mathcal{P}/\hbar} |\&\rangle = |\&\rangle$ for all $\mathbf{a}$, then $|\&\rangle = c|0\rangle$ for some complex constant c.

Let us introduce the unitary operator $U(\mathbf{a}) \equiv e^{i\mathbf{a}\cdot\mathcal{P}/\hbar}$. It then follows that

$$U(\mathbf{a})^\dagger \, \varphi(\mathbf{x}) \, U(\mathbf{a})$$
$$= \frac{1}{(2\pi)^{s/2}} \int [e^{i\mathbf{k}\cdot(\mathbf{x}+\mathbf{a})} a(\mathbf{k}) + e^{-i\mathbf{k}\cdot(\mathbf{x}+\mathbf{a})} a(\mathbf{k})^\dagger] \sqrt{\hbar/2} \, d^s k$$
$$= \varphi(\mathbf{x} + \mathbf{a}) \; .$$

In like manner it follows that

$$U(\mathbf{a})^\dagger \, \pi(\mathbf{x}) \, U(\mathbf{a}) = \pi(\mathbf{x} + \mathbf{a}) \; .$$

It is evident that $U(\mathbf{a})$ acts so as to generate finite spatial transformations of the fields. Consequently, $\mathcal{P}$ is referred to as the generator of space translations, and by expanding the two previous relations to first order in $\mathbf{a}$ we learn that

$$[\varphi(\mathbf{x}), \mathcal{P}] = -i\hbar \nabla \varphi(\mathbf{x}) \; ,$$
$$[\pi(\mathbf{x}), \mathcal{P}] = -i\hbar \nabla \pi(\mathbf{x}) \; .$$

In closing this discussion we observe that the space-translation generator may be given in component form simply as

$$\mathcal{P}_l \equiv \hbar \int a(\mathbf{k})^\dagger \, k_l \, a(\mathbf{k}) \, d^s k \; .$$

Another important set of operators is the family of self-adjoint *rotation generators* which we define as the antisymmetric tensor operator

$$\mathcal{J}_{lm} \equiv i\hbar \int a(\mathbf{k})^\dagger \left(\frac{\partial}{\partial k_l} k_m - \frac{\partial}{\partial k_m} k_l \right) a(\mathbf{k}) \, d^s k \, .$$

It immediately follows that

$$[a(\mathbf{k}), \mathcal{J}_{lm}] = i\hbar \left(\frac{\partial}{\partial k_l} k_m - \frac{\partial}{\partial k_m} k_l \right) a(\mathbf{k}) \, ,$$

$$[a(\mathbf{k})^\dagger, \mathcal{J}_{lm}] = -i\hbar \left(\frac{\partial}{\partial k_l} k_m - \frac{\partial}{\partial k_m} k_l \right) a(\mathbf{k})^\dagger \, .$$

Further properties of the rotation generators are most easily won if we first present a rather general commutation relation. Initially, let us introduce the following two general differential operators

$$\mathsf{A} \equiv \mathsf{A}(\mathbf{k}, i\hbar(\partial/\partial\mathbf{k})) \, , \qquad \mathsf{B} \equiv \mathsf{B}(\mathbf{k}, i\hbar(\partial/\partial\mathbf{k})) \, .$$

Then, based on the properties of the a and $a^\dagger$ operators, it follows quite generally that

$$[\int a(\mathbf{k})^\dagger \, \mathsf{A} \, a(\mathbf{k}) \, d^s k, \int a(\mathbf{k}')^\dagger \, \mathsf{B} \, a(\mathbf{k}') \, d^s k'] = \int a(\mathbf{k})^\dagger \, [\mathsf{A}, \mathsf{B}] \, a(\mathbf{k}) \, d^s k \, .$$

Let us next set

$$J_{lm} \equiv i\hbar \left(\frac{\partial}{\partial k_l} k_m - \frac{\partial}{\partial k_m} k_l \right) \, ,$$

and observe that

$$[J_{lm}, \hbar k_n] = i\hbar(\delta_{ln} \hbar k_m - \delta_{mn} \hbar k_l) \, ;$$

the double factors of $\hbar$ on the right-hand side become less peculiar if we employ the *momentum vector*

$$\mathbf{p} \equiv \hbar \mathbf{k}$$

in place of the *wave vector* $\mathbf{k}$; we shall find it convenient to use both $\mathbf{p}$ and $\mathbf{k}$ hereafter, as we have already done! In terms of $\mathbf{p}$ the preceding commutator reads

$$[J_{lm}, p_n] = i\hbar(\delta_{ln} p_m - \delta_{mn} p_l) \, ,$$

and the interpretation of this equation is that the momentum $\mathbf{p}$ transforms as a vector under rotations analogous to what we have already observed in the classical theory. Another commutation relation of great importance is given by

$$[J_{lm}, J_{rs}] = i\hbar(\delta_{lr} J_{ms} - \delta_{mr} J_{ls} - \delta_{ls} J_{mr} + \delta_{ms} J_{lr}) \, .$$

When we combine these commutation relations with the general commutation relation for operators given before, it follows that

$$[\mathcal{J}_{lm}, \mathcal{P}_n] = i\hbar(\delta_{ln} \mathcal{P}_m - \delta_{mn} \mathcal{P}_l) \, ,$$

$$[\mathcal{J}_{lm}, \mathcal{J}_{rs}] = i\hbar(\delta_{lr} \mathcal{J}_{ms} - \delta_{mr} \mathcal{J}_{ls} - \delta_{ls} \mathcal{J}_{mr} + \delta_{ms} \mathcal{J}_{lr}) \, .$$

In addition, the rotation operator evidently satisfies

$$\mathcal{J}_{lm}\,|0\rangle = 0\,, \qquad e^{-i\theta_{lm}\mathcal{J}_{lm}/\hbar}\,|0\rangle = |0\rangle\,;$$

however – and unlike the case of the space-translation generator – the no-particle state is *not* the only state invariant under rotations. As an example of another state invariant under arbitrary rotations consider

$$|\alpha\rangle \equiv \int \alpha(|\mathbf{k}|)\,a(\mathbf{k})^\dagger\,d^s k\,|0\rangle\,,$$

where we have deliberately chosen a smooth test function α that depends only on the magnitude $|\mathbf{k}|$ of the vector $\mathbf{k}$. It follows that

$$\mathcal{J}_{lm}|\alpha\rangle = \int \alpha(|\mathbf{k}|)\,[\mathcal{J}_{lm}, a(\mathbf{k})^\dagger]\,d^s k\,|0\rangle$$

$$= -i\hbar \int a(\mathbf{k})^\dagger \left(\frac{\partial}{\partial k_l}k_m - \frac{\partial}{\partial k_m}k_l\right)\alpha(|\mathbf{k}|)\,d^s k\,|0\rangle = 0\,.$$

The vanishing of this expression follows from integration by parts and recognizing that the differential operator involved makes an infinitesimal rotation of a rotation-invariant function, which evidently vanishes. More precisely, we note that the set of operators $\{\mathcal{J}_{lm}\}$ have a nontrivial subspace of $L^2(\mathbb{R}^s)$ on which all of them have the eigenvalue zero. As a consequence it follows that there exists a subspace of the Fock space on which all the operators $\mathcal{J}_{lm}$ will have the eigenvalue zero. For example, the unnormalized coherent state $|\psi\rangle$ defined for any smooth $\psi \in L^2(\mathbb{R}^s)$, such that $\psi = \psi(|\mathbf{k}|)$, i.e., the function is rotationally invariant, will satisfy $\mathcal{J}_{lm}\,|\psi\rangle = 0$.

By an appropriate Fourier transformation, it follows that

$$[\varphi(\mathbf{x}), \mathcal{J}_{lm}] = i\hbar\left(x_m\frac{\partial}{\partial x_l} - x_l\frac{\partial}{\partial x_m}\right)\varphi(\mathbf{x}) \equiv i\hbar J_{lm}\,\varphi(\mathbf{x})\,,$$

$$[\pi(\mathbf{x}), \mathcal{J}_{lm}] = i\hbar\left(x_m\frac{\partial}{\partial x_l} - x_l\frac{\partial}{\partial x_m}\right)\pi(\mathbf{x}) \equiv i\hbar J_{lm}\,\pi(\mathbf{x})\,.$$

If we introduce $U(\theta) \equiv \exp[-i\theta_{lm}\mathcal{J}_{lm}/\hbar]$, then it follows that

$$U(\theta)^\dagger\varphi(\mathbf{x})U(\theta) = \varphi(\mathbf{Rx})\,,$$

$$U(\theta)^\dagger\pi(\mathbf{x})U(\theta) = \pi(\mathbf{Rx})\,,$$

where

$$(\mathbf{Rx})_a = R_{ab}\,x_b \equiv (\exp[\theta_{lm}J_{lm}])_{ab}\,x_b\,,$$

namely a rotation of the vector $\mathbf{x}$ in the s-dimensional configuration space by an orthogonal transformation.

We can easily introduce still other operators that are defined with the help of different weightings of the integrand within the integral that defines the number operator. For example, let us next consider

$$\mathcal{H} = \int a(\mathbf{k})^\dagger \, \hbar\Omega(\mathbf{k}) \, a(\mathbf{k}) \, d^s k \, ,$$

where $\Omega(\mathbf{k}) > 0$ and is locally integrable, but is otherwise arbitrary for the moment. It is clear that $\mathcal{H}$ is a positive operator, i.e., $\mathcal{H} \geq 0$, which means that $\langle \psi | \mathcal{H} | \psi \rangle \geq 0$ for a general $|\psi\rangle$, and that

$$\mathcal{H} |0\rangle = 0 \, , \qquad e^{-i\mathcal{H}t/\hbar} |0\rangle = |0\rangle \, .$$

Furthermore, under the positivity condition on Ω, the state $|0\rangle$ is the only state up to a factor which is annihilated by the action of the operator $\mathcal{H}$. By analogy with the previous calculations we observe that

$$e^{i\mathcal{H}t/\hbar} a(\mathbf{k}) e^{-i\mathcal{H}t/\hbar} = e^{-i\Omega(\mathbf{k})t} \, a(\mathbf{k}) \, ,$$
$$e^{i\mathcal{H}t/\hbar} a(\mathbf{k})^\dagger e^{-i\mathcal{H}t/\hbar} = e^{i\Omega(\mathbf{k})t} \, a(\mathbf{k})^\dagger \, .$$

As the notation suggests, we have in mind that $\mathcal{H}$ represents the Hamiltonian for some system. We are now in position to demonstrate just how the operator $\mathcal{H}$ serves as a Hamiltonian for a "free field".

5.5 The dynamics of free fields

We shall define a free field as a field that satisfies a *homogeneous linear set of equations of motion*, and so it follows that the Hamiltonian for such a field is a homogeneous quadratic function of the momenta and coordinates. Of all the possible quadratic expressions for the classical Hamiltonian that could be considered, we shall concentrate on just those classical expressions of the form

$$H(f, g) = \tfrac{1}{2} \int [f(\mathbf{x})]^2 \, d^s x + \tfrac{1}{2} \int g(\mathbf{x}) \, W(\mathbf{x} - \mathbf{y}) \, g(\mathbf{y}) \, d^s x \, d^s y \, ,$$

for functions f and g and some distribution W chosen so that $0 \leq H(f, g) < \infty$, while $H(f, g) = 0$ holds only for $f = g \equiv 0$. This class of models includes, among others, the relativistic free field, and we shall discuss this important example below. For now it is convenient to discuss the wider class of examples for a general kernel W; the class of acceptable kernels is defined below.* An important fact about such Hamiltonians is that for any choice of W one of the equations of motion is invariably

$$\dot{g}(t, \mathbf{x}) \equiv \{g(t, \mathbf{x}), H(f, g)\} \equiv f(t, \mathbf{x}) \, .$$

---

* The important examples of so-called *generalized free fields* are developed in the Exercises at the conclusion of this chapter.

We generally assume that this dynamical relation between the canonical kinematical field variables must of necessity hold true in the quantum theory as well.

Since the integrals in question run over all space it is clear that the Hamiltonians under consideration all enjoy translation invariance. In other words, $H(f_\mathbf{a}, g_\mathbf{a}) = H(f, g)$, where $f_\mathbf{a}(\mathbf{x}) \equiv f(\mathbf{x} - \mathbf{a})$ and $g_\mathbf{a}(\mathbf{x}) \equiv g(\mathbf{x} - \mathbf{a})$. The expression for such Hamiltonians achieves a certain simplification in Fourier space. If we introduce

$$\tilde{f}(\mathbf{k}) \equiv \frac{1}{(2\pi)^{s/2}} \int e^{-i\mathbf{k}\cdot\mathbf{x}} f(\mathbf{x})\, d^s x \;,$$

$$\tilde{g}(\mathbf{k}) \equiv \frac{1}{(2\pi)^{s/2}} \int e^{-i\mathbf{k}\cdot\mathbf{x}} g(\mathbf{x})\, d^s x \;,$$

$$\Omega(\mathbf{k})^2 \equiv \frac{1}{(2\pi)^s} \int e^{-i\mathbf{k}\cdot\mathbf{x}} W(\mathbf{x})\, d^s x \;,$$

it follows that

$$H(f, g) = \tfrac{1}{2} \int [|\tilde{f}(\mathbf{k})|^2 + \Omega(\mathbf{k})^2 |\tilde{g}(\mathbf{k})|^2]\, d^s k \;.$$

Since we have required that $H(f, g) > 0$ for any $g \not\equiv 0$, we insist that $\Omega(\mathbf{k})^2 > 0$ for all $\mathbf{k} \in \mathbb{R}^s$, and so we may choose $\Omega(\mathbf{k}) = \sqrt{\Omega(\mathbf{k})^2} > 0$ as well as $[\Omega(\mathbf{k})]^{1/2} > 0$. Furthermore, if we introduce

$$\psi(\mathbf{k}) \equiv \frac{1}{\sqrt{2\hbar}} \{[\Omega(\mathbf{k})]^{1/2} \tilde{g}(\mathbf{k}) + i[\Omega(\mathbf{k})]^{-1/2} \tilde{f}(\mathbf{k})\} \;,$$

then it follows that

$$H(f, g) = \hbar \int \Omega(\mathbf{k}) |\psi(\mathbf{k})|^2\, d^s k \;.$$

The introduction of the factor $\hbar$ into the classical theory here is strictly for dimensional reasons, and in particular $\hbar\Omega$ has the dimensions of energy.

The expression for the Hamiltonian that we have obtained is an integral over *uncoupled* momentum dependent variables. The classical equations of motion expressed in these variables become

$$\dot{\psi}(t, \mathbf{k}) = \{\psi(t, \mathbf{k}), H(f, g)\} = -i\Omega(\mathbf{k})\, \psi(t, \mathbf{k}) \;,$$

and its complex conjugate, with the evident classical solution

$$\psi(t, \mathbf{k}) = e^{-i\Omega(\mathbf{k})t}\, \psi(\mathbf{k}) \;,$$
$$\psi(t, \mathbf{k})^* = e^{i\Omega(\mathbf{k})t}\, \psi(\mathbf{k})^* \;,$$

where we have chosen the initial condition as $\psi(0, \mathbf{k}) \equiv \psi(\mathbf{k})$. For each $\mathbf{k} \in \mathbb{R}^s$ we observe that the time evolution of the classical variable $\psi(\mathbf{k})$ and its complex conjugate derived here is identical to the time evolution of the quantum operator $a(\mathbf{k})$ and its adjoint. Thus we may tentatively adopt the quantum operator $\mathcal{H}$, as presented in the previous section, for a given function Ω as the quantum

Hamiltonian associated with the classical Hamiltonian $H(f,g)$. Let us confirm this tentative association.

We have agreed that the set of annihilation and creation operators a and $a^\dagger$ are irreducible and that all operators acting in the Fock space may be constructed out of them. We now define a family of field and momentum operators, one for each of the functions Ω introduced above, by the following transformations:

$$\varphi(\mathbf{x}) = \frac{1}{(2\pi)^{s/2}} \int [a(\mathbf{k})^\dagger e^{-i\mathbf{k}\cdot\mathbf{x}} + a(\mathbf{k})e^{i\mathbf{k}\cdot\mathbf{x}}] \sqrt{\frac{\hbar}{2\Omega(\mathbf{k})}}\, d^s k \;,$$

$$\pi(\mathbf{x}) = \frac{1}{(2\pi)^{s/2}} \int [ia(\mathbf{k})^\dagger e^{-i\mathbf{k}\cdot\mathbf{x}} - ia(\mathbf{k})e^{i\mathbf{k}\cdot\mathbf{x}}] \sqrt{\frac{\hbar\Omega(\mathbf{k})}{2}}\, d^s k \;.$$

The fields defined in this manner obviously depend on the choice of Ω; however, for convenience, we suppress that dependence in the notation. Based on the properties of a and $a^\dagger$, it is a simple exercise to show for any acceptable choice of Ω that these fields obey the desired canonical commutation relations, specifically

$$[\varphi(\mathbf{x}),\varphi(\mathbf{y})] = 0 \;, \qquad [\pi(\mathbf{x}),\pi(\mathbf{y})] = 0 \;,$$
$$[\varphi(\mathbf{x}),\pi(\mathbf{y})] = i\hbar\delta(\mathbf{x}-\mathbf{y}) \;,$$

which hold for all $\mathbf{x},\mathbf{y} \in \mathbb{R}^s$. But these facts are purely *kinematic;* what is also needed is to relate the fields *dynamically* as already demonstrated classically. To that end we examine the time evolution of the field operator as induced by the Hamiltonian $\mathcal{H}$. In particular, let us first examine

$$\dot\varphi(\mathbf{x}) \equiv -\frac{i}{\hbar}[\varphi(\mathbf{x}),\mathcal{H}]$$

$$= -\frac{i}{\hbar}\frac{1}{(2\pi)^{s/2}} \int \{[a(\mathbf{k})^\dagger,\mathcal{H}]e^{-i\mathbf{k}\cdot\mathbf{x}} + [a(\mathbf{k}),\mathcal{H}]e^{i\mathbf{k}\cdot\mathbf{x}}\} \sqrt{\frac{\hbar}{2\Omega(\mathbf{k})}}\, d^s k$$

$$= \frac{1}{(2\pi)^{s/2}} \int [ia(\mathbf{k})^\dagger e^{-i\mathbf{k}\cdot\mathbf{x}} - ia(\mathbf{k})e^{i\mathbf{k}\cdot\mathbf{x}}] \sqrt{\frac{\hbar\Omega(\mathbf{k})}{2}}\, d^s k$$

$$= \pi(\mathbf{x}) \;.$$

Thus, by using the *same* function Ω in the definition of the kinematical fields φ and π as in the Hamiltonian we have been able to connect the two basic kinematical variables in the sought-for dynamical fashion.

We are now in a position to offer the full dynamical solution for the free field according to the prescription

$$\varphi(t,\mathbf{x}) \equiv e^{i\mathcal{H}t/\hbar}\varphi(\mathbf{x})e^{-i\mathcal{H}t/\hbar}$$

$$= \frac{1}{(2\pi)^{s/2}} \int [a(\mathbf{k})^\dagger e^{-i\mathbf{k}\cdot\mathbf{x}+i\Omega(\mathbf{k})t} + a(\mathbf{k})e^{i\mathbf{k}\cdot\mathbf{x}-i\Omega(\mathbf{k})t}] \sqrt{\frac{\hbar}{2\Omega(\mathbf{k})}}\, d^s k \;.$$

In this form it is also evident that the time derivative of the field evaluated at time zero will reproduce the desired momentum only when the function Ω in the exponent multiplied by the time t is the same as the function Ω that weights the integration over $\mathbb{R}^s$. With the expression for the full space-time dependence of the field operator in hand, we are in a position to determine a number of correlation functions or expectation values of quantities of interest.

Most of the expectation values of interest can be readily deduced from the behavior of similar quantities given for one degree of freedom. Let us again introduce the self-adjoint smeared field operators

$$\varphi(f) \equiv \int \varphi(\mathbf{x}) f(\mathbf{x}) \, d^s x \,, \qquad \pi(g) \equiv \int \pi(\mathbf{x}) g(\mathbf{x}) \, d^s x \,,$$

defined in the present case for field operators that involve a general function Ω. Then, for example, we determine that

$$\langle 0| \exp\{i[\varphi(f) - \pi(g)]/\hbar\} |0\rangle$$
$$= \exp\{-(1/4\hbar) \int [\Omega(\mathbf{k})^{-1}|\tilde{f}(\mathbf{k})|^2 + \Omega(\mathbf{k})|\tilde{g}(\mathbf{k})|^2] \, d^s k\} \,,$$

and this expression is well defined for all f and g such that the exponent is finite, namely, whenever

$$\int [\Omega(\mathbf{k})^{-1}|\tilde{f}(\mathbf{k})|^2 + \Omega(\mathbf{k})|\tilde{g}(\mathbf{k})|^2] \, d^s k < \infty \,.$$

Observe that $(f,g) = \int \tilde{f}(\mathbf{k})^* \tilde{g}(\mathbf{k}) \, d^s k$ is also well defined for all such f and g thanks to the Schwarz inequality.

Coherent states

The important normalized coherent states for the present problem are given by

$$|f, g\rangle \equiv \exp\{i[\varphi(f) - \pi(g)]/\hbar\} |0\rangle \,,$$

defined for all f and g that constitute a dense subset of the final display equation of the last paragraph. (Remark: We note that these (normalized) coherent states are identical to the previous (unnormalized) coherent state after the latter are normalized, even though the argument has shifted from one complex function ψ to the two real functions f and g. This reemphasis on the argument exactly parallels that which occurs for a single degree of freedom; see [KlS 68, KlS 85].) The overlap of two such states may be deduced from the fact that

$$\langle f, g | f', g' \rangle = \exp\{i[(g, f') - (f, g')]/2\hbar\} \langle 0| f' - f, g' - g \rangle \,,$$

and the latter inner product follows from the one given in the previous paragraph. It follows from this relation that each such vector is normalized, i.e., $\langle f, g | f, g \rangle = 1$ for all f and g. *Expectations taken with respect to these normalized coherent states are of fundamental importance.* For instance, it follows that

$$\langle f, g | \varphi(\mathbf{x}) | f, g \rangle = g(\mathbf{x}) \,, \qquad \langle f, g | \pi(\mathbf{x}) | f, g \rangle = f(\mathbf{x}) \,,$$

and, even more interesting, that

$$\langle f, g | \boldsymbol{P} | f, g \rangle = -\int f(\mathbf{x}) \boldsymbol{\nabla} g(\mathbf{x}) \, d^s x \, ,$$

$$\langle f, g | \mathcal{J}_{ab} | f, g \rangle = \int f(\mathbf{x}) \, \mathsf{J}_{ab} \, g(\mathbf{x}) \, d^s x \, ,$$

$$\langle f, g | \mathcal{H} | f, g \rangle = \tfrac{1}{2} \int [f(\mathbf{x})]^2 \, d^s x + \tfrac{1}{2} \int g(\mathbf{x}) \, W(\mathbf{x} - \mathbf{y}) \, g(\mathbf{y}) \, d^s x \, d^s y \, .$$

These relations hold for any acceptable choice of W.

Weak correspondence principle

Observe carefully what these relations actually assert: *the* **classical** *generator in each case is given as the diagonal coherent-state expectation value of the* **quantum** *generator!* This association of quantum and classical generators goes under the name of the *weak correspondence principle* [Kl 67]. Note well that this association of classical and quantum quantities has *not* involved the limit $\hbar \to 0$; in point of fact, the value of $\hbar$ has not been changed at all. Here, in the weak correspondence principle, is the advertised **coexistence** *of the quantum and classical formalisms*, and an explicit mathematical expression of how the quantum and classical generators are related to each other. The validity of the weak correspondence principle has required the use of normal ordering, i.e., the operators $\boldsymbol{P}$, $\mathcal{J}_{ab}$ and $\mathcal{H}$ have been *normal ordered*, and are given in the three cases by

$$\boldsymbol{P} = -\int \, : \pi(\mathbf{x}) \, \boldsymbol{\nabla} \varphi(\mathbf{x}) : \, d^s x$$

$$= \hbar \int a(\mathbf{k})^\dagger \, \mathbf{k} \, a(\mathbf{k}) \, d^s k \, ;$$

$$\mathcal{J}_{ab} = \int \, : \pi(\mathbf{x}) \, \mathsf{J}_{ab} \, \varphi(\mathbf{x}) : \, d^s x$$

$$= i\hbar \int a(\mathbf{k})^\dagger \left(k_b \frac{\partial}{\partial k_a} - k_a \frac{\partial}{\partial k_b} \right) a(\mathbf{k}) \, d^s k \, ;$$

$$\mathcal{H} = \tfrac{1}{2} \int \!\! \int \, : [\pi(\mathbf{x}) \, \delta(\mathbf{x} - \mathbf{y}) \, \pi(\mathbf{y}) + \varphi(\mathbf{x}) \, W(\mathbf{x} - \mathbf{y}) \, \varphi(\mathbf{y})] : \, d^s x \, d^s y$$

$$= \hbar \int a(\mathbf{k})^\dagger \, \Omega(\mathbf{k}) \, a(\mathbf{k}) \, d^s k \, .$$

We emphasize that this association of a classical and a quantum expression *automatically* subtracts off the zero point energy. In Exercise 5.2 we stress the fact that – far from being accidental – the validity of the weak correspondence principle follows from general concepts. We will also encounter the weak correspondence principle in the next section when we take up relativistic models.

5.6 Relativistic free fields

The relativistic free field is determined by a proper choice of Hamiltonian. This is already true in the classical theory and it is no less true in the quantum theory. In the classical theory the Hamiltonian for the relativistic free field of mass m is given by

$$H = \tfrac{1}{2} \int \{[f(\mathbf{x})]^2 + [\boldsymbol{\nabla} g(\mathbf{x})]^2 + m^2 [g(\mathbf{x})]^2\}\, d^s x$$
$$\equiv \int H(\mathbf{x})\, d^s x \,.$$

Correspondingly, the quantum Hamiltonian operator is taken as

$$\mathcal{H} = \tfrac{1}{2} \int\, :\{[\pi(\mathbf{x})]^2 + [\boldsymbol{\nabla} \varphi(\mathbf{x})]^2 + m^2 [\varphi(\mathbf{x})]^2\}:\, d^s x$$
$$\equiv \int \mathcal{H}(\mathbf{x})\, d^s x \,.$$

The indicated expression for $\mathcal{H}$ is quite formal due to the fact that the field operators at points are not genuine operators and so their square is ill-defined. Nevertheless, the specifically indicated sum of the several ill-defined expressions leads to a meaningful operator that may, alternatively and properly, be defined as

$$\mathcal{H} = \int a(\mathbf{k})^\dagger \sqrt{\mathbf{p}^2 + m^2}\, a(\mathbf{k})\, d^s k \,,$$

where, as before, we have used $\mathbf{p} \equiv \hbar \mathbf{k}$. Thus we deal with an operator that belongs to the class extensively discussed above for which

$$\hbar^2 \Omega(\mathbf{k})^2 = \hbar^2 \omega(\mathbf{k})^2 \equiv \mathbf{p}^2 + m^2 \,.$$

Here we have introduced the function $\hbar^2 \omega(\mathbf{k})^2 \equiv \mathbf{p}^2 + m^2$ appropriate to a relativistic system of mass m; this relation is exactly the familiar relativistic dispersion law $E^2 = \mathbf{p}^2 + m^2$ in units where the velocity of light $c = 1$.

Hereafter, for the remainder of this chapter, we choose units such that $\hbar = 1$.

Observe that the relativistic choice of ω is rotationally invariant in $\mathbf{k}$-space. This implies that the kernel $W(\mathbf{x} - \mathbf{y})$ is in reality of the form $W(|\mathbf{x} - \mathbf{y}|)$, and so the relativistic Hamiltonians are not only translationally invariant but they are rotationally invariant as well. (Indeed, any Hamiltonian based on a function Ω that was only a function of the length of its vector argument would have both translational and rotational invariance.) As a consequence we have the Poisson bracket relations $\{J_{ab}, H\} = 0$ as well as the operator commutation relations $[\mathcal{J}_{ab}, \mathcal{H}] = 0$ for all pairs $a, b,\, a \neq b$.

In the case of a relativistic theory there are additional generators of interest that relate to *boosts*, namely the *relativity transformations*. Classically the generator is a vector quantity defined by

$$K_a(f,g) = -\int [f(\mathbf{x})\, t\, \partial_a\, g(\mathbf{x}) + x_a\, H(\mathbf{x})]\, d^s x \ .$$

Correspondingly, the quantum operator is given by

$$\mathcal{K}_a(f,g) = -\int :[\pi(\mathbf{x})\, t\, \partial_a\, \varphi(\mathbf{x}) + x_a\, \mathcal{H}(\mathbf{x})]:\, d^s x \ .$$

It is straightforward to verify that as Poisson brackets $\{P_l, K_m\} = \delta_{lm}H$ and $\{H, K_l\} = P_l$, as well as operator commutators $[\mathcal{P}_l, \mathcal{K}_m] = i\delta_{lm}\mathcal{H}$ and $[\mathcal{H}, \mathcal{K}_l] = i\mathcal{P}_l$, as required for the Lie algebra generators of the Poincaré group [ItZ 80]. Furthermore, we note that the boost generators also satisfy the weak correspondence principle inasmuch as

$$\langle f,g|\mathcal{K}_a|f,g\rangle = -\int [f(\mathbf{x})\, t\, \partial_a\, g(\mathbf{x}) + x_a\, H(\mathbf{x})]\, d^s x \ .$$

These relations effectively establish that the example under discussion refers to a relativistic free field.

Based on the extensive discussion already given for a general Ω we are able to immediately offer a variety of expectation values. For example, in terms of the fields φ and π smeared at any sharp time, e.g., $t = 0$, it follows for a relativistic free field of mass m that

$$\langle 0|\exp\{i[\varphi(f) - \pi(g)]\}|0\rangle$$
$$= \exp\{-(1/4)\textstyle\int[\omega(\mathbf{k})^{-1}|\tilde{f}(\mathbf{k})|^2 + \omega(\mathbf{k})|\tilde{g}(\mathbf{k})|^2]\, d^s k\} \ .$$

Instead, if we introduce the space-time smeared field

$$\varphi(h) \equiv \textstyle\int h(x)\varphi(x)\, d^n x \ ,$$

it follows that

$$\langle 0|\exp[i\varphi(h)]|0\rangle = \exp\{-(1/2)\textstyle\int h(x)\, \Delta_1(x - y)\, h(y)\, d^n x\, d^n y\} \ ,$$

where

$$\Delta_1(x - y) \equiv \frac{1}{2}\langle 0|[\varphi(x)\varphi(y) + \varphi(y)\varphi(x)]|0\rangle$$
$$= \frac{1}{(2\pi)^s}\int e^{i\mathbf{k}\cdot(\mathbf{x}-\mathbf{y})}\cos[\omega(\mathbf{k})(t_x - t_y)]\frac{d^s k}{2\omega(\mathbf{k})}$$
$$= \frac{1}{2(2\pi)^s}\int e^{ik(x-y)}\delta(p^2 + m^2)\, d^n k \ .$$

In Fourier space this expectation value reads

$$\langle 0|\exp[i\varphi(h)]|0\rangle = \exp\{-(\pi/2)\textstyle\int|\tilde{h}(k)|^2\delta(p^2 + m^2)\, d^n k\} \ ,$$

which implies that unless the test function h contains energy and momentum values "on the mass shell", i.e., that satisfy the relativistic dispersion relation, then the expectation value in question is unity.

Another expectation value of interest involves the time-ordering operator T and is given by

$$\langle 0|T \exp[i\varphi(h)]|0\rangle = \exp\{(i/2)\int h(x)\,\Delta_F(x-y)\,h(y)\,d^n x\,d^n y\}\,,$$

where

$$\Delta_F(x-y) \equiv i\langle 0|T\varphi(x)\varphi(y)|0\rangle$$

$$= \lim_{\epsilon \to 0} \frac{1}{(2\pi)^n} \int \frac{e^{ik(x-y)}}{p^2 + m^2 - i\epsilon}\,d^n k\,,$$

where ϵ is a small positive parameter that orients the contour of integration properly around the poles in the complex $k_0(=p_0)$ plane in a contour integration, and the limit $\epsilon \to 0^+$ is taken at the end of the calculation.

Expressions involving ϵ in the manner indicated above may be written in an alternative fashion as well. In particular, for $u \in \mathbb{R}$, we observe that

$$\lim_{\epsilon \to 0} \frac{1}{u - i\epsilon} = \lim_{\epsilon \to 0} \frac{u + i\epsilon}{u^2 + \epsilon^2}$$

$$= \mathsf{P}\frac{1}{u} + i\pi\delta(u)\,.$$

Here P denotes the *principal value* integral, while the second term is based on the fact that $\int [\epsilon/(u^2 + \epsilon^2)]\,du = \pi$ identically for all $\epsilon > 0$.

If we put this relation to use for us in Fourier space, then the time-ordered vacuum expectation value reads

$$\langle 0|T \exp[i\varphi(h)]\,|0\rangle = \lim_{\epsilon \to 0} \exp\left\{\frac{i}{2} \int \frac{|\tilde{h}(k)|^2}{p^2 + m^2 - i\epsilon}\,d^n k\right\}$$

$$= \exp\left\{\frac{i}{2} \int [\mathsf{P}\frac{1}{p^2 + m^2} + i\pi\delta(p^2 + m^2)]|\tilde{h}(k)|^2\,d^n k\right\}\,.$$

This result contains two terms: (1) a phase term that exists, in general, for any nonvanishing test function h, and (2) a damping term that arises only provided the test function h contains energy and momenta of the correct relativistic energy–momentum relationship.

It is important to note that what is here referred to from a mathematical point of view as a *test function* may also be called from a physical point of view an *external source*. In particular, if the source contains the appropriate values of energy and momenta, then it is possible for the external source to create particles out of the vacuum; see Exercise 5.7. If that is the case then the probability amplitude for the new state which includes a number of such created particles has an overlap with the vacuum that is less than unity. Indeed, the absolute square of the probability amplitude $\langle 0|T \exp[i\varphi(h)]|0\rangle$ is the probability of finding the vacuum in the new state. On the other hand, if the external source does *not* contain energy and momenta in the right relationship, then the probability amplitude is simply a phase factor and it has an absolute magnitude squared of unity, showing that the probability of making real particles from the vacuum is zero whenever the correct dispersion relation is not satisfied by the external source.

5.7 Functional integral formulation

Let us follow the discussion for path integrals presented in Chapter 4, and generalize that argument to the case of scalar free fields. First recall that for finitely many degrees of freedom, the formal path integral in the presence of an external source may be given by

$$\langle 0|\mathsf{T}\exp[i\int \Sigma\, h_k(t)Q_k(t)\,dt]|0\rangle$$

$$= \lim_{\epsilon\to 0}\mathcal{N}\int \exp\bigl(i\int\{\Sigma\, h_k(t)q_k(t) + \tfrac{1}{2}\Sigma\,[\dot{q}_k(t)^2 + i\epsilon q_k(t)^2]$$

$$- V(q(t))\}\,dt\,\bigr)\,\Pi\,\mathcal{D}q_k\,,$$

where the time integral in the exponent runs from $-\infty$ to ∞ and as boundary conditions we choose $\lim_{|t|\to\infty} q_k(t) = 0$. Thanks to this infinite integration range and to the presence of a small damping factor proportional to ϵ, all final states save for the ground state are projected out. The limit $\epsilon \to 0$ is reserved for the final step. To reduce the number of symbols, the presence of both the small damping factor and the ultimate limit that $\epsilon \to 0$ are frequently left implicit, and we shall adopt such a notational shorthand as well. From a formal point of view the extension of the path integral given above to the case of infinitely many degrees of freedom leads to exactly the same expression as for finitely many degrees of freedom! Of course, a limit is involved, and that limit may exist only under special conditions. For example, it may be enough to ensure that the coordinates of the external source h_k fall to zero sufficiently fast as $k \to \infty$. In that case the existence of the limit characterizes the allowed test sequences for the problem at hand. With this discussion in mind, we may simply reinterpret the formal path integral above as applying to the case of infinitely many degrees of freedom as well. Just as we were able to trade infinitely many coordinates for fields in the classical theory and in the operator formulation of quantum field theory, we may do so again within the path integral; in a certain sense such a trade amounts to little more than a formal change of variables within an integration. Since the expression in the exponent of the formal path integral represents the classical action integral as the integral of the Lagrangian, we can – maintaining the same level of formal manipulations – pass from one expression of the classical action to another. In the present case we can pass either to fields expressed in configuration space or to fields expressed in momentum space, and for the sake of the completeness we shall illustrate both cases. Let us begin with free fields.

For the moment, we revert to the more general dispersion relation as embodied in the symbol $\Omega(\mathbf{k})$. To ease the transition we initially adopt a particle-like language and concentrate on the expression

$$\langle 0|\mathsf{T}\exp[i\int h(x)\varphi(x)\,d^n x]|0\rangle$$

$$= \langle 0|\mathsf{T}\exp\{i\int[\tilde{h}(t,\mathbf{k})^* e^{-i\Omega(\mathbf{k})t}a(\mathbf{k})$$

$$+ \tilde{h}(t,\mathbf{k})e^{i\Omega(\mathbf{k})t}a(\mathbf{k})^\dagger]\,\frac{d^s k\,dt}{\sqrt{2\Omega(\mathbf{k})}}\,\}|0\rangle$$

$$= \mathcal{N}\int \exp\{i\int[\tilde{h}(t,\mathbf{k})^* q(t,\mathbf{k})$$

$$+ \tfrac{1}{2}|\dot{q}(t,\mathbf{k})|^2 - \tfrac{1}{2}\Omega(\mathbf{k})^2|q(t,\mathbf{k})|^2]\,d^s k\,dt\}\,\mathcal{D}q\,.$$

This formal expression is an immediate consequence of the corresponding path integral for a single degree of freedom and the fact that for each $\mathbf{k}$ value, the field at $\mathbf{k}$ is independent of the field for all other $\mathbf{k}$ values. Let us reexpress this final result in the language of space and time. In that case, and using the field notation, a Fourier transformation leads directly to the formal expression

$$\langle 0|\mathsf{T}\exp[i\int h(x)\varphi(x)\,d^n x]\,|0\rangle$$

$$= \mathcal{N}\int \exp\{i\int[h(t,\mathbf{x})\phi(t,\mathbf{x}) + \tfrac{1}{2}\dot{\phi}(t,\mathbf{x})^2]\,d^s x\,dt$$

$$- \tfrac{1}{2}i\int \phi(t,\mathbf{x})\,W(\mathbf{x}-\mathbf{y})\,\phi(t,\mathbf{y})\,d^s x\,d^s y\,dt\}\,\mathcal{D}\phi\,.$$

We may specialize to the case of a relativistic free particle by a proper choice of the kernel W. In particular, the relativistic free particle of mass m may be characterized by the functional integral (including the $i\epsilon$ factor but leaving the limit $\epsilon \to 0$ implicit)

$$\langle 0|\mathsf{T}\exp[i\int h(x)\varphi(x)\,d^n x]\,|0\rangle$$

$$= \mathcal{N}\int \exp(i\int\{h(x)\phi(x) - \tfrac{1}{2}[\partial_\mu \phi(x)]^2 - \tfrac{1}{2}(m^2 - i\epsilon)\phi(x)^2\}\,d^n x\,)\,\mathcal{D}\phi\,,$$

where the field is integrated over all space-time with the boundary condition that $\phi(x)$ vanishes at both spatial and temporal infinity. Although formal, this is an expression of fundamental importance.

Recall that we have discussed the general N-dimensional Gaussian integral in Chapter 4, leading to the result that

$$\int e^{i\Sigma b_k y_k - \frac{1}{2}\Sigma y_k A_{kl} y_l}\prod dy_k = \frac{(2\pi)^{N/2}}{\sqrt{\det(A)}}e^{-\frac{1}{2}\Sigma b_k A_{kl}^{(-1)} b_l}$$

provided, of course, that $A = \{A_{kl}\}$ is a symmetric $N \times N$ matrix such that the integral converges, a condition which also ensures that A has an inverse $A^{(-1)}$ defined so that $A^{(-1)}A = 1$. Observe that the final form of the exponent, $-\frac{1}{2}\Sigma b_k A_{kl}^{(-1)} b_l$, may be determined from the exponent of the integrand, $i\Sigma b_k y_k - \frac{1}{2}\Sigma y_k A_{kl} y_l$, evaluated at the value of y_l corresponding to the extremal point, namely for $ib_k = \Sigma A_{kl} y_l$. This is an extremely useful method to determine the exponent in a Gaussian integral.

We can immediately use this result to evaluate the functional integral for the relativistic free field of mass m. The result will be the exponential of a quadratic form that is given by the extremal value of the exponent of the integrand. That extremal occurs when

$$[\Box - (m^2 - i\epsilon)]\phi(x) = -h(x) \,,$$

where we have retained the necessary factor $(i\epsilon)$ in order for the integral to have converged in the first place. The solution to this equation consistent with its vanishing at infinity is

$$\phi(x) = -\int \frac{1}{[\Box - (m^2 - i\epsilon)]} \delta(x - y) \, h(y) \, d^n y$$

$$= \int \left\{ \frac{1}{(2\pi)^n} \int \frac{e^{ip(x-y)}}{p^2 + m^2 - i\epsilon} \, d^n p \right\} h(y) \, d^n y$$

$$\equiv \int \Delta_F(x - y) \, h(y) \, d^n y \,;$$

here, F stands for "Feynman". Substitution of this extremal ϕ into the exponent of the integrand leads to the result that

$$\langle 0 | \mathrm{T} \exp\left[i \int h(x) \varphi(x) \, d^n x \right] | 0 \rangle$$

$$= \exp\{ (i/2) \int h(x) \, \Delta_F(x - y) \, h(y) \, d^n x \, d^n y \} \,,$$

where the overall normalization has been fixed by the required value of unity when $h \equiv 0$. This result confirms the one previously obtained for the generating function of the time-ordered vacuum expectation values.

Up to this point our discussion has been a purely formal one and it is important to understand under what circumstances it is permissible to argue in strictly formal terms – and also when it is not. In the present case there is, underlying our formal discussion, a fully satisfactory proper mathematical meaning to the manipulations involved. Let us focus on the example of the relativistic free particle since the discussion in the other cases is entirely analogous. To give meaning to a formal functional integral of the kind under consideration it is necessary to employ some form of *regularization* designed to give meaning to an otherwise ill-defined expression. There are a great many ways to do so, but for the present let us focus on the so-called *lattice regularization* wherein one replaces the continuum of space-time by a (hyper)cubic lattice with a uniform lattice spacing $a > 0$. Initially, we also assume that the size of the entire lattice is finite so that we have in effect replaced the field theory with its infinite number of degrees of freedom by a system with a finite number, say N, of degrees of freedom. For this system with finitely many degrees of freedom we can formulate the functional integral as an ordinary integral for N variables. Ultimately, the answer of physical interest should be given by a limit in which $N \to \infty$. Let us see how this works out in practice for the relativistic free field.

As our regularized form for the path integral we consider

$$C_N \int \exp\{i[\Sigma h_k \phi_k a^n + \tfrac{1}{2}\Sigma(\phi_{k^\#} - \phi_k)^2 a^{n-2}$$
$$- \tfrac{1}{2}\Sigma(\phi_{k^\bullet} - \phi_k)^2 a^{n-2} - \tfrac{1}{2}(m^2 - i\epsilon)\Sigma\phi_k^2 a^n]\} \prod d\phi_k .$$

There are a number of notational points to clarify in this expression: C_N denotes a normalization factor adjusted so that the result is unity if $h_k = 0$ for all k; a^n represents the volume of a unit cell on the lattice; k denotes the label of a lattice site, i.e., $k = (k^0, k^1, \ldots, k^s)$, $k^j \in \mathbb{Z} \equiv \{0, \pm 1, \pm 2, \ldots\}$; $k^\# = (k^0+1, k^1, \ldots, k^s)$, namely the next future point in time, while, analogously, k^* denotes each one of the s nearest neighbors in the (positive) spatial directions, $k^1, \ldots, k^s$. The sums run over finitely many lattice sites and nearest neighbors as needed with (say) periodic boundary conditions. It is clear that the exponent of the integrand is nothing other than a Riemann sum approximation to the integral expression that holds in the continuum, and in which, for example, h_k denotes the average value of the test function h over the lattice cell centered at the lattice site k, and correspondingly for ϕ_k. Clearly, with a lattice regularization present, only a finite number (N) of ordinary integrals are involved, and the desired answer for the generating functional of the time-ordered vacuum expectation values is formally given as the limit $N \to \infty$, at least in principle. For the case of the free field, the result of those N integrations can be readily determined by the general formula for Gaussian integrals, and so the answer in the continuum limit is obtained by taking the limit $N \to \infty$ of the expression obtained after doing the Gaussian integrations. In turn, the expression that results from this operation is recognized as the same as that obtained by directly solving for the continuum form of the exponent as we have already done, and this observation provides a justification for the procedure outlined above for determining the result of the functional integral.

5.8 Euclidean-space functional integral formulation

An appropriate analytic continuation in the "time" variable takes one from Minkowski to Euclidean space. In so doing, the integrand of the functional integral changes from an oscillatory to a positive one, but even for this real functional integral it is generally necessary to still introduce a lattice regularized formulation. Thus we may as well go directly from some form of regularization such as the Minkowski lattice formulation to the Euclidean lattice formulation by a simple analytic procedure. Let one of the lattice spacing "a" terms in the lattice formulation be identified as referring to the time direction, and analytically continue *that a* (and not any other) from a to $-ia$, a rotation of $-\pi/2$ in the complex plane. This has the effect of changing the Minkowski lattice formulation into the Euclidean expression

$$N_N \int \exp[\Sigma h_k \phi_k a^n - \tfrac{1}{2}\Sigma(\phi_{k^*} - \phi_k)^2 a^{n-2} - \tfrac{1}{2}m^2\Sigma\phi_k^2 a^n] \, \Pi \, d\phi_k \, .$$

Note that for the *time* increment, the (temporal) a in question is in the *denominator*, while for the *space* increments, the (temporal) a in question lies in the *numerator*. Thus the time- and space-increment terms respond differently to the complex rotation of the variable. In the expression for the integrand, the sum over k^* implicitly covers all n dimensions, i.e., both the former time direction and all s spatial directions, and they all enter with a common sign characteristic of a Euclidean (rather than a Minkowski) metric. This is a very important formulation for the relativistic free field, and we shall soon extend it to interacting field models as well. If we formally take the continuum limit and interchange it with the multidimensional integrations, then we are led to a formal functional integral in the case of the Euclidean field theory given by

$$S\{h\} = \mathcal{N} \int \exp(\int \{h(x)\phi(x) - \tfrac{1}{2}[(\nabla\phi)(x)^2 + m^2\phi(x)^2]\} \, d^n x \,) \, \mathcal{D}\phi \, .$$

Here we have begun to use the symbol $\mathcal{N}$ as a formal normalization constant; although its specific value may actually change from one equation to another, the manner of its definition will generally be clear from the context. The formal expression $S\{h\}$ may be regarded in either of two ways. On the one hand, it may be viewed as just a formal short-hand notation for the lattice regularized expression followed by the continuum limit. On the other hand, there is in fact a probability measure μ_F (F for "free") on distribution-valued fields such that

$$S\{h\} = \int e^{\phi(h)} \, d\mu_F(\phi)$$

holds in a well-defined mathematical sense [GeV 64, Sk 74]. In the Euclidean case under discussion, the measure μ_F is a proper measure (i.e., countably additive) unlike the case for the Minkowski functional integral where the "measure" that one often assumes is not a proper measure (i.e., it is only finitely additive); these two contrasting types of measures will be illustrated by a simple example in Chapter 6. For the present, suffice it to say that the Euclidean situation offers a level of mathematical characterization that is unavailable in the Minkowski formulation, namely that of a genuine *stochastic process*. Let us denote the ensemble average of some quantity $(\cdot)$ by $\langle(\cdot)\rangle \equiv \int(\cdot) \, d\mu_F$, and let us use the same symbol ϕ to denote the stochastic variable. In that case the Euclidean-space moment generating functional may be written in the form

$$S(h) = \langle e^{\phi(h)} \rangle = \exp\{\tfrac{1}{2}\langle [\phi(h)]^2 \rangle\}$$
$$= \exp\{\tfrac{1}{2}\int h(x) \, \langle \phi(x)\phi(y) \rangle \, h(y) \, d^n x \, d^n y \} \, .$$

Consequently, the entire expression is determined completely by the two-point correlation function $\langle \phi(x)\phi(y) \rangle$, just as the case for any mean-zero Gaussian random variable.

The form of the Euclidean-space two-point function is given by the integral expression

$$C(x - y) \equiv \langle \phi(x)\phi(y) \rangle = \frac{1}{(2\pi)^n} \int \frac{e^{ip \cdot (x-y)}}{p^2 + m^2} d^n p .$$

This integral may be explicitly evaluated in terms of modified Bessel functions [JaR 80]. However, we are not especially interested in the exact functional form of this correlation function; rather we are interested in *the behavior of $C(x - y)$ when x and y are close to one another.* In our units the distance $|x - y|$ is small when $m|x - y| \ll 1$, in which case, when $n \geq 3$, we may use the approximation that

$$C(x - y) \simeq C_0(x - y) \equiv \frac{1}{(2\pi)^n} \int \frac{e^{ip \cdot (x-y)}}{p^2} d^n p$$
$$\propto |x - y|^{-(n-2)} , \qquad n \geq 3 ,$$

where the indicated functional dependence for small $|x - y|$ is essentially determined simply on dimensional grounds. This short-distance behavior is characteristic of all relativistic free fields in dimensions $n \geq 3$ and is independent of the mass m. In two space-time dimensions ($n = 2$) the mass parameter cannot be eliminated from the denominator, for otherwise the integral would diverge at zero momentum as well as infinite momentum. The short-distance behavior in the case of two space-time dimensions involves the mass m and (with $p \equiv mr$) is given by

$$C(x - y) = \frac{1}{(2\pi)^2} \int \frac{e^{imr \cdot (x-y)}}{r^2 + 1} d^2 r$$
$$\propto -\ln(m|x - y|) , \qquad n = 2 .$$

The short-distance behavior for the two-point correlation functions that we have arrived at will be of great significance as we turn our attention to the construction of local powers of the field. We shall find the proper proportionality constant for each of the cases later.

Local products

As a preliminary to the study of additional scalar field models we take up the question of *local powers* of the stochastic variable $\phi(x)$. A brief argument will convince us that we cannot use the conventional notion of local power. First of all, we note that for any $n \geq 2$ it already follows that $\langle \phi(x)^2 \rangle = C(0) = \infty$. For some random variables this condition may simply imply the divergence of a moment, but for a mean-zero Gaussian variable it actually implies that $\phi(x)^2 =$

∞ almost everywhere.* Let us instead focus on *normally-ordered local products*, which, because of the similarity with the definition in the case of operator fields, we denote by $: \phi(x)^p :$, $p \in \{1, 2, 3, \ldots\}$, and which we formally define by the generating function

$$: e^{c\phi(x)} : \equiv \frac{e^{c\phi(x)}}{\langle e^{c\phi(x)} \rangle}$$

$$\equiv e^{c\phi(x) - \frac{1}{2}c^2\langle\phi(x)^2\rangle} .$$

Properly, the calculation of normal-ordered powers proceeds with the argument of the generating function a smeared field, then expanding in powers of c, after which the smearing is removed. However, from a formal point of view the same result arises if we simply expand both sides of the formal generating function in powers of the parameter c. For example, it formally follows that

$$: \phi(x) : = \phi(x) ,$$
$$: \phi(x)^2 : = \phi(x)^2 - \langle\phi(x)^2\rangle ,$$
$$: \phi(x)^3 : = \phi(x)^3 - 3\phi(x)\langle\phi(x)^2\rangle ,$$
$$: \phi(x)^4 : = \phi(x)^4 - 6\phi(x)^2\langle\phi(x)^2\rangle + 3\langle\phi(x)^2\rangle^2 ,$$

etc. In these expressions observe that $\langle\phi(x)^2\rangle = C(0) = \langle\phi(0)^2\rangle$ is a divergent constant that is independent of the point x. To see that such expressions make sense as distributions we first observe – without proof for the moment – that

$$\langle [\int h(x) : \phi(x)^p : d^n x]^2 \rangle = p! \int h(x) \langle\phi(x)\phi(y)\rangle^p h(y) \, d^n x \, d^n y .$$

This expression converges at infinity because we are dealing with test functions that fall off at infinity as fast as we like. On the other hand, convergence (or lack of it) when $x \approx y$ is quite another matter. For $n = 2$ the short distance behavior has a logarithmic divergence $-\ln(m|x - y|)$, and so we may conclude that

$$\int_{m|x-y|<1} [-\ln(m|x - y|)]^p \, d^n x < \infty$$

holds for all $p \geq 0$. For $n \geq 3$, the convergence depends on a specific relationship of p and n. Since the short distance behavior of the correlation function behaves

---

* Let X denote a mean-zero Gaussian random variable. Then it follows that

$$\langle e^{-\frac{1}{2}X^2} \rangle = \frac{1}{(2\pi)^{1/2}} \int e^{-\frac{1}{2}t^2} \langle e^{itX} \rangle \, dt$$

$$= \frac{1}{(2\pi)^{1/2}} \int e^{-\frac{1}{2}(1+\langle X^2\rangle)t^2} \, dt = \frac{1}{\sqrt{1 + \langle X^2 \rangle}} .$$

Thus if $\langle X^2 \rangle = \infty$ it follows that $\langle \exp[-\frac{1}{2}X^2] \rangle = 0$, which can only hold provided $X^2 = \infty$ almost everywhere.

as $|x - y|^{-(n-2)}$, we will have convergence when $x \approx y$ provided that

$$\int_{m|x-y|<1} \frac{1}{|x - y|^{p(n-2)}} d^n x < \infty \, ,$$

and this criterion demands that $p(n - 2) < n$. That is, we have convergence whenever

$$p < \frac{n}{n - 2} \, ,$$

a relation that holds for all $n \geq 2$, assuming the proper interpretation in the case $n = 2$. The finiteness of such a moment implies that the stochastic variable

$$: \phi^p : (h) \equiv \int h(x) : \phi(x)^p : d^n x \, , \qquad p < n/(n - 2)$$

is almost everywhere finite. Therefore, $: \phi^p : (h)$ is necessarily well defined almost everywhere. These normal ordered powers will serve as the local powers, especially when we seek to go beyond free theories.

We now return to the proof that the second moment of the stochastic variable $: \phi^p : (h)$ has the indicated form. The basic arguments in the proof are already present in the simplest case where $p = 2$. Because the stochastic process under consideration is Gaussian with a zero mean, it follows that

$$\langle \phi(u)\phi(v)\phi(x)\phi(y) \rangle = \langle \phi(u)\phi(v) \rangle \langle \phi(x)\phi(y) \rangle$$
$$+ \langle \phi(u)\phi(x) \rangle \langle \phi(v)\phi(y) \rangle + \langle \phi(u)\phi(y) \rangle \langle \phi(v)\phi(x) \rangle \, .$$

Now we specialize this expression by setting $u = x$ and $v = y$. Thus we are led to the relation that

$$\langle \phi(x)^2 \phi(y)^2 \rangle = 2\langle \phi(x)\phi(y) \rangle^2 + \langle \phi(x)\phi(x) \rangle^2 \, .$$

The term on the left-hand side and the last factor on the right-hand side are divergent, and we now move the two divergent terms to the same (left-hand) side of the equation to learn that

$$\langle : \phi(x)^2 :: \phi(y)^2 : \rangle = 2\langle \phi(x)\phi(y) \rangle^2 \, .$$

In like manner it follows that all divergences from coinciding points are subtracted off in forming the higher order, normally ordered local powers, thus confirming the quoted relation for the moments.

5.9 Interacting scalar fields

As a typical example of an interacting scalar quantum field theory let us focus on the formal functional integral given by

$$\mathcal{C} \int \exp(\int \{h(x)\phi(x) - \tfrac{1}{2}[\nabla \phi(x)]^2 - \tfrac{1}{2}m^2 \phi(x)^2 - g : \phi(x)^4 : \} d^n x \,) \, \mathcal{D}\phi \, .$$

Let us set about trying to give some mathematical meaning to this formal expression. The constant term that is part of the normal ordering is already included in the overall normalization factor $\mathcal{C}$, which may always be determined at the end of the calculation by an appropriate normalization condition.

If we wish to give better meaning to the formal functional integral we may invoke the stochastic process that describes the free field of mass m. Once again we denote by μ_F the measure on fields that describes the free field. Thus a somewhat better formulation of the formal path integral is given by

$$\mathcal{C} \int \exp\{\int [h(x)\phi(x) - g : \phi(x)^4 :] d^n x\} \, d\mu_F(\phi) \, .$$

This expression is still formal and thus unsatisfactory, however, because the local density $: \phi(x)^4 :$ leads to a stochastic variable only when integrated against a test function and, in particular, not the function "one" for all space-time. A closer look at the test function criterion reveals that the function "one" defined over a compact region, and zero elsewhere, would be an acceptable test function in the present case. Thus we modify our formal path integral once again to read

$$S(h) = \lim_{\Lambda \to \infty} C_\Lambda \int \exp\{\int_\Lambda [h(x)\phi(x) - g : \phi(x)^4 :] d^n x\} \, d\mu_F(\phi) \, ,$$

where by Λ we mean a finite volume of space-time which we can take to be a (hyper)cube centered on the origin of coordinates, and by $\Lambda \to \infty$ we mean that the volume approaches infinity in a natural and symmetric way. Actually this formulation is entirely satisfactory whenever the locally integrated normal-ordered power is a proper stochastic variable, and that situation occurs, as derived above, whenever $p < n/(n-2)$, or specifically for $p = 4$ when $n < 8/3$, i.e., $n = 2$. Thus the formulation of the Euclidean φ_2^4 field theory offered above is a perfectly acceptable one and the theory has been successfully analyzed from this starting point [GlJ 87, FeFS 92]. Unfortunately, it is not a suitable starting point for higher space-time dimensions.

Perturbation theory

Let us take up the important question of a perturbation analysis in a power series in the coupling constant g. The perturbation approach to study the question of the existence and the (approximate) calculation of $S(h)$ is an important one and can be used for any space-time dimension n that one chooses. To a certain extent some features of the calculation are common to any dimension, and so let us discuss those initially. For the sake of convenience, we employ a formal notation suppressing the volume Λ and its associated limit. Thus, first of all, let us consider the formal power series given by

$$S(h) = \sum_{q=0}^{\infty} \frac{(-g)^q}{q!} C' \langle e^{\phi(h)} [\int : \phi(x)^4 : d^n x]^q \rangle$$

$$= \sum_{q=0,\, p=0}^{\infty} \frac{(-g)^q}{q!p!} C' \langle [\phi(h)]^p [\int : \phi(x)^4 : d^n x]^q \rangle \,.$$

In the last line the problem has been reduced, in effect, to a sum of correlation functions of the field in the Gaussian ensemble of the relativistic free field with mass m. In turn, this leads to a multitude of contributions each of which involves sums of products of the two-point correlation function of the Gaussian ensemble. The factor C' is to be chosen so that the overall normalization $S(0) = 1$. In practice, the effect of this normalization can be most simply incorporated within perturbation theory by confining attention to those contributions that are entirely connected to the external terms $\phi(h)$ [ItZ 80, Ka 93, PeS 95].

As we have discussed in Chapter 4 and indicated earlier in this chapter for the four-point function, the correlation functions of a mean-zero Gaussian ensemble are entirely determined by the two-point function

$$C(x - y) \equiv \langle \phi(x)\phi(y) \rangle = \frac{1}{(2\pi)^n} \int \frac{e^{ip\cdot(x-y)}}{p^2 + m^2} d^n p$$

$$\simeq \frac{k_n}{|x-y|^{n-2}}, \qquad m|x-y| \ll 1, \quad n \geq 3 \,,$$

$$\simeq -k_2 \ln(m|x-y|), \qquad m|x-y| \ll 1, \quad n = 2 \,,$$

where the vector square and vector inner product are taken here in the Euclidean metric. The last two lines are devoted to the short distance behavior in the two cases $n \geq 3$ and $n = 2$. Here the constants k_n may be determined by the following relatively simple argument. For $n \geq 3$ we consider the distributional relation

$$\frac{k_n}{|x|^{n-2}} \equiv \frac{1}{(2\pi)^n} \int \frac{e^{ip\cdot x}}{p^2} d^n p$$

which holds simply on the grounds of a change of variables in the integration. We now integrate both sides of this expression with a Gaussian test function leading to

$$\int \frac{k_n}{|x|^{n-2}} e^{-x^2/2} d^n x = \frac{1}{(2\pi)^n} \int \frac{e^{ip\cdot x}}{p^2} e^{-x^2/2} d^n p \, d^n x$$

$$= \frac{1}{(2\pi)^{n/2}} \int \frac{1}{p^2} e^{-p^2/2} d^n p \,.$$

Let us introduce $u^2 \equiv x^2/2$ and $v^2 \equiv p^2/2$, where $0 < u < \infty$ as well as $0 < v < \infty$. Since the angular integrations on both sides cancel we are left with just one-dimensional radial integrals leading to

$$k_n \int_0^\infty e^{-u^2} u \, du = \frac{1}{4(\pi)^{n/2}} \int_0^\infty e^{-v^2} v^{n-3} \, dv \,,$$

from which we conclude that

$$k_n = \frac{\Gamma(n/2 - 1)}{4(\pi)^{n/2}} \,, \qquad n \geq 3 \,.$$

For $n = 2$ we can use the fact that as $m \to 0$ the natural logarithm holds over an ever increasing range of space-time. Let us again integrate with a Gaussian test function leading to

$$\int e^{-x^2/2} \left[-k_2 \ln(m|x|) \right] d^2x \simeq \frac{1}{(2\pi)} \int \frac{e^{-p^2/2}}{p^2 + m^2} \, d^2p \,,$$

which as m becomes smaller and smaller we may separate into

$$-k_2 \ln(m) \int e^{-x^2/2} \, d^2x + l.o.t.$$

$$= \frac{1}{(2\pi)} \int_{|p|^2 < 1} \frac{1}{p^2 + m^2} \, d^2p$$

$$+ \frac{1}{(2\pi)} \int_{|p|^2 < 1} \frac{[e^{-p^2/2} - 1]}{p^2 + m^2} \, d^2p + l.o.t. \,,$$

where *l.o.t.* means lower order terms. It follows directly from the leading divergences as $m \to 0$ that

$$k_2 = \frac{1}{2\pi} \,.$$

Space-time dimension $n = 3$

For $n = 3$, as we shall observe below, the naturally defined formal theory has one ultraviolet divergence which requires an infinite mass renormalization. This means that besides normal ordering it is necessary to add an additional, divergent multiple of : $\phi(x)^2$:. The need for such a modification may be readily demonstrated. For the present we proceed formally, and consider the two-point function given by

$$\langle \phi(h)^2 \exp\{-g \int : \phi(x)^4 : d^3x\} \rangle \,.$$

We expand the exponent in a power series in g and focus in particular on the second-order term

$$\tfrac{1}{2} g^2 \int \langle \phi(h)^2 : \phi(x)^4 :: \phi(y)^4 : \rangle \, d^3x \, d^3y \,.$$

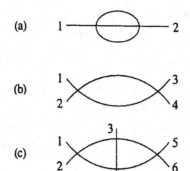

Fig. 5.1. Several contributions to the generating functional represented graphically. The free ends labelled 1, 2, 3, etc., each denote a separate factor $\phi(h)$. The crosses resulting from the intersection of two lines represent terms of the form $: \phi(\cdot)^4 :$ arising from the interaction. Each intersection is accompanied by one power of the coupling constant g.

Among the several contributions this integral leads to, we single out a particular term, conveniently represented by Fig. 5.1, part (a), for further consideration. The term in question is analytically given by*

$$96\,g^2 \int \langle\phi(h)\phi(x)\rangle\langle\phi(x)\phi(y)\rangle^3 \langle\phi(y)\phi(h)\rangle\,d^3x\,d^3y$$

$$\equiv 96\,g^2 \int \overline{h}(x)\,C(x-y)^3\,\overline{h}(y)\,d^3x\,d^3y$$

$$\simeq [96/(4\pi)^3]\,g^2 \int \overline{h}(x)\,|x-y|^{-3}\,\overline{h}(y)\,d^3x\,d^3y$$

$$= (3/2\pi^3)\,g^2 \int \overline{h}(u+v/2)\,|v|^{-3}\,\overline{h}(u-v/2)\,d^3u\,d^3v\;,$$

where $\overline{h}(x) \equiv \langle\phi(h)\phi(x)\rangle$. Due to the rapid fall off of $\overline{h}(x)$ for large $|x|$, the integrals over u and v may be considered to be restricted to a finite domain of $\mathbb{R}^3$, say $|v| \leq 1$. On the other hand, there is a divergence in the integral that arises at $v = 0$. Observe that if a Taylor expansion is made about u of either term $\overline{h}$, then the extra v terms that appear render the integral over v convergent at the origin. Thus the divergence is equivalent to that of

$$(3/2\pi^3)\,g^2 \int \overline{h}(u)\,|v|^{-3}\,\overline{h}(u)\,d^3u\,d^3v\;.$$

This term is first made finite by regularization, say by replacing $|v|^{-3}$ by $(|v| + \epsilon)^{-3}$ for a very small $\epsilon > 0$. However, we also may note that

$$[4\pi|\ln(\epsilon)|]^{-1}(|v| + \epsilon)^{-3}$$

---

* The prefactor $96 = (4!)^2/3!$ is a combinatorial factor representing the number of similar terms arising from the previous expression. For rules on such factors see, e.g., [ItZ 80].

converges to the three-dimensional delta function as $\epsilon \to 0$. Thus we observe that we can renormalize our regularized expression by introducing

$$\lim_{\epsilon \to 0} (3/2\pi^3)\, g^2 \int \overline{h}(x)\{(|x-y|+\epsilon)^{-3} - [4\pi|\ln(\epsilon)|]\, \delta(x-y)\}\overline{h}(y)\, d^3x\, d^3y$$

which now has the virtue of being finite. Note that the effect of the $\delta(x-y)$ term is identical to that which would follow from an additional interaction proportional to $\frac{1}{2} : \phi(x)^2 :$. Therefore, if we instead reconsider the problem stated in the form

$$\lim_{\epsilon \to 0} \langle \phi(h)^2 \exp\{-\textstyle\int [g : \phi_\epsilon(x)^4 : + c_\epsilon\, g^2 : \phi(x)^2 :]\, d^3x\}\rangle \, ,$$

where $\phi_\epsilon(x)$ is a field with the regularized two-point function,

$$\langle \phi_\epsilon(x)\phi_\epsilon(0)\rangle = (1/4\pi)(|x|+\epsilon)^{-1}\, , \qquad |x| \ll 1\, ,$$

and

$$c_\epsilon \equiv (3/\pi^2)|\ln(\epsilon)|\, ,$$

then we would have eliminated the indicated divergence to order g^2. It is a triumph of combinatorics and renormalization theory that in this case this is the *only* counterterm needed to render the expansion finite to all orders of the coupling constant g. Although by itself this fact is suggestive, it no way proves that this regularization and renormalization provide a meaningful regularization and renormalization procedure *outside* of perturbation theory. Nevertheless, this is the case. In a triumph of mathematical physics (see, e.g., [FeFS 92]) it has been shown that the expression

$$\lim_{\Lambda \to \infty} \lim_{\epsilon \to 0} C_\Lambda \langle \exp\{\textstyle\int_\Lambda [h(x)\phi(x) - g : \phi_\epsilon(x)^4 : - c_\epsilon\, g^2 : \phi(x)^2 :]\, d^3x\}\rangle$$

converges to a nontrivial result and that a perturbation theory in the coupling constant g is asymptotic to that result.

Space-time dimension $n = 4$

For $n = 4$ the naturally defined formal theory has three distinct categories of ultraviolet divergences which require a mass renormalization, a field strength renormalization, and a coupling constant renormalization. In fact, it is necessary to add counterterms for each of the three basic contributions in the formal action. Let us initially argue for the first two types of counterterms. The start of our analysis follows the discussion in the case $n = 3$, and we again consider the two-point function given by

$$\langle \phi(h)^2 \exp\{-g\textstyle\int : \phi(x)^4 : d^4x\}\rangle \, .$$

Let us expand the exponent in a power series in g and focus again, in particular, on the second-order term

$$\tfrac{1}{2}g^2 \int \langle \phi(h)^2 : \phi(x)^4 :: \phi(y)^4 : \rangle\, d^4x\, d^4y \, .$$

Among the several contributions this integral leads to, we single out the following term [see Fig. 5.1, part (a), again]:

$$96 \, g^2 \int \langle \phi(h)\phi(x)\rangle \langle \phi(x)\phi(y)\rangle^3 \langle \phi(y)\phi(h)\rangle \, d^4x \, d^4y$$

$$\equiv 96 \, g^2 \int \overline{h}(x) \, C(x-y)^3 \, \overline{h}(y) \, d^4x \, d^4y$$

$$\simeq [96/(4\pi^2)^3] \, g^2 \int \overline{h}(x) \, |x-y|^{-6} \, \overline{h}(y) \, d^4x \, d^4y$$

$$= (3/2\pi^6) \, g^2 \int \overline{h} \, (u+v/2)|v|^{-6} \, \overline{h}(u-v/2) \, d^4u \, d^4v \, ,$$

where the integrals over u and v may be considered to be restricted to a finite domain of $\mathbb{R}^4$, say $|v| \leq 1$. On the other hand, there is a divergence in the integral that arises at $v = 0$. Observe if a Taylor expansion is made of either term $\overline{h}$, then after two such terms, the extra v terms that appear render the integral over v convergent at the origin. Thus the divergence is equivalent to that for

$$(3/2\pi^6) \, g^2 [\int \overline{h}(u) \, |v|^{-6} \, \overline{h}(u) \, d^4u \, d^4v$$

$$- (1/4) \int \overline{h}_{,j} \, (u) \, v^j \, |v|^{-6} \, v^k \, \overline{h}_{,k} \, (u) \, d^4u \, d^4v$$

$$+ (1/4) \int \overline{h}_{,jk} \, (u) \, v^j \, |v|^{-6} \, v^k \, \overline{h}(u) \, d^4u \, d^4v] \, ;$$

a possible term involving only one derivative vanishes, on regularization, by symmetry. We shall regularize the divergences by replacing the factor $|v|$ in the denominator by $(|v| + \epsilon)$ just as before. Let us consider the two contributions separately. The first term becomes $\int (|v| + \epsilon)^{-6} \, d^4v = \epsilon^{-2}\pi^2$, which may be renormalized by the introduction of a mass counterterm of the form $(3/4\pi^4) \, g^2\epsilon^{-2} : \phi(x)^2 :$. In the two other factors only those terms for which $j = k$ contribute due to symmetry, each such term has the same divergence, and therefore the regularized form of such terms is also given, with summation on j implied, by

$$-(3/16\pi^6) \, g^2 \int \overline{h}_{,j} \, (u)(|v| + \epsilon)^{-4} \overline{h}_{,j} \, (u) \, d^4u \, d^4v \, .$$

The divergence here is given by $\int_0^1 (|v| + \epsilon)^{-4} \, d^4v = 2\pi^2 \, |\ln(\epsilon)|$, which in turn may be canceled by the counterterm $(3/16\pi^4) \, g^2 \, |\ln(\epsilon)| : (\nabla\phi)(x)^2 :$. This kind of renormalization is referred to as field strength renormalization.

To deal with the coupling constant renormalization we focus next on another expression of the form

$$\tfrac{1}{2}g^2 \int \langle \phi(h)^4 : \phi(x)^4 :: \phi(y)^4 : \rangle \, d^4x \, d^4y \, .$$

Of all the terms to which this gives rise we consider further the term [see Fig. 5.1, part (b)] [note $288 = (4!)^2/2!$]

$$288\, g^2 \int \overline{H}(x)\, C(x-y)^2\, \overline{H}(y)\, d^4x\, d^4y$$

$$\simeq [288/(4\pi^2)^2]\, g^2 \int \overline{H}(x)\, |x-y|^{-4}\, \overline{H}(y)\, d^4x\, d^4y\ .$$

Here $\overline{H}(x) \equiv \langle \phi(h)\phi(x) \rangle^2 = \overline{h}(x)^2$ which is a smooth function that we may assume falls to zero rapidly as $|x|$ becomes large. Thus, as before, the divergence arises when $x = y$. The term in question, therefore, is equivalent to

$$(18/\pi^4)\, g^2 \int \overline{H}(u + v/2)\, |v|^{-4}\, \overline{H}(u - v/2)\, d^4u\, d^4v\ ,$$

and the divergence is given by

$$(18/\pi^4)\, g^2 \int \overline{H}(u)\, |v|^{-4}\, \overline{H}(u)\, d^4u\, d^4v\ .$$

Regularization leads to the integral $\int_0^1 (|v|+\epsilon)^{-4}\, d^4v = (2\pi^2)|\ln(\epsilon)|$, which may be compensated by the counterterm $(3/2\pi^2)\, g^2|\ln(\epsilon)| : \phi(x)^4 :$. Such a modification amounts to a renormalization of the quartic interaction coupling constant.

If we included the three renormalization counterterms then we would have eliminated the indicated divergence to order g^2. However, and unlike the case of $n = 3$, when higher-order terms in the perturbation series are considered additional divergences arise, but, fortunately, they all may be subsumed into modifications of the mass, coupling constant, and the coefficient of the gradient-squared term in the formal Lagrangian. Thus the coefficients of these factors become power series in the coupling constant, all terms of which have a common characteristic divergence behavior. This behavior is typical of theories that are called strictly renormalizable. Although by itself this fact is suggestive, it no way proves that this kind of regularization and renormalization provide a meaningful regularization and renormalization procedure *outside* of perturbation theory. In fact, it is now widely believed that this is *not* the case, namely, that the full theory treated outside of perturbation theory yields a qualitatively different answer from that given by the perturbation solution. Indeed, outside of perturbation theory, the quartic self-coupled scalar quantum field theory is believed to be a *free* theory, i.e., to lead to a free field theory or perhaps to a generalized free field theory [FeFS 92, Ca 88]. It is of considerable interest that the results of a perturbation treatment, on the one hand, and of a nonperturbative analysis, on the other hand, can have such diverse answers.

Fortunately, this dilemma seems to have a resolution. A careful examination of the perturbation series, especially including a partial resummation of certain selected contributions, leads one to believe that a resummed perturbation series also may well lead to a free theory result. Let us give a qualitative picture of such

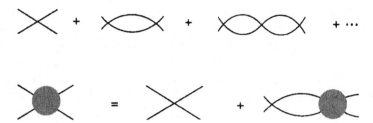

Fig. 5.2. A graphical representation for the sum of an infinite number of terms of a special family of contributions. The top line illustrates the separate graphs each of which has the interpretation presented for Fig. 5.1. The bottom line graphically represents an equation, in fact an integral equation, whose solution leads, simply by iteration, to the complete sum of the contributions represented in the top line.

an argument. First consider Fig. 5.2. This figure represents an integral equation for the summation of an infinite number of contributions thereby leading to an expression beyond any finite order of perturbation theory. The qualitative structure of such a partial summation is of the form

$$\frac{gC}{(1 + g|\ln(\epsilon)|C)},$$

where C denotes a finite, positive contribution the details of which are not important for the present discussion (an explicit example of this type will be given in Chapter 9). The essential fact to note is that as the regularization is removed, i.e., as $\epsilon \to 0$, the denominator diverges sending the overall term to zero. In other words, although order by order the contributions to the interaction involve divergences, when selected contributions are resummed it follows that the overall contribution of an infinite number of terms actually forces the nonlinear coupling constant to vanish; this kind of argument is said to be due to a "Landau pole" (or "Landau ghost") [ItZ 80]. Here, then, in the context of perturbation theory is an argument, heuristic to be sure, that the "complete" theory including the nonlinear interaction corresponds to a noninteracting theory. This is some measure of support for the triviality of the conventionally formulated ϕ_4^4, the truth of which has been all but rigorously established [FeFS 92]. We will revisit the ϕ_4^4 theory again in Chapter 11.

Space-time dimension $n \geq 5$

For $n = 5$ the naturally defined formal theory has an unlimited number of distinct categories of ultraviolet divergences which require not only a mass renormalization, a field strength renormalization, and a coupling constant renormalization, but require the introduction and renormalization of an infinite number of additional self interactions. In other words, the formal starting Lagrangian that contains a quartic interaction ϕ^4 must be supplemented by an unending series of interactions of the form $: \phi^{2p} :$ where $p = 3, 4, 5, \ldots$ (as well as other counterterms involving derivatives). The divergent part of the coefficients of these auxiliary interactions is determined so as to cancel those divergences that arise from lower order interactions. However, that leaves open the question of any finite part remaining for these interactions, and the arbitrariness in these factors, as the argument is generally made, renders such theories useless for any predictive purposes, inasmuch as an infinite number of measurements need to be made first so as to fix the finite parts of these needed counterterm interactions. Such theories are called nonrenormalizable, and it is clear from this qualitative account that they lead to essentially insurmountable difficulties. Our purpose here is not to offer any miracle cure for such "sick" theories, but rather to indicate how the need for such an endless chain of higher-order interactions arises. (We shall revisit these theories in Chapter 11.) For convenience, we shall confine our discussion to the case $n = 5$.

To assess the specific need for renormalizations and various counter-terms for such a theory, we accept the lessons we have learned from the analysis in space-time dimensions $n = 3$ and $n = 4$. In particular, for $n = 4$ we found the need for a coupling constant renormalization. In the present case let us examine the term

$$(1/4!)g^4 \langle \phi(h)^6 \, [\int : \phi(x)^4 : d^5x]^4 \rangle \, .$$

Observe that this term is higher-order in g than we have previously considered. Among the several contributions to which this factor gives rise, we focus on [see Fig. 5.1, part (c)] [note $144 = (4!)^2/(2!)^2$]

$$144\,g^4 \int \overline{H}(w)\,\overline{H}(x)\,\overline{h}(y)\,\overline{h}(z)\,C(x-y)\,C(x-z)\,C(y-z)$$
$$\times\, C(w-y)\,C(w-z)\,d^5w\,d^5x\,d^5y\,d^5z \, ,$$

where $\overline{h}$ and $\overline{H}$ have their previous meanings. As was the case before, the integration domains are effectively restricted for large arguments. For short distances $C(x-y) \simeq (1/8\pi^2)|x-y|^{-3}$, and therefore we are faced with estimating

$$144(1/8\pi^2)^5\,g^4 \int \overline{H}(w)\,\overline{H}(x)\,\overline{h}(y)\,\overline{h}(z)\,|x-y|^{-3}\,|x-z|^{-3}\,|y-z|^{-3}$$
$$\times\, |w-y|^{-3}\,|w-z|^{-3}\,d^5w\,d^5x\,d^5y\,d^5z \, .$$

We may estimate this expression by carrying out the integrations over both w and x which lead to a result proportional to

$$g^4 \int \overline{h}(y)\,|y-z|^{-5}\,\overline{h}(z)\,d^5y\,d^5z \, .$$

A change of variables leads to

$$g^4 \int \overline{h}\,(u+v/2)|v|^{-5}\,\overline{h}(u-v/2)\,d^5u\,d^5v \, ,$$

and much as before the divergent part is given by

$$g^4 \int \overline{h}(u)\, |v|^{-5}\, \overline{h}(u)\, d^5u\, d^5v \;,$$

which is logarithmically divergent, i.e., proportional to $|\ln(\epsilon)|$. For present purposes it is sufficient to determine that the integral in question is divergent, and we do not need to explicitly evaluate the coefficient. The counterterm for this divergence is proportional to $g^4|\ln(\epsilon)| : \phi(x)^6 :$, which is seen to be an entirely new local interaction. In turn, this nonvanishing interaction may be used in the evaluation of even further contributions. For example, we may next examine

$$(g^7/3!)\, |\ln(\epsilon)| \, \langle \phi(h)^8 \textstyle\int : \phi(w)^6 : d^5w\, [\int : \phi(x)^4 : d^5x]^3 \rangle$$

which, among others, contributes a term proportional to

$$g^7\, |\ln(\epsilon)| \, \int \overline{H}(w)\, \overline{H}(x)\, \overline{h}(y)^3\, \overline{h}(z)\, C(x-y)\, C(x-z)\, C(y-z)$$
$$\times\, C(w-y)\, C(w-z)\, d^5w\, d^5x\, d^5y\, d^5z \;,$$

with an integral which is just as divergent as was the previously considered case. This particular divergence requires the introduction of a counterterm proportional to $g^7\,|\ln(\epsilon)|^2 : \phi(x)^8 :$. And so it goes *ad infinitum* with the need to introduce counterterms of an ever higher order.

One may ask the question of how these conclusions obtained on the basis of perturbation theory relate to conclusions that can be drawn from nonperturbative analyses. In fact, a nonperturbative analysis of such models has been made on the basis of a Euclidean lattice model formulation (see next section), and it has been shown [FeFS 92] that models such as ϕ_5^4 rigorously tend to free or generalized free theories in the limit that the regularization is removed. In other words, while the perturbation analysis requires an unending set of higher-order counterterms for its definition order by order in the coupling constant, the nonperturbative analysis leads to a theory in which all nonlinear interactions effectively vanish. One may again argue – as was the case for $n = 4$ – that a partial summation of suitable contributions leads to the suppression of any nonlinear interaction, and so in this heuristic sense the results of the nonperturbative study may be qualitatively understood.

We note without proof that for the scalar quantum field models ϕ_n^4, with $n \geq 6$, the results are qualitatively the same as for $n = 5$ and all such theories are nonrenormalizable.

Coupling constant dimensionality

It is useful to observe that the qualitative characteristics of the results obtained above for the properties of ϕ_n^4 models can also be determined simply on the basis of coupling constant dimensionality. In units where $\hbar = 1$, or alternatively regarded as dimensionless, and all other quantities measured in mass dimensions, it follows that $\dim(x) = -1$, $\dim(\phi) = (n-2)/2$, and $\dim(g) = 4-n$. Now in any perturbation series divergences occur at short distances or equivalently at high momentum, regularized by a cutoff we shall generically call Λ. We assume that Λ, which has mass dimension one, is huge and is the dominant dimensional term outside of the coupling constant. Thus a general expression – for example, the counterterm for some particular interaction term – will involve a series of terms such as $\sum_{r,s}^{\infty} C_{rs}\, g^r \Lambda^s$, where C_{rs} denote dimensionless constants and $r \geq 0$. If $s = 0$ the indicated term is to be interpreted as $\ln(\Lambda)$, while if $s < 0$, that factor is taken as unity. Hereafter, the argument is based simply on dimensional grounds. If the mass dimension of g is positive, as it is for $n \leq 3$, then only a finite number of r-terms and s-terms will contribute. If the mass dimension of g is zero, as it is for $n = 4$, then arbitrarily many r-terms may appear, although only a limited number of s-terms. However, if the mass dimension of g is negative, as it is for $n \geq 5$, then arbitrarily many r-terms and a related number of s-terms will appear in the generic sum of this kind. In short, a theory is superrenormalizable if its coupling constant has positive mass dimension, is strictly renormalizable if its coupling constant has zero mass dimension, and it is nonrenormalizable if its coupling constant has negative mass dimension.

Indeed, this characterization of interacting models goes beyond the quartic interaction that has been our main concern. Consider, more generally, scalar quantum field theory models ϕ_n^p, namely, those with a pth-power self interaction in an n-dimensional space-time. The dimensionality of the associated coupling constant is determined to be $p + n(1 - p/2)$. On the basis of dimensionality alone we are led to conclude that for a superrenormalizable theory, $p < 2n/(n-2)$, for a strictly renormalizable theory, $p = 2n/(n-2)$, and for a nonrenormalizable theory, $p > 2n/(n-2)$. In point of fact, detailed calculations, very much along the lines outlined above for the special case $p = 4$, confirm these assignments.

We shall have more to say about issues of renormalizability and nonrenormalizability in subsequent chapters.

5.10 Euclidean-space lattice regularization

An alternative regularization provides another approach to self-interacting scalar quantum field theories, and it is a natural generalization of the lattice formulation already studied in the case of the relativistic free theories. In particular, let us consider

$$S(h) = C \int \exp\{\Sigma[h_k\, \phi_k\, a^n - \tfrac{1}{2}(\phi_{k^*} - \phi_k)^2 a^{n-2}$$
$$- \tfrac{1}{2}m_0^2\, \phi_k^2\, a^n - g : \phi_k^4 : a^n\,]\}\, \Pi\, d\phi_k\,.$$

In this expression we assume that the total number of lattice sites $N \equiv L^n$, say, is finite, and recall that the continuum limit means not only that $a \to 0$, but that $N \to \infty$ in such a way that eventually the integration covers all of $\mathbb{R}^n$. Guided by the continuum theory, we have included normal ordering for the quartic interaction term. For $n = 2$ and $n = 3$, and allowing for an appropriate definition of the bare parameters, it has been demonstrated that the continuum limit of this lattice formulation generates a nontrivial Euclidean-space quantum field theory, which in addition satisfies the appropriate axioms to permit a transformation back to a Minkowski theory, and which yields a theory that is equivalent to the one defined asymptotically by its perturbation series in the coupling constant g. On the other hand, for $n \geq 5$, and quite probably for $n = 4$, the so defined theory leads to a Gaussian – often called trivial – result [FeFS 92]. We will touch upon these results in Chapter 11.

Exercises

5.1 We have shown that Fock space is spanned by nonnormalized coherent states of the form

$$|\psi\rangle \equiv e^{\int \psi(\mathbf{k})a(\mathbf{k})^\dagger \, d^s k}|0\rangle = e^{a^\dagger(\psi)}|0\rangle \, ,$$

for all *complex* $\psi \in L^2(\mathbb{R}^s)$. Show that the same space is spanned if the functions $\psi(\mathbf{k})$ are restricted to be *real* functions. This fact establishes that the set of coherent states for strictly real arguments is also a total set for Fock space.

5.2 Assume that the operator $\mathcal{P}$ represents the space-translation generator which is endowed with the fundamental properties (for $\hbar = 1$) that

$$i[\mathcal{P}, \phi(\mathbf{x})] = -\boldsymbol{\nabla}\phi(\mathbf{x}) \, , \qquad i[\mathcal{P}, \pi(\mathbf{x})] = -\boldsymbol{\nabla}\pi(\mathbf{x}) \, .$$

Let $|f, g\rangle \equiv \exp\{i[\phi(f) - \pi(g)]\}|0\rangle$ denote the usual coherent states, and assume that $\mathcal{P}|0\rangle = 0$. Show that these properties suffice to establish that

$$\langle f, g|\mathcal{P}|f, g\rangle = -\int f(\mathbf{x}) \, \boldsymbol{\nabla} g(\mathbf{x}) \, d^s x$$

follows as an *identity*, whatever form $\mathcal{P}$ may have as an operator. Since this expectation value coincides with the classical space-translation generator, this result shows the general validity of the weak correspondence principle for this operator.

Repeat this problem for the space-rotation operators $\mathcal{J}_{ab}$ for a space of dimension 2 or greater.

5.3 Consider $J < \infty$ independent free fields as defined in Section 5.5, each with its own distinct function $\Omega_j(\mathbf{k})$, $1 \le j \le J$. These fields satisfy

$$[\varphi_j(\mathbf{x}), \pi_k(\mathbf{y})] = i\delta_{jk}\delta(\mathbf{x} - \mathbf{y}) ,$$

where $1 \le j, k \le J$, and all other commutators vanish. Let $\{f_j\}$ and $\{g_j\}$ denote two sets of real test functions, and $\varphi_j(f_j) = \int \varphi_j(\mathbf{x}) f_j(\mathbf{x}) \, d^s x$, and likewise for $\pi_j(g_j)$ (summation implied). Then it follows that

$$\langle 0|e^{i[\varphi_j(f_j) - \pi_j(g_j)]}|0\rangle$$
$$= \exp\{-\tfrac{1}{4}\int \Sigma_j[\Omega_j(\mathbf{k})^{-1}|\tilde{f}_j(\mathbf{k})|^2 + \Omega_j(\mathbf{k})|\tilde{g}_j(\mathbf{k})|^2] \, d^s k\} .$$

The states

$$|\{f, g\}\rangle \equiv e^{i[\varphi_j(f_j) - \pi_j(g_j)]}|0\rangle$$

denote the normalized coherent states for the several fields, and it follows, for example, that

$$\langle\{f, g\}|\mathcal{P}|\{f, g\}\rangle = -\int f_j(\mathbf{x})\,\boldsymbol{\nabla} g_j(\mathbf{x}) \, d^s x ,$$
$$\langle\{f, g\}|\mathcal{H}|\{f, g\}\rangle = \tfrac{1}{2}\int \Sigma_j[|\tilde{f}_j(\mathbf{k})|^2 + \Omega_j(\mathbf{k})^2|\tilde{g}_j(\mathbf{k})|^2] \, d^s k ,$$

which are further examples of the weak correspondence principle.

Now consider "collapsing" the field, by which we mean choosing the hitherto independent test functions $\{f_j\}$ and $\{g_j\}$ as no longer independent. In particular, for fixed f and g, let

$$f_j(\mathbf{x}) \equiv \sqrt{\rho_j}\, f(\mathbf{x}) , \qquad g_j(\mathbf{x}) \equiv \sqrt{\rho_j}\, g(\mathbf{x}) ,$$

where the set of *nonnegative numbers* $\{\rho_j\}$ is at our disposal. With summation implied, let us define

$$\varphi(\mathbf{x}) \equiv \sqrt{\rho_j}\, \varphi_j(\mathbf{x}) , \qquad \pi(\mathbf{y}) \equiv \sqrt{\rho_j}\, \pi_j(\mathbf{y}) .$$

We first assume that $\Sigma\rho_j = 1$. In this case show that

$$[\varphi(\mathbf{x}), \pi(\mathbf{y})] = i\delta(\mathbf{x} - \mathbf{y}) ,$$

i.e., that these fields satisfy the conventional single-field canonical commutation relations. Let $|f, g\rangle \equiv \exp\{i[\varphi(f) - \pi(g)]\}|0\rangle$ denote coherent states in the present case. Show that, in general, the set of coherent states $\{|f, g\rangle\}$ does *not* span the same Hilbert space as spanned by the set of states $\{|\{f, g\}\rangle\}$. Show also that

$$\langle 0|f, g\rangle = \exp\{-\tfrac{1}{4}\int \Sigma_j\rho_j[\Omega_j(\mathbf{k})^{-1}|\tilde{f}(\mathbf{k})|^2 + \Omega_j(\mathbf{k})|\tilde{g}(\mathbf{k})|^2] \, d^s k\}$$
$$\equiv \exp\{-\tfrac{1}{4}\int [A(\mathbf{k})|\tilde{f}(\mathbf{k})|^2 + B(\mathbf{k})|\tilde{g}(\mathbf{k})|^2] \, d^s k\} ,$$

where

$$A(\mathbf{k}) \equiv \Sigma_j\rho_j[\Omega_j(\mathbf{k})]^{-1} ,$$
$$B(\mathbf{k}) \equiv \Sigma_j\rho_j[\Omega_j(\mathbf{k})] .$$

Note well that it follows that $A(\mathbf{k})B(\mathbf{k}) \geq 1$. Excluding equality of any of the Ω functions, then $AB = 1$ holds if and only if $\rho_j = \delta_{j1}$; if two or more ρ_j are nonvanishing, then $AB > 1$. In addition, show that

$$\langle f, g | \mathcal{P} | f, g \rangle = -\int f(\mathbf{x}) \, \nabla g(\mathbf{x}) \, d^s x \, ,$$

$$\langle f, g | \mathcal{H} | f, g \rangle = \tfrac{1}{2} \int [|\tilde{f}(\mathbf{k})|^2 + M(\mathbf{k})^2 |\tilde{g}(\mathbf{k})|^2] \, d^s k \, ,$$

$$M(\mathbf{k})^2 \equiv \Sigma_j \rho_j [\Omega_j(\mathbf{k})]^2 \, .$$

5.4 Generalize the previous problem to a *continuous* distribution of distinct Ω terms, i.e., taking suitable limits so that $\Omega_j(\cdot) \to \Omega(\cdot, u)$ and $\rho_j \to \rho(u)$, $0 \leq u < \infty$, subject to $\int \rho(u) \, du = 1$.

5.5 Specialize the previous two problems to the case of *relativistic* generalized free fields, specifically, by choosing $\Omega_j(\mathbf{k})^2 = \mathbf{p}^2 + m_j^2$, or more precisely $\Omega(\mathbf{k}, m) = \mathbf{p}^2 + m^2$, where m plays the role of u. Determine that $M(\mathbf{k})^2 = \mathbf{p}^2 + m_0^2$, where $m_0^2 \equiv \int m^2 \rho(m) \, dm$.

5.6 For the case of relativistic generalized free fields, determine that the time-ordered two-point function is given (for $\hbar = 1$) by

$$\langle 0 | T\varphi(x)\varphi(y) | 0 \rangle = \frac{-i}{(2\pi)^n} \int \frac{e^{ip(x-y)}}{p^2 + m^2 - i\epsilon} \, d^n p \, \rho(m) \, dm$$

$$= \frac{1}{(2\pi)^s} \int \frac{e^{i\mathbf{p}\cdot(\mathbf{x}-\mathbf{y}) - i\omega(\mathbf{p})|t_x - t_y|}}{2\,\omega(\mathbf{p})} \, d^s p \, \rho(m) \, dm \, ,$$

where $\omega(\mathbf{p}) \equiv \sqrt{\mathbf{p}^2 + m^2}$. Discuss the singularity structure of this function for $x \simeq y$ in the case that $\int \rho(m) \, dm = 1$. Repeat this part of the problem for Euclidean fields.

If $|t_x - t_y|$ is replaced by $(t_x - t_y)$ above, the result is the usual two-point function, not the time-ordered one. Determine how the singularity for $x \simeq y$ changes in that case.

We now extend the scope of the original definition of $\rho(m)$ and no longer require that $\int \rho(m) \, dm < \infty$. Discuss the singularity structure of the time-ordered two-point function when $x \simeq y$ in those cases where, for large m,

$$\rho(m) \propto m^{2r-1} \, , \qquad 0 \leq r < \infty \, .$$

Find the range of r for which smooth test functions of compact support exist such that $\langle 0 | T\varphi(h)^2 | 0 \rangle < \infty$, where as usual $\varphi(h) = \int \varphi(x) \, h(x) \, d^n x$. For r values larger than this it is necessary to modify the test function space to exclude any coincident points [OsS 73, OsS 75]. Repeat this part of the problem for Euclidean fields, and in Minkowski space for the usual two-point function.

The important nonnegative function, more precisely a distribution, $\rho(m)$ is called the *spectral weight function*.

5.7 For a fixed function h and for φ a relativistic free field of mass m, find the probability amplitude that the state

$$\mathsf{T}\, e^{i\int \varphi(x)h(x)\, d^n x}\, |0\rangle$$

has a number N of particles, $N \in \{0, 1, 2, \ldots\}$. Show that the sum of the probabilities is unity independently of h; see, e.g., [Th 58].

6

Expanding the Data Base

WHAT TO LOOK FOR

In the previous chapters, we have developed a rather standard view toward classical and quantum theory illustrating the three canonical quantization procedures, namely those of Schrödinger, Heisenberg, and Feynman. In this chapter we want to return to Hilbert space and analyze several new issues as well as some old issues in greater depth than was previously the case. These issues will include operators and their domains, bilinear forms, infinite-product representations, irreducible representations of the canonical operators, as well as the important notion of "tags" (unitary invariants of operator representations). Some general remarks on measures, probability distributions, characteristic functions, and infinitely divisible distributions are also included here. Additionally, a few comments about Brownian motion are included. We close with some general remarks regarding canonical quantum scalar fields. The properties developed in this chapter will find application in our study of model quantum field theories in subsequent chapters.

6.1 Hilbert space and operators, revisited

In Chapter 3 we already introduced and discussed at some length properties of Hilbert spaces suitable for quantum mechanical applications. In all earlier applications we have used the elegant notation of Dirac, and this notation is generally appropriate for most applications. Unfortunately, the Dirac notation is awkward to use for certain general questions, especially related to operator adjoints, and as a consequence we shall also introduce and use a common mathematical notation for the vectors in a Hilbert space $\mathfrak{H}$, say as lower-case Greek letters ψ, ϕ, ξ, etc., and especially an inner product denoted by $(\psi, \phi) \; [= (\phi, \psi)^*]$ that is linear in the second element (ϕ) and antilinear in the first element (ψ).* The Hilbert space vectors form a complex linear vector space, with 0 denoting the zero element with the property that $\psi + 0 \equiv \psi$ for all $\psi \in \mathfrak{H}$. If $a \in \mathbf{C}$, then $a\psi \in \mathfrak{H}$ whenever $\psi \in \mathfrak{H}$. It follows from these relations that $(0, \phi) = 0$ for any $\phi \in \mathfrak{H}$. The norm of a vector is given by $\|\psi\| \equiv +\sqrt{(\psi, \psi)}$ with $\|\psi\| = 0$ if and only if $\psi = 0$. Only vectors for which $\|\psi\| < \infty$ belong to the Hilbert space, which as usual is assumed to be separable and complete (more later). The Schwarz inequality reads $|(\phi, \psi)| \leq \|\phi\| \|\psi\|$, while the triangle inequality is given by $\|\phi + \psi\| \leq \|\phi\| + \|\psi\|$. It is convenient to regard $\|\psi - \phi\|$ as the distance between the indicated vectors in $\mathfrak{H}$. There are two common notions of convergence of vectors. Weak convergence of a sequence of vectors $\{\phi_n\}_{n=1}^{\infty}$ to a vector ϕ holds if and only if for all $\lambda \in \mathfrak{H}$, it follows that $\lim_{n \to \infty} (\lambda, \phi_n - \phi) = 0$. Strong convergence of a sequence $\{\phi_n\}_{n=1}^{\infty}$ to a vector ϕ, on the other hand, means that $\lim_{n \to \infty} \|\phi_n - \phi\| = 0$. Equivalently, we may say that the sequence $\{\phi_n\}_{n=1}^{\infty}$ is a *Cauchy sequence* if for any $\epsilon > 0$ there exists a $P(\epsilon) < \infty$ such that $\|\phi_n - \phi_m\| < \epsilon$ holds provided $m > P$ and $n > P$. Hilbert space is complete in the sense that it already contains the limits of all Cauchy sequences. When we speak of the convergence of vectors without further qualification, we always have strong convergence in mind.

It is convenient at this point to introduce the concept of a *dense set* of vectors. A set $\mathcal{D}$ is a dense set of vectors if for any $\epsilon > 0$ and any $\lambda \in \mathfrak{H}$, there exists a vector $\phi \in \mathcal{D}$ such that $\|\lambda - \phi\| < \epsilon$. Intuitively, this means that for any vector in the Hilbert space (λ) there is a vector in the dense set (ϕ) that is arbitrarily close to it. (The analogy of the rationals being dense in the reals, or the set of continuous functions being dense in the space $L^2([0, 1])$ should help convey the meaning of this concept.) We are primarily interested in operators A for which the domain $\mathfrak{D}(A)$ is dense in the Hilbert space $\mathfrak{H}$, and let us restrict attention initially to that case. We observe that if $(\lambda, \phi) = 0$ for all ϕ in a dense set, then it follows by continuity that $(\lambda, \xi) = 0$ for all $\xi \in \mathfrak{H}$, and hence $\lambda = 0$. For completeness we again define a *total set*. A set $\mathcal{T}$ is a total set if for all $\tau \in \mathcal{T}$, $(\tau, \psi) = 0$ implies that $\psi = 0$. (A total set is also commonly called a "complete set"; we prefer to use the phrase "total set" since in many applications $\mathcal{T}$ contains an *un*countable number of vectors.) Note that the finite linear span (the set of vectors given by all finite linear combinations) of a total set leads to a dense set.

Operators will generally (but not universally) be denoted by upper-case Roman letters, e.g., A, B, C, etc., and the domain of an operator, say A, is denoted by $\mathfrak{D}(A)$. We shall restrict attention only to *linear operators*, namely operators that have the property that $A(a\phi + b\psi) = aA\phi + bA\psi$ for all $a, b \in \mathbf{C}$ and all

---

* As we reexamine Hilbert space somewhat more closely in this chapter it will be unavoidable that some of the basic groundwork will be covered again. An excellent introductory text is [RiS 55].

$\phi, \psi \in \mathfrak{D}(A)$. Linear Hilbert space operators may be divided into two families: bounded and unbounded operators. For a bounded linear operator, say B, it follows that $\|B\psi\| \leq b\|\psi\|$ for some $0 \leq b < \infty$. Since in this case $\|B0\| = 0$ it follows that $B0 = 0$ for all bounded B. The operator norm of a bounded operator is defined by $\|B\| \equiv \sup \|B\psi\|/\|\psi\| < \infty$ where the supremum is over all $\psi \neq 0$. For any bounded operator B, the domain $\mathfrak{D}(B) = \mathfrak{H}$. For an unbounded operator, say A, $\|A\psi\|$ is not finite for all $\psi \in \mathfrak{H}$. In fact, $\|A\psi\| < \infty$ only for vectors $\psi \in \mathfrak{D}(A)$, although it is sometimes convenient if $\mathfrak{D}(A)$ does not consist of all such vectors. We note that the zero vector is always in the domain of any operator, and $A0 \equiv 0$.

Two notions of the convergence of a sequence of operators are common. On one hand, a sequence of operators B_n converges weakly to an operator B whenever $\lim \langle \phi | (B_n - B) | \psi \rangle = 0$ for all $|\phi\rangle$ and $|\psi\rangle$ in a common dense domain. On the other hand, a sequence of operators C_n converges strongly to an operator C provided that $\lim \|(C_n - C)|\psi\rangle\| = 0$ for all $|\psi\rangle$ in a common dense domain. In both cases, if the sequence of operators is uniformly bounded, e.g., if $\|B_n\| \leq b < \infty$ for all n, then it suffices to show convergence for $|\phi\rangle$ and $|\psi\rangle$ in a total set of vectors. This procedure is an important and extremely practical criterion to show convergence, and it is a procedure that we will use frequently. Because of its importance, we restate one of our most important tools in the following form:

Let $\{B_n\}$ be a sequence of uniformly bounded operators, $\|B_n\| \leq b$, for all $n \in \{1, 2, 3, \ldots\}$. Let $\mathcal{T}$ and $\mathcal{T}'$ denote two total sets of vectors (which may be the same set). Assume that

$$\langle \tau | B_n | \tau' \rangle \equiv B_n(\tau; \tau') \to B(\tau; \tau'),$$

that is, that the indicated matrix elements converge as complex numbers for every pair of vectors $\tau \in \mathcal{T}$ and $\tau' \in \mathcal{T}'$. Then the sequence $\{B_n\}$ of uniformly bounded operators converges weakly to a bounded operator B which is fully characterized by the fact that

$$\langle \tau | B | \tau' \rangle \equiv B(\tau; \tau').$$

We leave the proof of this important fact to Exercise 6.1.

Hilbert spaces are either finite dimensional or countably infinite dimensional. The former are like finite-dimensional complex Euclidean spaces, and all operators (or matrices, if you prefer) are bounded operators, and one's intuition is almost always correct. In infinite dimensional Hilbert spaces, on the other hand, very unusual things can transpire that occasionally defy common sense!

As linear operators we might expect that

$$\lim_{n \to \infty} A \psi_n = A \lim_{n \to \infty} \psi_n$$

for a sequence of vectors $\{\psi_n\}_{n=1}^{\infty}$ such that each $\psi_n \in \mathfrak{D}(A)$ as well as $\lim \psi_n \in \mathfrak{D}(A)$. Some operators satisfy this equality while others do not. That further divides unbounded operators into two classes, namely the *closable operators* for which this equality always holds and the *nonclosable operators* for which it sometimes fails. Rather than deal with all possible vector sequences, it is sufficient to examine those situations where $\lim_{n \to \infty} \psi_n \to 0$. For a closable operator, $\lim_{n \to \infty} A\psi_n$ either converges to 0 or it fails to converge in $\mathfrak{H}$. For a nonclosable operator, on the other hand, besides these possibilities, it may happen that $\lim_{n \to \infty} A\psi_n$ converges to a vector in the Hilbert space that is different from 0! A simple example will illustrate this phenomenon. Let us consider the sequence space l^2 composed of square-summable sequences $\{\phi^k\}_{k=1}^{\infty}$; we put the index on the top this time to make room for the sequence label. Choose the operator $A = \{A_{kl}\}$ with entries $A_{kl} = l\delta_{k1}$, namely the sequence $1, 2, 3, \ldots$ along the first row and zero otherwise. This matrix defines an operator for which the domain consists of all sequences for which $\Sigma_{k=1}^{\infty} k\phi^k$ is finite. However, consider the sequence of vectors $\phi_n = \{\phi_n^k\}$ where $\phi_n^k = \delta_{nk}/n$. It follows that $\phi_n \to 0$ but $A\phi_n \to \xi$, where $\xi^k = \delta_{k1}$ which is not the zero vector! In brief, we have illustrated a simple operator and a sequence of vectors such that

$$0 \neq \xi = \lim_{n \to \infty} A\,\phi_n \neq A \lim_{n \to \infty} \phi_n = A\,0 = 0\,.$$

The reader can well imagine the difficulties such operators may cause, and although we will encounter such operators in passing, it is reassuring to learn that all operators of physical interest will not have such bizarre properties. For the most part we will confine our attention to closable operators for which this kind of situation will never occur.

The adjoint of an operator A may be uniquely defined when $\mathfrak{D}(A)$ is dense. In particular, the adjoint $A^\dagger$ of A is defined as follows. For a suitable $\psi \in \mathfrak{H}$, let ξ be a vector such that $(\xi, \phi) = (\psi, A\phi)$ holds for all vectors $\phi \in \mathfrak{D}(A)$. Then $\psi \in \mathfrak{D}(A^\dagger)$ and $A^\dagger \psi = \xi$. Note that it is necessary for the domain of A to be dense in order for ξ to be uniquely determined. At the very least the vector $0 \in \mathfrak{D}(A^\dagger)$ and $A^\dagger 0 = 0$ as required. For some operators $A^\dagger$, 0 is the *only* vector in $\mathfrak{D}(A^\dagger)$.* On the other hand, such an operator is not especially useful, and we often actually refer to it as a "nonoperator" or "not an operator", a terminology based on the essentially total lack of vectors on which it is defined.

---

* It is not too difficult to devise such an operator. Consider a sequence Hilbert space with operators represented as semi-infinite square matrices. Start with the unit matrix and divide it into consecutive blocks along the main diagonal of 1×1, 2×2, 3×3, etc. Take the 2×2 block and shift it all the way to the left so that it gets one entry in the first column. Repeat such a shift all the way to the left for the 3×3 block so that it gets one entry in the first column, and continue this process *ad infinitum*. The result is an operator for which 0 is the only vector in its domain!

A symmetric operator (which is called a Hermitian operator in the physics literature) is one for which $(A\phi, \psi) = (\phi, A\psi)$ holds for all $\phi, \psi \in \mathfrak{D}(A)$. It follows that $A^\dagger = A$ on $\mathfrak{D}(A)$, from which we learn that $\mathfrak{D}(A^\dagger) \supseteq \mathfrak{D}(A)$ [simply because $\mathfrak{D}(A^\dagger)$ is definitely not smaller than $\mathfrak{D}(A)$]. The special case of equality holds for the very important case of self-adjoint operators; i.e., if $A^\dagger = A$ and $\mathfrak{D}(A^\dagger) = \mathfrak{D}(A)$ then we say that the operator A is *self adjoint*. Notice that a self-adjoint operator is necessarily symmetric, but the converse is not necessarily true. In particular, a symmetric operator may uniquely determine a self-adjoint operator; in this case the operator is called essentially self adjoint. An example of such an operator is the differential operator $P = -i(\partial/\partial x)$ acting on the space $L^2((-\infty, \infty))$ with $\mathfrak{D}(P) = \mathfrak{D}_o(P) = \{\psi : \psi \in \mathcal{S}\}$, where $\mathcal{S}$ is the space of functions with continuous derivatives of arbitrary order which all fall to zero at infinity faster than any power, or $\mathfrak{D}(P) \equiv \{\psi : \int [|\psi'(x)|^2 + |\psi(x)|^2] \, dx < \infty\}$. The second domain is the maximal domain on which $\mathfrak{D}(P^\dagger) = \mathfrak{D}(P)$, and one speaks of extending the original smaller domain $\mathfrak{D}_o(P)$ to the maximal domain $\mathfrak{D}(P)$. For some symmetric operators it is possible to extend the operator by extending its domain so that it becomes self adjoint, but that extension is not unique and therefore various distinct self-adjoint operators may arise. An example of this type is for the operator $P = -i(\partial/\partial x)$ acting this time on the space $L^2([0, 1])$. In this case we choose as the initial domain $\mathfrak{D}_o(P) = \{\psi : \psi' \in L^2([0, 1]), \psi(0) = \psi(1) = 0\}$. Now $\mathfrak{D}_o(P^\dagger) = \{\psi : \psi' \in L^2([0, 1])\}$ without any need for boundary conditions. We can extend the original domain of P (thereby shrinking the domain of $P^\dagger$) and arrive at a self-adjoint operator where $\mathfrak{D}(P^\dagger) = \mathfrak{D}(P) = \{\psi : \psi' \in L^2([0, 1]), \psi(1) = e^{i\alpha}\psi(0)\}$, where $\alpha \in \mathbb{R}$ is a needed new parameter to define the extension. Each different α (mod 2π) leads to a distinct operator, say P_α, with a spectrum depending on the parameter α. In the third category are symmetric operators that can never be extended to self-adjoint operators. An example of such an operator is $P = -i(\partial/\partial x)$ acting on the Hilbert space $L^2([0, \infty))$ with $\mathfrak{D}(P) = \{\psi : \psi' \in L^2([0, \infty)), \psi(0) = 0\}$. In this case $\mathfrak{D}(P^\dagger) = \{\psi : \psi' \in L^2([0, \infty))\}$ without a boundary condition required. A rapid way to see the category into which a symmetric operator fits is the following: consider the distributional solutions of the two equations $P^\dagger \psi_\pm(x) = \pm i \psi_\pm(x)$. Let (n_+, n_-) denote the number of linearly independent, square-integrable solutions to these two equations; the numbers $n_\pm$ are called the deficiency indices. If $n_+ = n_- = 0$, the symmetric operator is essentially self adjoint. If $n_+ = n_- > 0$, the operator admits self-adjoint extensions, but n_+^2 real parameters are needed to specify the operator. If $n_+ \neq n_-$, no self-adjoint operator exists. In the examples for P considered in this paragraph, $\psi_\pm(x) = e^{\mp x}$ and the deficiency indices (n_+, n_-) are $(0, 0)$ for $L^2((-\infty, \infty))$; $(1, 1)$ for $L^2([0, 1])$; and $(1, 0)$ for $L^2([0, \infty))$.

Recall that a unitary operator U is defined by the fact that $UU^\dagger = U^\dagger U = \mathbf{1}$. Now it is of the utmost importance that every unitary operator U is given by $U = e^{iA}$ and $U^\dagger = e^{-iA}$ for some self-adjoint operator A, and conversely, every self-adjoint operator A generates a unitary operator and its adjoint by way of $e^{iA} \equiv U$ and $e^{-iA} \equiv U^\dagger$. All this seems familiar save for the question of how e^{iA} is to be defined as an operator. We may include in the domain of every self-adjoint operator a dense set of *analytic vectors*. A vector ψ is an analytic vector for an operator A provided that $\Sigma_{n=0}^{\infty} \|A^n\psi\|/n! < \infty$. For such vectors we define e^{iA} by means of its power series expansion, namely

$$e^{iA}\,\psi \equiv \sum_{n=0}^{\infty} \frac{i^n A^n}{n!}\,\psi\,,$$

which is well defined because the series converges strongly to a vector in $\mathfrak{H}$ for all analytic vectors. On the analytic vectors it follows that $\|e^{iA}\,\psi\| = \|e^{-iA}\,\psi\| = \|\psi\|$. Since the set of analytic vectors is dense we can extend the definition of e^{iA} to all vectors in $\mathfrak{H}$ by continuity, specifically for an arbitrary vector ϕ by $e^{iA}\phi = \lim e^{iA}\,\psi_n$ where each ψ_n is an analytic vector for A and the sequence is chosen so that $\lim \|\phi - \psi_n\| = 0$.

A *unitary one-parameter group* is a set of operators $\{U(t)\}_{t\in\mathbb{R}}$ that satisfies, for all $t, s \in \mathbb{R}$, the following four properties: (1) $U(0) = \mathbf{1}$; (2) $U(t)U(s) = U(t+s)$; (3) $U^\dagger(t) = U(-t)$; and (4) $U(t)$ is weakly continuous, which means, for all $\psi, \phi \in \mathfrak{H}$, that $(\psi, U(t)\phi)$ is a continuous function of t. If a set of operators fulfills these four conditions then it follows that there exists a self-adjoint operator A such that $U(t) = e^{itA}$. This rule is extremely useful to determine whether an operator is self adjoint or not, and we will have many occasions to use it. The relation of these conditions to representations of an operator A is straightforward. In particular, if we diagonalize A then it follows for each pair of vectors ψ, ϕ that there exists a complex, countably additive measure (hereafter simply called a measure) $\sigma_{\psi,\phi}$ such that

$$(\psi, U(t)\phi) = \int e^{ita}\, d\sigma_{\psi,\phi}(a)\,.$$

If $\psi = \phi$ then the measure $\sigma_{\phi,\phi}$ is nonnegative, and if ϕ is normalized, then the measure is a probability measure. In the latter case the function

$$C(t) \equiv \int e^{ita}\, d\mu_\phi(a)$$

defines the characteristic function for a probability measure $\mu_\phi \equiv \sigma_{\phi,\phi}$ with $(\phi, \phi) = 1$. As a characteristic function $C(t)$ is continuous whether the measure μ is discrete, absolutely continuous, or singular continuous (see below). Thus we see one aspect of how continuity becomes an important feature of the definition of a one-parameter unitary group.

As a second aspect of the continuity condition let us see what can emerge if we omit that requirement. Consider the example where A is represented by multiplication by y such that $U(t) = e^{ity}$ and $U^\dagger(t) = e^{-ity}$ hold for all $y \in \mathbb{R}$. Clearly the first three conditions for a one-parameter group are fulfilled, and for some particular normalized state ω we identify

$$F(t) \equiv (\omega, U(t)\omega) \equiv \lim_{Y \to \infty} (2Y)^{-1} \int_{-Y}^{Y} e^{ity} \, dy .$$

It is evident that $F(0) = 1$ while $F(t) = 0$ for all $t \neq 0$. Thus the putative characteristic function $F(t)$ is *not* continuous. As a consequence it follows that the vectors $\omega(t) \equiv U(t)\omega$ are mutually orthogonal, $(\omega(s), \omega(t)) = 0$, for every pair such that $s \neq t$. Hence, in this representation there are an *uncountable* number of orthogonal vectors, and this is the characteristic situation in a *nonseparable* Hilbert space. If one attempts to introduce a "measure" to represent this inner product in the usual manner,

$$F(t) = (\omega, U(t)\omega) \equiv \int e^{ity} \, d\tau(y) ,$$

one would find that τ is only *finitely* additive rather than having the very important property of being *countably* additive. To illustrate this point we observe that if $\chi(y)$ denotes a nonnegative function of compact support with the property that $\Sigma_{n=-\infty}^{\infty} \chi(y - n) \equiv 1$ provides a decomposition of unity, then it follows that

$$0 = \sum_{n=-\infty}^{\infty} \int \chi(y - n) \, d\tau(y) < \int \sum_{n=-\infty}^{\infty} \chi(y - n) \, d\tau(y) = 1 ,$$

which for the right choice of χ demonstrates a violation of countable additivity. As a second consequence of this fact we observe that

$$0 = \lim_{c \to 0+} \int e^{-cy^2} \, d\tau(y) < \int \lim_{c \to 0+} e^{-cy^2} \, d\tau(y) = 1 ,$$

a relation that violates Fatou's Lemma for a nonnegative measure m which states, for a sequence of nonnegative functions $f_n(y)$ and when both limits exist, that $\lim_{n \to \infty} \int f_n(y) \, dm(y) \geq \int \lim_{n \to \infty} f_n(y) \, dm(y)$ [KoF 57, ReS 80].

We conclude our discussion of nonseparable Hilbert spaces; the reader should note that there is no evidence that the description of any physical process requires the use of nonseparable spaces.

6.2 Canonical commutation relations, revisited

In the light of the discussion of the preceding section, let us revisit the question of the representation (for a single degree of freedom) of the pair of irreducible, self-adjoint momentum and coordinate operators, P and Q, that satisfy not only the Heisenberg commutation relation $[Q, P] = i\hbar \mathbb{1}$ but also the (stronger) Weyl form of the commutation relation expressed in terms of the unitary family of operators

$$U[p, q] \equiv e^{(ipQ - iqP)/\hbar} ,$$

for all $(p, q) \in \mathbb{R}^2$, which fulfill

$$U[p, q]U[p', q'] = e^{i(pq' - qp')/2\hbar} U[p + p', q + q'] .$$

As is well known there is only *one* representation of the Weyl commutation relations up to unitary equivalence; von Neumann [vN 31] has given an elegant proof of this fact, but there is a more elementary way to come to the same conclusion. In Chapter 3 we introduced the annihilation and creation operators – say $a = (Q + iP)/\sqrt{2\hbar}$ and $a^\dagger = (Q - iP)/\sqrt{2\hbar}$ – that satisfy the commutation relation $[a, a^\dagger] = \mathbb{1}$. Let us assume that the symmetric operator $\mathsf{N} = a^\dagger a = (P^2 + Q^2 - \hbar)/2$ is essentially self adjoint on the domain $\mathfrak{D}(P^2) \cap \mathfrak{D}(Q^2)$. Physically, this is equivalent to the assumption that the harmonic oscillator Hamiltonian is an observable. Mathematically, it is equivalent to the assumption that the basic kinematical operators P and Q satisfy the Weyl form of the commutation relations and not simply the Heisenberg form of the commutation relations. With that assumption it follows for every $s > 0$ and an arbitrary vector $\phi \in \mathfrak{H}$ that $e^{-s\mathsf{N}}\phi$ is an analytic vector for a and for $a^\dagger$, and as a consequence the relations

$$e^{za^\dagger - z^* a} \left(e^{-s\mathsf{N}} \phi \right) = e^{-\frac{1}{2} z^* z} e^{za^\dagger} e^{-z^* a} \left(e^{-s\mathsf{N}} \phi \right) = e^{\frac{1}{2} z^* z} e^{-z^* a} e^{za^\dagger} \left(e^{-s\mathsf{N}} \phi \right)$$

hold directly by resumming the appropriate power series expansion. Furthermore, the operator $e^{za^\dagger - z^* a}$ is unitary on the analytic vectors and through continuity it preserves that property on all of $\mathfrak{H}$. The previous equations imply the Weyl commutation relations when we observe that

$$U[p, q] = e^{(ipQ - iqP)/\hbar} = e^{[(q+ip)a^\dagger/\sqrt{2\hbar} - (q-ip)a/\sqrt{2\hbar}]} .$$

An interesting discussion has been given of a pair of canonical degrees of freedom that satisfy the Heisenberg commutation relations on the chosen domain but for which N is not essentially self adjoint and for which the Weyl commutation relations do not hold [Re 88, Ar 92].

6.3 Annihilation and creation operators; bilinear forms

In this section we wish to reexamine the usual annihilation and creation field operators one more time. In particular, let us consider the local operators $a(\mathbf{k})$ and $a(\mathbf{k})^\dagger$, for all $\mathbf{k} \in \mathbb{R}^s$, subject to the familiar commutation relation $[a(\mathbf{k}), a(\mathbf{l})^\dagger] = \delta(\mathbf{k} - \mathbf{l})$ with all other commutators vanishing. We have introduced (un)normalized coherent states for a single degree of freedom in Chapters 3 and 4, and (un)normalized coherent states for infinitely many degrees of freedom, i.e., for a field, in Chapter 5. Field coherent states may be loosely defined by the condition that $a(\mathbf{k})|\psi\rangle = \psi(\mathbf{k})|\psi\rangle$, leading to our earlier definition for the unnormalized coherent states that $|\psi\rangle = \exp[\int \psi(\mathbf{k}) a(\mathbf{k})^\dagger d^s k] |0\rangle$ for all $\psi \in L^2(\mathbb{R}^s)$. Indeed, for the smeared annihilation operator we have

$$a(\alpha^*) |\psi\rangle \equiv \int \alpha(\mathbf{k})^* a(\mathbf{k}) d^s k |\psi\rangle = \int \alpha(\mathbf{k})^* \psi(\mathbf{k}) d^s k |\psi\rangle ,$$

which is then well defined for all $\alpha \in L^2(\mathbb{R}^s)$. This much is either explicit or at least implicit in earlier chapters.

We now wish to observe that for *each* $\mathbf{k} \in \mathbb{R}^s$ the local operator $a(\mathbf{k})$ is *actually an operator!* Indeed, to demonstrate that $a(\mathbf{k})$ is an operator it suffices to define it on a suitable domain, and for that purpose we consider a *subset* of coherent states $|\zeta\rangle = \exp[\int \zeta(\mathbf{k}) \, a(\mathbf{k})^\dagger \, d^s k] \, |0\rangle$ for which the function ζ is *continuous*. We observe that the set of such coherent states forms a total set. For each such coherent state it follows that

$$a(\mathbf{k}) \, |\zeta\rangle = \zeta(\mathbf{k}) \, |\zeta\rangle \,,$$

which is well defined for each $\mathbf{k}$ since ζ is continuous and thus pointwise uniquely defined. The operator $a(\mathbf{k})$ is also well defined on finite linear combinations of this subset, which means that we have a dense set of vectors in Fock space on which $a(\mathbf{k})$ is well defined for each $\mathbf{k} \in \mathbb{R}^s$. Next we show that *each of these operators is nonclosable*. To show that $a(\mathbf{k})$ is nonclosable it suffices to exhibit a sequence of vectors $|\phi_n\rangle$ such that $\lim_{n\to\infty} |\phi_n\rangle = 0$, that is $\lim_{n\to\infty} \langle \phi_n | \phi_n \rangle = 0$, while $\lim_{n\to\infty} a(\mathbf{k}) \, |\phi_n\rangle = |\theta\rangle \neq 0$ for an arbitrary $\mathbf{k} \in \mathbb{R}^s$. To that end let $|\phi_n\rangle = |\zeta_n\rangle - |0\rangle$, where the function $\zeta_n(\mathbf{q}) \equiv \exp[-n(\mathbf{q} - \mathbf{k})^2]$, in which case we conclude that

$$0 \neq |0\rangle = \lim a(\mathbf{k}) \, |\phi_n\rangle \neq a(\mathbf{k}) \lim \, |\phi_n\rangle = a(\mathbf{k}) \, 0 = 0 \,,$$

demonstrating that the operator is nonclosable.

We next inquire as to the consequence of this fact for the operator $a(\mathbf{k})^\dagger$ at a given point $\mathbf{k} \in \mathbb{R}^s$. The answer is that $a(\mathbf{k})^\dagger$ has only the vector 0 in its domain [or as we loosely say $a(\mathbf{k})^\dagger$ is "not an operator"]. By definition, $a(\mathbf{k})^\dagger \, 0 = 0$, so that, as usual, $0 \in \mathfrak{D}(a(\mathbf{k})^\dagger)$ for each $\mathbf{k} \in \mathbb{R}^s$. However, for any *nonzero* vector $|\psi\rangle$, necessarily with $\langle \psi | \psi \rangle > 0$, it follows from the commutation relation that

$$\langle \psi | a(\mathbf{k}) a^\dagger(\mathbf{k}) | \psi \rangle = \delta(\mathbf{0}) \langle \psi | \psi \rangle + \langle \psi | a^\dagger(\mathbf{k}) a(\mathbf{k}) | \psi \rangle = \infty$$

and so $|\psi\rangle$ is not contained in $\mathfrak{D}(a^\dagger(\mathbf{k}))$. On the other hand, if we consider the smeared operator $a(\alpha^*)$ introduced above then it follows that this operator *is* closable for any $\alpha \in L^2$, and furthermore both $a(\alpha^*)$ and its adjoint $a(\alpha^*)^\dagger \equiv a^\dagger(\alpha)$ are densely defined operators.

With $a(\mathbf{k})$ for each $\mathbf{k} \in \mathbb{R}^s$ an operator it is easy to show that powers of this operator, $a(\mathbf{k})^p$, $p \geq 2$, are also well defined, and indeed the set of coherent states $|\zeta\rangle$ with continuous function arguments and their linear combinations provide a dense domain of definition for such operators. However, there is a strong difference between the first power $(p = 1)$ and higher powers $(p \geq 2)$. For an arbitrary $\alpha \in L^2$ let us consider the smeared field operator

$$a^p(\alpha^*) \equiv \int \alpha(\mathbf{k})^* \, a(\mathbf{k})^p \, d^s k \,.$$

When $p = 1$ this operator is closable and has a densely defined adjoint. On the other hand, when $p \geq 2$, this operator, even though it is smeared, is *not* closable and the domain of its adjoint contains only the single vector 0! The proof of these remarks is similar for all $p \geq 2$, and so we will outline the proof only in the case $p = 2$. Let

$$|\psi_{2,n}\rangle = (1/\sqrt{2}) \int \phi_n(\mathbf{k}, \mathbf{l}) \, a(\mathbf{k})^\dagger \, a(\mathbf{l})^\dagger \, d^s k \, d^s l \, |0\rangle$$

denote a two-particle state. Choose $\alpha \in L^2(\mathbb{R}^s)$ and let

$$\phi_n(\mathbf{k}, \mathbf{l}) = \alpha(\mathbf{k}) \exp[-\mathbf{k}^2 - \mathbf{l}^2 - n(\mathbf{k} - \mathbf{l})^2] \,.$$

Then it follows that $\lim |\psi_{2,n}\rangle = 0$ while

$$\lim a^2(\alpha^*) |\psi_{2,n}\rangle = 2 \int \exp(-2\mathbf{k}^2) |\alpha(\mathbf{k})|^2 \, d^s k \, |0\rangle \neq 0$$

whenever $\alpha \not\equiv 0$. The proof that 0 is the only vector in $\mathfrak{D}(a^2(\alpha^*)^\dagger)$ is straightforward and is left to the reader; see Exercise 6.3.

In summary, we have established that for each $\mathbf{k} \in \mathbb{R}^s$, $a(\mathbf{k})$ is an operator and therefore automatically a local operator as well. On the other hand, $a(\mathbf{k})^\dagger$ is a local operator but not an operator. Additionally, for each $\alpha \in L^2$ and $p \geq 2$, $a^p(\alpha^*)$ is an operator, but $a^{p\dagger}(\alpha) \equiv [a^p(\alpha^*)]^\dagger$ is not an operator. We shall have occasion to appeal to this theorem in Chapter 10 to help us decide for certain model problems just which expressions may be used to construct relevant Hamiltonian operators and which expressions may not.

Bilinear forms

It sometimes occurs that an expression does not qualify as an operator but it does qualify as a kind of suboperator known as a *bilinear form*, often called a quadratic form. Dealing with an operator O generally means that the expression $(\phi, O\psi)$ is well defined for all $\psi \in \mathfrak{D}(O)$ (which may be less than all of $\mathfrak{H}$) and for all $\phi \in \mathfrak{H}$. On the other hand, a form F is defined only for a limited set of both vectors involved. To help understand this concept, we return to and extend the Dirac notation for Hilbert spaces so that $\langle \lambda | F | \psi \rangle$ also applies to a form and is well defined provided both $|\psi\rangle$ *and* $|\lambda\rangle$ belong to special vector sets determined by the form domain $\mathfrak{D}_{form}(F)$ consisting of ordered pairs. As an example of a form we offer $a(\mathbf{k})^\dagger$ for any $\mathbf{k} \in \mathbb{R}^s$. As the form domain we choose both vectors from the set of coherent states with continuous function argument, vectors which we called $|\zeta\rangle$ above. In that case

$$\langle \zeta | a(\mathbf{k})^\dagger | \zeta' \rangle = \zeta(\mathbf{k})^* \, \langle \zeta | \zeta' \rangle \,.$$

This formula may be extended to finite linear combinations of both vectors without difficulty. Another example with the same form domain is given by

$$\langle \zeta | [a(\mathbf{k})^\dagger + a(\mathbf{k})] | \zeta' \rangle = [\zeta(\mathbf{k})^* + \zeta'(\mathbf{k})] \, \langle \zeta | \zeta' \rangle \,.$$

Now let us pass to some alternative examples of forms in configuration space.

Consider the free fields of Chapter 5 based on a suitable function $\Omega(\mathbf{k})$ and the local field operator defined (for $\hbar = 1$) by

$$\varphi(\mathbf{x}) = \frac{1}{(2\pi)^{s/2}} \int [a(\mathbf{k})^\dagger \, e^{-i\mathbf{k}\cdot\mathbf{x}} + a(\mathbf{k}) \, e^{i\mathbf{k}\cdot\mathbf{x}}] \frac{d^s k}{\sqrt{2\Omega(\mathbf{k})}} \ .$$

It follows as a form that

$$\langle \zeta | \varphi(\mathbf{x}) | \zeta' \rangle = \gamma(\mathbf{x}) \, \langle \zeta | \zeta' \rangle \ ,$$

where

$$\gamma(\mathbf{x}) \equiv \frac{1}{(2\pi)^{s/2}} \int [\zeta(\mathbf{k})^* \, e^{-i\mathbf{k}\cdot\mathbf{x}} + \zeta'(\mathbf{k}) \, e^{i\mathbf{k}\cdot\mathbf{x}}] \frac{d^s k}{\sqrt{2\Omega(\mathbf{k})}} \ .$$

Indeed, in this case we can even say more, namely

$$\langle \zeta | : e^{a\varphi(\mathbf{x})} : | \zeta' \rangle = e^{a\gamma(\mathbf{x})} \, \langle \zeta | \zeta' \rangle \ ,$$

for all a, which leads in particular to an expression such as

$$\langle \zeta | \int : \varphi^4(\mathbf{x}) : \, d^s x \, | \zeta' \rangle = \int \gamma^4(\mathbf{x}) \, d^s x \, \langle \zeta | \zeta' \rangle \ ,$$

corresponding, at least in appearance, to continuous coherent state matrix elements of a putative interaction term in the Hamiltonian.

While such expressions look intriguing and even physically relevant as forms, it must be appreciated that in general they cannot be extended to operators, i.e., operators with more than the zero vector in their domain. Since an operator O takes any vector from a selected set (its domain) into a vector in Hilbert space, it follows that the expression $(O\psi, O\psi)$ is finite for any vector $\psi \in \mathfrak{D}(O)$. This expression is quadratic in the operator, and may be reexpressed as $(O\psi, O\psi) = \Sigma_{n=1}^\infty |(\beta_n, O\psi)|^2$ for any complete orthonormal basis $\{\beta_n\}_{n=1}^\infty$. On the other hand, the particular expression for a form generally does not permit an operator to be defined because the sum in question would not converge (for a nonzero vector ψ) for any basis set. As a simple example consider the $\delta(0)$-form acting in $L^2(\mathbb{R})$. For two continuous elements $u, v \in L^2$, we set

$$(u, \delta(0)v) \equiv \int u^*(x)\delta(x)v(x) \, dx = u^*(0)v(0) \ ,$$

which is well defined as a form for continuous functions u and v with arbitrary values at the origin; however, it could never be extended to an operator for this would require that continuous functions v exist with arbitrary values at the origin for which $(\delta(0)v, \delta(0)v) < \infty$, and no such functions exist.

In studying quantum field theory, and especially the Hamiltonian operator appropriate to some model system, it is common to initially set up the Hamiltonian as a form between two sets of well-behaved vectors in the Hilbert space. This kind of construction is relatively easy to do, just as was the case illustrated above in our discussion of the form $\int : \phi^4(\mathbf{x}) : \, d^s x$ for a proposed interaction term. But a form construction is not enough; it is necessary for the Hamiltonian to be an operator – and a self-adjoint operator at that. *It is fair to say that all* the effort expended in regularization and renormalization in the context of quantum field theory is an effort to convert the expression for the Hamiltonian from a bilinear form to a self-adjoint operator! The difficulty in doing so in general is reflected in the familiar difficulties found in generating well-defined quantum field theories.

6.4 Product representations

In this section we discuss a family of very simple but nevertheless instructive quantum mechanical problems dealing with an infinite number of degrees of freedom. These examples have Hamiltonians of the form

$$\overline{\mathcal{H}} = \sum_{n=1}^{\infty} \mathcal{H}(P_n, Q_n) \, ,$$

where $[Q_n, P_m] = i\delta_{nm}\mathbb{1}$ (with $\hbar = 1$) is the only nonvanishing commutator. It follows that such models are nothing but an independent replication of the same Hamiltonian for a single degree of freedom to a countably infinite number of degrees of freedom. Of course, care must be taken in defining $\mathcal{H}$ so that the sum defining $\overline{\mathcal{H}}$ converges. Let us see how we can build a suitable Hilbert space for this problem. First, if $\phi \equiv \phi^0$ denotes the ground state of $\mathcal{H}$ for a single degree of freedom, then it is clear that $\otimes_{n=1}^{\infty}\phi_n$ denotes the ground state of $\overline{\mathcal{H}}$. On the other hand – as we have already observed in Chapter 5 in relation to a similar question for the harmonic oscillator – it is necessary that the ground state energy of each $\mathcal{H}$ be zero, i.e., $\mathcal{H}\phi = 0$, or otherwise the ground state energy of $\overline{\mathcal{H}}$ would diverge. Granting that, we have now found one acceptable vector in a proposed Hilbert space! We may find others as follows, and, for convenience, let us assume that $\mathcal{H}$ has only a nondegenerate, discrete spectrum. Let ϕ^k, $k = 0, 1, 2, \ldots$, denote the complete set of discrete, normalized eigenstates of any given $\mathcal{H}$ each of which has associated energy levels E^k that fulfill $0 = E^0 < E^1 < E^2 < \ldots$. Then some additional states in the proposed Hilbert space are given by product vectors of the form

$$\phi_{[k_1, k_2, \ldots]} \equiv \otimes_{n=1}^{\infty} \phi_n^{k_n}$$

provided that $k_n = 0$ for all n sufficiently large, or otherwise the associated state would have an infinite energy. The inner product of two such states is given by

$$(\phi_{[k_1, k_2, \ldots]}, \phi_{[l_1, l_2, \ldots]}) = \prod_{n=1}^{\infty}(\phi^{k_n}, \phi^{l_n}) = \prod_{n=1}^{\infty} \delta_{k_n, l_n} \, ,$$

i.e., either 1 or 0, as is to be expected. Next we invoke linearity to introduce states

$$\Phi \equiv \sum_{m=1}^{M} a_m \, \phi_{[k_1^m, k_2^m, \ldots]} \, ,$$

where $M < \infty$, and these states have an inner product that follows by linearity from the given inner product. Finally, we complete the Hilbert space by taking limits as $M \to \infty$ including all vectors obtained as the limit points of all Cauchy sequences.

This construction may seem belabored but there is nevertheless a point to be made about it. It should be clear from the construction that the resultant Hilbert space depends heavily on the particular ground state ϕ chosen and therefore on the particular Hamiltonian $\mathcal{H}$. If we change a parameter in $\mathcal{H}$ however slightly, leading say to $\mathcal{H}'$, we in turn change the ground state ϕ slightly as well, say to ϕ', so that $|(\phi', \phi)| < 1$. As a consequence the resultant Hilbert spaces for the two systems will have no vector in common save the zero vector! This result holds because

$$\prod_{n=1}^{\infty} (\phi'_n, \phi_n) = 0$$

as the infinite power of a complex number of magnitude strictly less than one, no matter how close to one it may be. No amount of excited states, superposition, and/or completion of the space will ever put a nonvanishing vector from one Hilbert space into the other Hilbert space. (Indeed, following von Neumann [vN 38], this complete orthogonality is taken to hold even if the state ϕ and ϕ' are identical except for a nonvanishing phase factor; i.e., one interprets $e^{i\infty}$ as 0.) Thus it is entirely natural to label the resultant Hilbert space by the (ground state) vector on which it is based, and so we identify $\mathfrak{H}_\phi$ as the Hilbert space constructed in the manner outlined above. Of course, we can introduce a unitary mapping between the vectors in one Hilbert space $\mathfrak{H}_\phi$ and another Hilbert space $\mathfrak{H}_{\phi'}$ given by $\overline{V} = \otimes_{n=1}^{\infty} V_n$, where $V_n \phi_n^k = \phi_n'^k$ for all k, and defined by linearity and continuity on the rest of the Hilbert space. On the other hand, the existence of this unitary operator merely expresses the equivalence of all separable Hilbert spaces to each other, and it should not be taken to signify that any unitary equivalence of the basic kinematical operators necessarily holds.

6.5 Inequivalent representations and tags

In the context of the kind of problem discussed in the preceding section let us examine under what condition sets of the basic kinematic operators $\{P_n\}, \{Q_n\}$ and $\{P'_n\}, \{Q'_n\}$, appropriate to the two systems, can be shown to be unitarily equivalent to each other. In fact, let us study this question for the associated Weyl operators. To that end we introduce the operators F_N and F'_N, $N < \infty$, defined for all $(p, q) \in \mathbb{R}^2$, by

$$F_N \equiv N^{-1} \sum_{n=1}^{N} e^{i(pQ_n - qP_n)/\hbar} \,, \qquad F'_N \equiv N^{-1} \sum_{n=1}^{N} e^{i(pQ'_n - qP'_n)/\hbar} \,.$$

For $N > n$ it readily follows that

$$[P_n, F_N] = \frac{p}{N} e^{i(pQ_n - qP_n)/\hbar} , \qquad [P'_n, F'_N] = \frac{p}{N} e^{i(pQ'_n - qP'_n)/\hbar} ,$$

$$[Q_n, F_N] = \frac{q}{N} e^{i(pQ_n - qP_n)/\hbar} , \qquad [Q'_n, F'_N] = \frac{q}{N} e^{i(pQ'_n - qP'_n)/\hbar} .$$

Thus as $N \to \infty$, we see that $F = \lim F_N$ and $F' = \lim F'_N$ *commute* with all the variables $\{P_n\}, \{Q_n\}$ and $\{P'_n\}, \{Q'_n\}$, respectively. As irreducible operator sets it follows that F and F' are each proportional to the identity operator, specifically $F = f(p, q)\mathbf{1}$ and $F' = f'(p, q)\mathbf{1}'$, for some c-number functions $f(p, q)$ and $f'(p, q)$ that depend on the free parameters p and q. We can readily evaluate these c-number functions by using any normalized vector in each Hilbert space, and for convenience we choose the ground state introduced in the preceding section, which leads to

$$f(p, q) = \lim_{N \to \infty} N^{-1} \sum_{n=1}^{N} (\phi_n, e^{i(pQ_n - qP_n)/\hbar} \phi_n) = (\phi, e^{i(pQ - qP)/\hbar} \phi) ,$$

where the last term refers to the expectation for any single degree of freedom (the index of which is then suppressed). It follows likewise that $f'(p, q) = (\phi', e^{i(pQ' - qP')/\hbar} \phi')$.

Now suppose that the operator sets $\{P_n\}, \{Q_n\}$ and $\{P'_n\}, \{Q'_n\}$ are related by a unitary transformation in the sense that

$$\overline{U} e^{i(pQ_n - qP_n)/\hbar} \overline{U}^\dagger = e^{i(pQ'_n - qP'_n)/\hbar}$$

holds for all $n \geq 1$ and all $(p, q) \in \mathbb{R}^2$. If this is the case, it follows that $\overline{U} F_N \overline{U}^\dagger = F'_N$ holds for all N, and so too in the limit $N \to \infty$ which implies that $f(p, q) = f'(p, q)$ must hold for all p and q. In other words, *the two representations are unitarily equivalent if and only if the two representation "tags" f and f' coincide for all free parameters p and q.* Here, in this expression, we define a "tag" as a c-number unitary invariant derived from an algebra of operators.* However, since the representation of each P and Q is irreducible for each individual degree of freedom, the equality $f(p, q) = f'(p, q)$ can hold if and only if $\phi' = e^{i\alpha}\phi$, for some real α, namely if the two states are identical apart from a possible phase factor. Conversely, if the two states differ by more than merely an overall phase factor, then the two sets of Weyl operators are *not* unitarily equivalent [KlMW 66]! Since there are uncountably many distinct normalized Hilbert space vectors that are not in the same ray (not related by a phase factor), it follows that there are uncountably many inequivalent irreducible representations of the canonical commutation relations in the case of an infinite number of degrees of freedom. This situation could not stand in greater contrast to the situation,

---

* Tags were discussed in [KlM 65]. Tags are rather analogous to Casimir operators that are often used to distinguish among irreducible representations in group theory [Gi 74].

expressed in Weyl form, of a single, unique irreducible representation up to unitary equivalence in the case of any finite number of canonical degrees of freedom!

Why field theories have divergences

In the case of product representations we have determined a representation of the canonical operators such that

$$\overline{\mathcal{H}} = \sum_{n=1}^{\infty} [\tfrac{1}{2}(P_n^2 + Q_n^2) + gQ_n^4 - c], \qquad g > 0$$

is well defined provided that the ground state φ_n for each degree of freedom has zero energy. While the Hamiltonian is well defined as written, it is important to realize that it is *not* well defined if one tries to write it in the form

$$\overline{\mathcal{H}} = \overline{\mathcal{H}}_0 + g\overline{\mathcal{V}},$$

$$\overline{\mathcal{H}}_0 \equiv \sum_{n=1}^{\infty} \tfrac{1}{2}(P_n^2 + Q_n^2 - c_0),$$

$$\overline{\mathcal{V}} \equiv \sum_{n=1}^{\infty} (Q_n^4 - c_1),$$

for any c_0 and c_1 such that $c = c_0 + gc_1$. In particular, if the representation for the canonical operators is chosen correctly for $\overline{\mathcal{H}}$ then it is incorrect for both $\overline{\mathcal{H}}_o$ and $\overline{\mathcal{V}}$, neither of which is an operator under these circumstances; if instead, the operator representation is chosen so that $\overline{\mathcal{H}}_o$ is well defined, then neither $\overline{\mathcal{V}}$ nor $\overline{\mathcal{H}}$ is an operator. To make a division of the Hamiltonian as indicated has given rise to two of the three terms that do not correspond to operators since they would have only the zero vector in their domain. In point of fact, the "+" sign that combines $\overline{\mathcal{H}}_o$ and $g\overline{\mathcal{V}}$ is strictly formal and should not be read as the sum of two well defined expressions. The situation is not significantly changed if normal ordering is used; see Exercise 6.4. To expand a Hamiltonian operator into two terms joined by such a formal plus sign is a sure method to court trouble!*

Although most problems of physical interest do not involve product representations of the canonical operators, the principle nevertheless holds rather generally true that the plus (or minus) signs that enter into a formal Hamiltonian expression generally do not join two or more well-defined operators when an infinite number of degrees of freedom are involved. Any explicit or implicit assumption to the contrary leads to a computation in which divergences arise. In short,

---

* The example of this section is an elementary illustration of what is generally called Haag's Theorem [Ha 55, Ha 96].

divergences arise when one attempts to use a representation of the canonical operators that is not ideally matched to the particular operator(s) in question.

Approaching infinity as a limit

If the operator representation of an infinite number of degrees of freedom has such esoteric properties, then let us try backing off to a large but finite number of degrees of freedom, say N, and taking the limit $N \to \infty$ as the final step in the calculation. This approach is evidently a kind of cutoff which for the models under discussion in this section is a very natural one. As a concrete example let us continue with a quartic anharmonic oscillator; other problems may be treated in a similar fashion. Thus we are led to consider

$$\overline{\mathcal{H}}_N = \sum_{n=1}^{N} [\tfrac{1}{2}(P_n^2 + Q_n^2) + gQ_n^4 - c] \,, \qquad g > 0 \,,$$

where $c = c(g)$ is chosen so that the ground state has zero energy. Of course, for the present example it is not difficult to see how the limit $N \to \infty$ may be taken, but our goal is also to gain insight into a similar limiting procedure for more complicated problems for which the various degrees of freedom do not act independently of each other. In particular, let us try to study our problem by means of perturbation theory from the free model defined by $g = 0$. A simple, but relatively typical, calculation will illustrate the type of behavior that may be expected in the general situation. For example, let us consider the identity

$$\exp\left\{ -iT \, \Sigma_{n=1}^N [\tfrac{1}{2}(P_n^2 + Q_n^2) + gQ_n^4 - c]/\hbar \right\}$$
$$= \exp\left\{ -iT \, \Sigma_{n=1}^N [\tfrac{1}{2}(P_n^2 + Q_n^2) - c]/\hbar \right\}$$
$$- (ig/\hbar) \int_0^T dt \, \exp\left\{ -i(T - t) \, \Sigma_{n=1}^N [\tfrac{1}{2}(P_n^2 + Q_n^2) - c]/\hbar \right\}$$
$$\times (\Sigma_{n=1}^N Q_n^4) \exp\left\{ -it \, \Sigma_{n=1}^N [\tfrac{1}{2}(P_n^2 + Q_n^2) + gQ_n^4 - c]/\hbar \right\} \,.$$

It is clear that one may iterate the right-hand side by repeatedly inserting the definition from the left-hand side, and this iteration leads to a series expansion that represents the interacting theory in terms of properties of the noninteracting theory. In particular, the series expansion would involve various powers of the interaction term each of which is separated by propagators appropriate to the free theory with $g = 0$. In this regard the development is exactly like that which occurs in quantum mechanics for a finite and fixed number of degrees of freedom. So long as $N < \infty$, the perturbation series makes sense order by order, and in principle, we can imagine evaluating the successive terms in the perturbation expansion represented by the iterated right-hand side of this equation, adding them up, and determining the left-hand side, namely the theory with $g > 0$. This is the ideal goal of a perturbation analysis, and with simple enough examples, or with suitably controlled approximations in more difficult cases, it is actually possible to approach interacting theories on the basis of calculations with suitably cutoff theories. Once the interacting theory has been calculated, either exactly or with sufficient accuracy, then for the case at hand it becomes clear that the limit $N \to \infty$ may be taken. The only ambiguity may be to find the constant c designed to zero-out the ground state energy, but this can be determined by a consistency condition when needed.

In such an approach the correct answers may be determined by working with a system of finitely many degrees of freedom, solving the problem at that level, reducing all the questions of interest to expectation values, i.e., to complex numbers, and taking the limit as the number of degrees of freedom approaches infinity – namely, removing the cutoff. The virtue of this point of view is that exotic issues such as uncountably many inequivalent irreducible representations of the basic kinematical operators are completely side stepped; these issues still exist, but they are not central to the analysis. The approach via cutoffs and only removing them as the final step of the calculation has certain pragmatic advantages. On the other hand, this pragmatic approach is limited in one important aspect. *In particular, it assumes that the desired limit may be obtained through this particular sequence of models.* Natural as this assumption may seem, it nevertheless must be understood that it is indeed an assumption, and therefore may not always hold. *The several models studied in subsequent chapters in this book are devoted to illustrating cases where this assumption fails, and in fact, fails dramatically.*

6.6 Probability distributions and characteristic functions

Probability distributions for a single real variable arise from *measures*. In general, a probability measure is a nonnegative, nondecreasing, normalized, right continuous function $\mu(x)$, $x \in \mathbb{R}$. The properties that μ must fulfill are: (1) $\mu(-\infty) = 0$, (2) $\mu(\infty) = 1$, (3) $\mu(x) \le \mu(x+y)$, $0 < y$, and (4) $\lim_{\delta \to 0+} \mu(x - \delta) \le \mu(x) = \lim_{\delta \to 0+} \mu(x+\delta)$, for any x [KoF 57]. There are three qualitatively distinct types of functions with these properties. Typical of the first is the fact that μ is not a continuous function, a good example being given by the Heaviside function $H(x)$ defined so that $H(x) = 0$, $x < 0$, and $H(x) = 1$, $x \ge 1$. In this case the measure $\mu(x) = H(x)$ corresponds to a formal distribution given by $\delta(x)$, namely a Dirac delta function. Such measures are called *discrete*. Typical of another kind is that μ is a continuous function and furthermore that $\mu(x) = \int_{-\infty}^{x} \rho(y)\, dy$, where ρ is a nonnegative integrable function normalized so that $\mu(\infty) = 1$. Such measures are called *absolutely continuous*. Observe what may seem evident, that in this case μ is the integral of its derivative. Finally, there is a third kind of measure characterized by the fact that μ is a continuous function, but μ is *not* the integral of its derivative! Indeed, the derivative of μ is almost everywhere zero so that its integral would be zero. Such measures are called *singular continuous*. We will give an implicit example of a singular continuous measure in a moment. A general measure is given as the convex linear combination of measures of the three kinds discussed.

It is also very convenient to describe measures by their Fourier transforms, referred to as *characteristic functions* [Lu 70], and written as

$$C(t) \equiv \int e^{itx}\, d\mu(x) \,,$$

for each of the three probability measures or any suitable convex combination. Every characteristic function is continuous, and this is sufficiently important for us to demonstrate that fact. It follows that

$$|C(t) - C(s)| \leq \int |e^{itx} - e^{isx}|\, d\mu(x)$$

$$= \left(\int_{x<B} + \int_{B \leq x < A} + \int_{A \leq x} \right) |e^{itx} - e^{isx}|\, d\mu(x)$$

$$\leq \int_{B \leq x < A} |e^{itx} - e^{isx}|\, d\mu(x) + 2[\mu(B) + 1 - \mu(A)] \,,$$

where in the last line we have used $|e^{itx} - e^{isx}| \leq 2$. Now for any preassigned $\epsilon > 0$, we can choose B finite but sufficiently negative and A finite but sufficiently positive so that the last factor is less than $\epsilon/2$. Next we can bring s and t sufficiently close so that the first term is also less than $\epsilon/2$, which implies that the function C is continuous for any measure.

On the other hand, the three different kinds of measures give rise to characteristic functions with rather different properties. To see this, let us also introduce

$$C_T(t) \equiv \frac{1}{T} \int_t^{T+t} C(s)\, ds = \frac{1}{T} \int_0^T C(s+t)\, ds \,,$$

for some T, $0 < T < \infty$. It then follows that a purely absolutely continuous measure is characterized by the fact that $\lim_{t \to \infty} C(t) = 0$ as well as $\lim_{t \to \infty} C_T(t) = 0$. In turn, a purely discrete measure is characterized by the fact that $\lim_{t \to \infty} C(t) \neq 0$, e.g., as may be shown for a single sequence $t_n \to \infty$, as well as $\lim_{t \to \infty} C_T(t) \neq 0$. Finally, a singular continuous measure is characterized by the fact that although $\lim_{t \to \infty} C(t) \neq 0$, we nevertheless have $\lim_{t \to \infty} C_T(t) = 0$. We offer three characteristic functions as examples of pure distributions of the three types. Discrete: $C(t) = e^{it}$; absolutely continuous: $C(t) = e^{-|t|}$; and singular continuous: $C(t) = \Pi_{n=1}^{\infty} \cos(t/3^n)$. In the case of the latter, this expression implicitly defines a singular continuous measure as well.

It is noteworthy that one can begin completely with characteristic functions and derive the measures. Let a function $C(t)$ defined on the real line be continuous, normalized so that $C(0) = 1$, and positive definite in the sense that

$$\sum_{m,n=1}^{N} z_n^* z_m C(t_m - t_n) \geq 0$$

holds for all complex coefficients $\{z_n\}$, all sets of points $\{t_n\}$, and all $N < \infty$. Then it follows that $C(t)$ is a characteristic function and admits the representation

$$C(t) = \int e^{itx} \, d\mu(x)$$

for a uniquely defined probability measure μ [Lu 70].

If one deals with a sequence of distributions $\{\mu_n\}$ then one may ask whether such a sequence converges in some sense as $n \to \infty$. In many cases, convergence of a sequence of characteristic functions is an easier question to consider than convergence of the distributions themselves. In particular, if $C_n(t)$ denotes a sequence of characteristic functions associated with the measures μ_n, and the functions $C_n(t)$ converge to a continuous function $C(t)$, then it follows that the function $C(t)$ is a characteristic function and admits the representation

$$C(t) \equiv \int e^{itx} \, d\mu(x)$$

for some probability measure μ. In this case one says that the measures μ_n have converged weakly to the measure μ. The weak convergence of measures can change their type. For example, a sequence of absolutely continuous distributions may converge to a singular distribution. A sequence of purely discrete measures may converge to a singular continuous distribution, or to an absolutely continuous distribution, etc.

As discussed in Chapters 3 and 4, and especially in the present chapter, a self-adjoint operator X determines a one-parameter unitary group, and, in turn, a one-parameter group along with a normalized vector determines the expression $\langle \psi | e^{itX} | \psi \rangle$. Evidently, this function satisfies the criteria to qualify as a characteristic function, and therefore there exists a probability measure μ such that

$$C(t) = \int e^{itx} \, d\mu(x) = \langle \psi | e^{itX} | \psi \rangle \, .$$

A sequence of characteristic functions can then be generated by either a sequence of self-adjoint operators X_n or a sequence of normalized states $|\psi_n\rangle$, or a combination of both.

The extension of the previous discussion to probability distributions and characteristic functions for a finite number of variables is straightforward. When it comes to an infinite number of variables, the nature of probability distributions and their associated characteristic functionals is rather more involved, and we shall content ourselves with citing some relevant references [GeV 64, Sk 74].

6.7 Infinitely divisible distributions

It is straightforward to show that the product of two characteristic functions yields another characteristic function. In particular, if $C_1(t)C_2(t)$ is the product of two characteristic functions, then the only nonimmediate feature of showing that the result is a characteristic function is that the product satisfies the positive-definite function requirement. To establish this requirement, first consider that

$$\sum_{m,n=1}^{N} \sum_{r,s=1}^{N} z_n^* z_m w_r^* w_s \, C_1(t_m - t_n) \, C_2(t_s - t_r) \geq 0 \,,$$

which by superposition over the products $z_m w_s \rightarrow \Sigma_j z_m^j w_s^j = u_{m,s}$, and similarly for $z_n^* w_r^*$, implies that

$$\sum_{m,n=1}^{N} \sum_{r,s=1}^{N} u_{n,r}^* u_{m,s} \, C_1(t_m - t_n) \, C_2(t_s - t_r) \geq 0 \,,$$

which, further, by means of suitable limits, extends to arbitrary complex coefficients $\{u_{m,s}\}$. Finally, specialization to the case $u_{m,s} = \delta_{ms} v_m$ leads to the desired property that

$$\sum_{m,n=1}^{N} v_n^* v_m \, C_1(t_m - t_n) \, C_2(t_m - t_n) \geq 0 \,.$$

As is well known, the measure belonging to this product involves a convolution and is given by

$$C(t) = C_1(t)C_2(t) = \int e^{it(x+y)} \, d\mu_1(x) \, d\mu_2(y)$$
$$= \int e^{itb} \, d\mu(b) \,,$$

where $\mu(b) \equiv \int \mu_1(b - y) \, d\mu_2(y)$, which evidently satisfies the properties to be a probability measure. Of course, the two measures μ_1 and μ_2, or equivalently the two characteristic functions C_1 and C_2, may be identical. It follows in fact that if $C(t)$ is a characteristic function, then $C(t)^p$ is also a characteristic function for any $p \in \{1, 2, 3, \ldots\}$.

Rather than taking an arbitrary power of a characteristic function to make a new one, let us consider the subclass of characteristic functions which, if we take the qth *root* of the characteristic function,

$$C(t)^{1/q} \,,$$

still leads to a characteristic function for all $q \in \{1, 2, 3, \ldots\}$. These are the characteristic functions of *infinitely divisible distributions*. It follows that $C(t)^{p/q}$ determines a characteristic function for all rational values p/q, and by continuity of the limit we conclude that $C(t)^r$ is also a characteristic function for all real r, $0 < r < \infty$. Such characteristic functions admit an important canonical representation [Lu 70] that we now proceed to derive.

Since $C(t) = [C(t)^{1/m}]^m$ for any m and $C(t)^{1/m}$ is a characteristic function, it follows from $C(0) = 1$ and continuity, and for sufficiently large m, that $C(t)^{1/m} \neq$

0. As a consequence $C(t) \neq 0$ for all $t \in \mathbb{R}$. Hence, we may set

$$C(t) = e^{-L(t)}$$

for all t. Clearly $L(0) = 0$ and $\Re(L(t)) \geq 0$; here $\Re$ denotes the "real part". Now for every $q > 0$,

$$C_q(t) \equiv C(t)^{1/q} = e^{-L(t)/q} \equiv \int e^{itx}\, d\rho_q(x) \,.$$

In turn, we see that

$$\begin{aligned}
L(t) &= \lim_{q\to\infty} q[1 - e^{-L(t)/q}] = \lim_{q\to\infty} \int [1 - e^{itx}]\, q\, d\rho_q(x) \\
&= \lim_{q\to\infty} \int \{[1 + itx/(1+x^2) - e^{itx}] - itx/(1+x^2)\}\, q\, d\rho_q(x) \\
&= \lim_{q\to\infty} (\int_{|x|\leq 1/q} + \int_{|x|>1/q})[1 + itx/(1+x^2) - e^{itx}]\, q\, d\rho_q(x) \\
&\quad - it \lim_{q\to\infty} \int x/(1+x^2)\, q\, d\rho_q(x) \\
&= -it \lim_{q\to\infty} \int x/(1+x^2)\, q\, d\rho_q(x) \\
&\quad + \lim_{q\to\infty} \int_{|x|\leq 1/q}[1 + itx/(1+x^2) - e^{itx}]\, q\, d\rho_q(x) \\
&\quad + \lim_{q\to\infty} \int_{|x|>1/q}[1 + itx/(1+x^2) - e^{itx}]\, q\, d\rho_q(x) \\
&\equiv -iat + \tfrac{1}{2}bt^2 + \int_{|x|>0}[1 + itx/(1+x^2) - e^{itx}]\, d\sigma(x) \,,
\end{aligned}$$

where $a \in \mathbb{R}$, $b \geq 0$, and σ is a nonnegative measure for which

$$\int [x^2/(1+x^2)]\, d\sigma(x) < \infty \,, \qquad |x| > 0 \,.$$

This final representation for $L(t)$ is called the Levy canonical representation. If $\sigma \equiv 0$, then the resultant distribution is a Gaussian with mean a and variance b. If $b = 0\,(=a)$ and $\sigma \neq 0$, then we generically refer to such a distribution as a Poisson distribution. More specifically, if $\int d\sigma(x) < \infty$, we are dealing with a compound Poisson process, while if $\int d\sigma(x) = \infty$, we are dealing with a generalized Poisson process [DeF 75]. This latter case will be of great significance for our model studies!

Besides the Gaussian distribution, there are comparatively few other infinitely divisible distributions that are explicitly known. One of the most important examples is the Cauchy distribution given by

$$\begin{aligned}
C(t) = e^{-c|t|} &= \int e^{itx} \frac{c}{\pi(x^2+c^2)}\, dx \\
&= \exp\{-\int [1 - \cos(tx)](c/\pi x^2)\, dx\} \,,
\end{aligned}$$

where $c > 0$ is an arbitrary parameter of the distribution. Observe in this case that the characteristic function $C_q(t) = C(t)^{1/q}$ is equivalently obtained simply by changing the parameter c to c/q.

6.8 Basic properties of Brownian motion

A stochastic process describes a set of random functions of (say) time governed by some distribution. We denote averages in the ensemble by either $\langle(\cdot)\rangle$ or $\int(\cdot)\,d\rho$, where ρ denotes a suitable probability measure on random functions.

An important special case is that of a standard Wiener process, which is a stochastic process with a distribution characterized by four special properties. If $W(t)$, $t \geq 0$, denotes a random function, then the standard Wiener ensemble is characterized by the following four properties:

$$W(0) = 0 \,,$$

$$\langle W(t)\rangle = 0 \,,$$

$$\langle W(t)W(u)\rangle = \min(t,u) \,,$$

the distribution is Gaussian .

A general Wiener process X is related to a standard Wiener process W by

$$X(t) = \sqrt{k}\,W(t-t') + x' \,, \qquad t \geq t' \,,$$

and differs by a rescaling ($\sqrt{k}$), by a time shift ($t \to t - t'$), and by an offset at the initial time (x'). The four defining properties of a general Wiener process evidently follow from those of a standard Wiener process.

All four defining properties of a standard Wiener process are embodied in the *single* formula

$$\langle \exp[i\int s(t)W(t)\,dt]\rangle \equiv \exp[-\tfrac{1}{2}\int s(t)\min(t,u)\,s(u)\,dt\,du] \,,$$

where s is a (generalized) function defined on $0 \leq t < \infty$, which may be chosen as a suitable distribution (especially including δ-functions) for which the right-hand side is well defined. This particular expectation, along with the use of the expression

$$\delta(y) = \frac{1}{2\pi}\int e^{ixy}\,dx \,,$$

enables many quantities of interest to be obtained. For convenience we next list several expressions with an associated comment:

$$\langle[X(t) - X(u)]^2\rangle = k\,|t-u| \,,$$

from which it follows (see Chapter 8) that the time derivative of Brownian motion paths is nowhere defined, i.e., $|\dot{X}(t)| = \infty$ for all $t > t'$ with probability one;

$$\langle[X(t) - X(u)]^4\rangle = 3k^2\,|t-u|^2 \,,$$

from which it follows that Brownian motion paths $X(t)$ are continuous functions with probability one;

$$p(x,t) \equiv \langle \delta(x - W(t)) \rangle$$

$$= \frac{e^{-x^2/2t}}{\sqrt{2\pi t}} \,, \qquad 0 < t \,,$$

where $p(x,t)$ is the probability density for finding standard Brownian motion paths at time t at position x;

$$p(x,t; y, u) \equiv \frac{\langle \delta(x - X(t)) \, \delta(y - X(u)) \rangle}{\langle \delta(y - X(u)) \rangle}$$

$$= \frac{e^{-(x-y)^2/2k(t-u)}}{\sqrt{2\pi k(t-u)}} \,, \qquad t' < u < t \,,$$

where $p(x,t; y, u)$ denotes the probability density of finding general Brownian motion paths at time t and position x given the fact that all the paths must pass through the position y at the earlier time u;

$$p(x,t|0,1) \equiv \frac{\langle \delta(W(1)) \, \delta(x - W(t)) \rangle}{\langle \delta(W(1)) \rangle}$$

$$= \frac{e^{-x^2/2t(1-t)}}{\sqrt{2\pi t(1-t)}} \,, \qquad 0 < t < 1 \,,$$

where $p(x,t|0,1)$ is the probability density of finding standard Brownian motion paths at time t at position x conditioned on the fact that all paths pass through zero at the later time $t = 1$. This process is often called a *Brownian bridge*; and

$$p(x,t|x'', T)$$
$$\equiv \frac{\langle \delta(x'' - X(T)) \, \delta(x - X(t)) \rangle}{\langle \delta(x'' - X(T)) \rangle} \,, \qquad 0 = t' < t < T \,,$$
$$= \frac{1}{\sqrt{2\pi k\, t(1 - t/T)}} \exp\left\{ -\frac{[x - x'(1 - t/T) - x''t/T]^2}{2k\, t(1 - t/T)} \right\} \,,$$

where $p(x,t|x'', T)$ denotes the probability density (for $t' = 0$) of finding general Brownian motion paths at the time t and position x given the fact that all the paths pass through the position x'' at the later time T. This particular probability density will be used in Chapter 8.

For additional features of Brownian motion paths, and therefore of Wiener measures, the reader may consult [Ro 96].

6.9 General features of canonical models

Suppose we consider a scalar field model whose sharp-time field $\varphi(\mathbf{x})$ is a local self-adjoint operator. That is, $\varphi(f) = \int f(\mathbf{x})\, \varphi(\mathbf{x})\, d^s x$ defines a self-adjoint operator for a suitably wide class of real test functions f; we take up this very question of the class of test functions in the next paragraph. Let $|0\rangle$ be a distinguished normalized state – typically the ground state of the relevant Hamiltonian – and define the states

$$|f\rangle \equiv e^{i\varphi(f)} |0\rangle$$

for all f. The important functional

$$E(f) \equiv \langle 0| e^{i\varphi(f)} |0\rangle \equiv \langle 0|f\rangle \,,$$

normalized so that $E(0) = 1$, is known as the expectation functional for the field φ relative to the state $|0\rangle$, or with φ and $|0\rangle$ understood, simply as the *expectation functional*. It follows that

$$\langle f|f'\rangle = \langle 0|f' - f\rangle = \langle f' - f|0\rangle^* = \langle -f| - f'\rangle^* = E(f' - f) \,.$$

Maximum test function space

The expectation functional satisfies a minimal degree of continuity. In particular, let f_1 be such that $\varphi(f_1)$ is self adjoint. Then for this fixed f_1 the operator $\exp[it_1\varphi(f_1)]$ fulfills the conditions to be a one-parameter unitary group and it follows that the function $C(t_1) \equiv E(t_1 f_1)$ is continuous for $t_1 \in \mathbb{R}$. In like manner, the function

$$C(t_1, t_2, \ldots, t_N) \equiv E(t_1 f_1 + t_2 f_2 + \cdots + t_N f_N)$$

is jointly continuous for $(t_1, t_2, \ldots, t_N) \in \mathbb{R}^N$ for arbitrary $N < \infty$ provided that $\varphi(f_n)$ is self adjoint for all n, $1 \leq n \leq N$. However, one can even say more. Consider the functional

$$d\{f; f'\}^2 \equiv \frac{1}{\sqrt{4\pi}} \int_{-\infty}^{\infty} \{1 - E(t[f - f'])\} e^{-t^2/4} \, dt$$

$$= \langle 0|[1 - e^{-\varphi(f-f')^2}]|0\rangle \,.$$

It may be shown [HeK 70] that the functional $d\{f; f'\}$ serves to define a *metric* on test function space, i.e., the space of functions f, and that one may complete the test function space by including the limit points of Cauchy sequences in the metric d. Furthermore, for each and every element f of the completed test function space, the operator $\varphi(f)$ is self adjoint. Thus, the maximally allowed set of test functions for a given system is defined and determined by the expectation functional itself. Other equivalent metrics may also be used, and in general it is not really necessary to extend the test function space to its maximum size. Instead, it generally suffices to work with a convenient but dense set of test functions, and we shall implicitly do so hereafter.

As an example of the use of this metric consider the simple example of the free field introduced in Chapter 5 for which

$$E(f) = \exp[-(1/4)\int\Omega(\mathbf{k}) |\tilde{f}(\mathbf{k})|^2 \, d^s k] \,.$$

In this case

$$d\{f;0\}^2 = [1 - 1/\sqrt{1 + \int \Omega(\mathbf{k}) \,|\tilde{f}(\mathbf{k})|^2 \, d^s\kappa}] \,,$$

which is equivalent to the metric

$$d_\Omega\{f;0\}^2 \equiv \int \Omega(\mathbf{k}) \,|\tilde{f}(\mathbf{k})|^2 \, d^s k \,.$$

Thus, in this case, it follows that all functions for which $d_\Omega\{f;0\} < \infty$ are admissible test functions.

Araki relations

Besides the field $\varphi(\mathbf{x})$, we also assume there is a symmetric (i.e., Hermitian) canonical momentum $\pi(\mathbf{x})$ that satisfies the canonical commutation relations in the form (with $\hbar = 1$)

$$[\varphi(\mathbf{x}), \pi(\mathbf{y})] = i\,\delta(\mathbf{x} - \mathbf{y}) \,,$$

and also in the stronger form

$$e^{i\varphi(f)}\,\pi(\mathbf{x})\,e^{-i\varphi(f)} = \pi(\mathbf{x}) - f(\mathbf{x}) \,.$$

We next assume that the distinguished state $|0\rangle$ introduced above is *time-reversal invariant*, i.e., $T|0\rangle = |0\rangle$. In addition, recall that

$$T^\dagger a\,\varphi(\mathbf{x})T = a^*\varphi(\mathbf{x}) \,,$$
$$T^\dagger b\,\pi(\mathbf{x})T = -b^*\pi(\mathbf{x}) \,,$$

where a and b are complex numbers, and recall that the time-reversal transformation is anti-unitary and therefore $(T\lambda, T\chi) = (\chi, \lambda)$. For convenience, we also introduce an alternative notation for which $\Omega = |0\rangle$ (N.B. This is a different use of Ω from above!) and

$$\langle f|\mathcal{O}|f'\rangle \equiv (e^{i\varphi(f)}\Omega, \mathcal{O}\,e^{i\varphi(f')}\Omega)$$

for a general operator $\mathcal{O}$.

Time-reversal invariance can help us determine certain matrix elements [Ar 60a, Ar 60b; see also Chapter 4]. Let us consider the matrix elements of π, and for this purpose it is necessary to assume that the vectors $\{|f\rangle\}$ are in the form domain of $\pi(\mathbf{x})$ for sufficiently smooth f. In particular, and using the fact that $T\Omega = \Omega$,

$$
\begin{aligned}
\langle f|\pi(\mathbf{x})|f'\rangle &= (e^{i\varphi(f)}T\Omega, \pi(\mathbf{x})e^{i\varphi(f')}T\Omega) \\
&= -(Te^{-i\varphi(f)}\Omega, T\pi(\mathbf{x})e^{-i\varphi(f')}\Omega) \\
&= -(e^{-i\varphi(f')}\Omega, \pi(\mathbf{x})e^{-i\varphi(f)}\Omega) \\
&= -(e^{i\varphi(f)}\Omega, [\pi(\mathbf{x}) - f(\mathbf{x}) - f'(\mathbf{x})]e^{i\varphi(f')}\Omega) \\
&= -\langle f|[\pi(\mathbf{x}) - f(\mathbf{x}) - f'(\mathbf{x})]|f'\rangle \,,
\end{aligned}
$$

which directly leads to the fundamental relation

$$\langle f|\pi(\mathbf{x})|f'\rangle = \tfrac{1}{2}[f(\mathbf{x}) + f'(\mathbf{x})]\langle f|f'\rangle \,.$$

Let us now turn to some general remarks about dynamics. Assume that the Hamiltonian $\mathcal{H}$ is formally given by

$$``\mathcal{H} = \tfrac{1}{2}\int \pi(\mathbf{x})^2\, d^s x + \mathcal{V}(\varphi)\text{''},$$

and that $|0\rangle$ is the ground state of the Hamiltonian adjusted so that it satisfies $\mathcal{H}|0\rangle = 0$. Since the Hamiltonian satisfies $\mathcal{H}(-\pi, \varphi) = \mathcal{H}(\pi, \varphi)$, it is time-reversal invariant, and it follows that our requirement for $|0\rangle$ to be time-reversal invariant is consistent. Additionally, we learn that

$$\pi(\mathbf{x}) = i[\mathcal{H}, \varphi(\mathbf{x})] \,,$$

which provides an important connection between the Hamiltonian, the field, and the momentum. Indeed, it is this connection which is important, even more so than the presumed form for $\mathcal{H}$. Matrix elements for the Hamiltonian follow immediately, provided that we assume that the states $\{|f\rangle\}$ are in the form domain of $\mathcal{H}$. First observe that

$$\begin{aligned}
\tfrac{1}{2}[f(\mathbf{x}) + f'(\mathbf{x})]\langle f|f'\rangle &= \langle f|\pi(\mathbf{x})|f'\rangle \\
&= i\langle f|\,[\mathcal{H}, \varphi(\mathbf{x})]|f'\rangle \\
&= [\delta/\delta f(\mathbf{x}) + \delta/\delta f'(\mathbf{x})]\,\langle f|\mathcal{H}|f'\rangle \,.
\end{aligned}$$

The solution to this first-order functional differential equation is given by

$$\langle f|\mathcal{H}|f'\rangle = \tfrac{1}{2}\int f(\mathbf{x})f'(\mathbf{x})\, d^s x \,\langle f|f'\rangle + G\{f - f'\} \,,$$

where G denotes the "constant" of integration. Insisting that $\langle f|\mathcal{H}|0\rangle = 0$ tells us that $G\{f\} \equiv 0$. Consequently, we have obtained a second important set of matrix elements, namely,

$$\langle f|\mathcal{H}|f'\rangle = \tfrac{1}{2}(f, f')\,\langle f|f'\rangle \,.$$

The matrix elements of the momentum operator $\pi(\mathbf{x})$ and of the Hamiltonian $\mathcal{H}$ in the set of states $\{|f\rangle\}$ derived here are subsequently referred to as the "Araki relations".

In certain circumstances the set of states $\{|f\rangle\}$ for all f, or a suitably dense set, forms a total set of states. That is, the states $\{|f\rangle\}$ span the Hilbert space. In favorable cases, this even means that the given matrix elements may uniquely determine the local momentum operator $\pi(\mathbf{x})$ and/or the Hamiltonian $\mathcal{H}$ in terms of the expectation functional $E(f)$. This is an important fact when it occurs. This unique determination holds, for example, for any of the so-called free field theories discussed in Chapter 5. On the other hand, for any of the (nonfree) *generalized* free field theories discussed in the Exercises at the end of Chapter 5 that admit canonical commutation relations, the states $\{|f\rangle\}$ do *not* span the Hilbert space, and while the matrix elements that are given by the Araki relations are nevertheless correct, these matrix elements, by themselves, *cannot* in general determine the canonical momentum or the Hamiltonian operators.

Exercises

6.1 Let $\{B_n\}$ be a sequence of uniformly bounded operators, $\|B_n\| \leq b$, for all $n \in \{1, 2, 3, \ldots\}$. Let $\mathcal{T}$ and $\mathcal{T}'$ denote two total sets of vectors (which may be the same set). Assume that

$$\langle \tau | B_n | \tau' \rangle \equiv B_n(\tau; \tau') \to B(\tau; \tau') \,,$$

i.e., that the indicated matrix elements converge as complex numbers for every fixed pair of vectors $\tau \in \mathcal{T}$ and $\tau' \in \mathcal{T}'$. In this case, show that the sequence $\{B_n\}$ of uniformly bounded operators converges weakly to a bounded operator B which is fully characterized by the fact that

$$\langle \tau | B | \tau' \rangle \equiv B(\tau; \tau') \,.$$

6.2 Consider the expression

$$\int f(x)\, e^{-ikx}\, dx = \int g(x)^*\, e^{-ikx}\, h(x)\, dx = \int \tilde{g}(q)^*\, \tilde{h}(q+k)\, dq \,,$$

where $f \in L^1$ and both g and h are in L^2. It is clear that $|e^{ikx}| = 1$ and hence the corresponding transformation is uniformly bounded. Thus as $k \to \infty$, according to the results of the previous Exercise, we need only establish the limit for a total set of functions $\tilde{g}$ and $\tilde{h}$. Pick a convenient total set of functions and establish the *Riemann–Lesbesgue Lemma*, that

$$\lim_{k \to \pm\infty} \int_{-\infty}^{\infty} e^{-ikx}\, f(x)\, dx = 0$$

whenever $\int |f(x)|\, dx < \infty$.

6.3 For any $\alpha \in L^2(\mathbb{R}^s)$, show that the operator

$$\int \alpha(\mathbf{k}) a(\mathbf{k})^{\dagger\, 2}\, d^s k$$

has only the zero vector in its domain.

6.4 Consider the Hamiltonian composed of a countable number of noninteracting degrees of freedom and given by

$$\overline{\mathcal{H}} = \sum_{n=1}^{\infty} [\tfrac{1}{2} : (P_n^2 + Q_n^2) : + g : Q_n^4 : -c']$$

with c' chosen so that the ground state energy of $\overline{\mathcal{H}}$ vanishes. Here, normal ordering can be taken with respect to any state you like. Show that although there is an irreducible representation of the canonical operators satisfying $[Q_n, P_m] = i\delta_{mn}$ for which the full Hamiltonian is a well-defined operator, it is not possible to define the Hamiltonian operator as the sum of a well-defined, free, quadratic Hamiltonian plus a well-defined interaction potential whenever $g > 0$. Indeed, confirm that for no choice of irreducible representation of the canonical operators is it possible to have well-defined total kinetic energy and total interaction potential operators. Discuss the kind of difficulties that can arise if one nevertheless proceeds formally and expands certain expressions (e.g., the propagator) into a power series in the interaction potential.

6.5 Discuss the product representation of the operators $\{P_n, Q_n\}$ in which the Hamiltonian $\overline{\mathcal{H}} \equiv \Sigma_{n=1}^{\infty} \mathcal{H}_n$, where

$$\mathcal{H}_n = \tfrac{1}{2}(P_n^2 + Q_n^2) + g_n Q_n^4 - c_n \,,$$

is well defined. Here, the coupling constants $g_n \geq 0$, but otherwise form an arbitrary sequence, and the constants c_n are chosen appropriately. Show how the representation of the canonical operators depends on the sequence $\{g_n\}$.

6.6 Consider the operators $P_n = (-i\partial/\partial x)^n$, $n \in \{1, 2, 3, \ldots\}$, on the space $L^2([0,1])$ with $\mathfrak{D}(P_n) = \{\psi : \int [\,|\psi(x)|^2 + |\psi^{(n)}(x)|^2\,]\,dx < \infty, \ \psi(0) = \psi(1) = 0\}$. Show that the deficiency indices in this case are (n, n), which then entails n^2 real parameters to obtain a self-adjoint extension. For $n = 2$, establish the domain of the self-adjoint operators and find the dependence of the spectrum on the additional parameters.

7

Rotationally Symmetric Models

WHAT TO LOOK FOR

Symmetry is a powerful tool in studying and sometimes even solving classical and especially quantum mechanical problems. The models chosen for this chapter are rich with symmetry. Although the classical theory of each model is well defined and nontrivial, the conventional formulation of the quantum theory leads to trivial results. However, by expanding the starting point slightly we are able to offer a fully satisfactory solution to these quantum models, and symmetry plays a key role in finding that solution outside the conventional framework. In the present chapter we generally choose natural units for which $\hbar = 1$.

7.1 Classical formulation

We describe the models by means of their classical Hamiltonian. As we shall observe, these models admit a variety of equivalent formulations. The first formulation is based on a coordinate approach involving an infinite number of momentum and coordinate variables, $p \equiv \{p_n\}_{n=1}^{\infty}$ and $q \equiv \{q_n\}_{n=1}^{\infty}$. Each model Hamiltonian is well defined for all $p \in l^2$ and $q \in l^2$, specifically, when $\Sigma p_n^2 < \infty$ and $\Sigma q_n^2 < \infty$, and the classical Hamiltonian on which we shall focus is explicitly given by

$$H(p,q) = \tfrac{1}{2}\Sigma\,(p_n^2 + m_o^2\,q_n^2) + \lambda[\,\Sigma\,q_n^2\,]^2\;.$$

Here the sums in question run from 1 to ∞. (We shall consider the same model for truncated summations that run from 1 to $N < \infty$ as well.) The constant $\lambda \geq 0$ is a coupling constant for the quartic nonlinear interaction. If $\lambda = 0$ we deal with a linear and noninteracting theory, while if $\lambda > 0$ the theory is nonlinear. Moreover, $H(p,q) \geq 0$ with $H(p,q) = 0$ holding only for $p = q = 0$. The important feature to observe about the model is a *rotational symmetry* among all the coordinates and momenta, which is a simple consequence of the

145

fact that the Hamiltonian is a function of only two expressions, $p^2 \equiv \Sigma p_n^2$ and $q^2 \equiv \Sigma q_n^2$. In other words, the symmetry group is $O(\infty)$ for the complete model [or $O(N)$ for the model with a truncated summation]. The transformations of this group are real, square, semi-infinite, orthogonal matrices $\{O_{nm}\}_{n,m=1}^{\infty}$, and each transformation in question is given by

$$\{p_n\} \to \{\Sigma_m O_{nm}\, p_m\}\,, \qquad \{q_n\} \to \{\Sigma_m O_{nm}\, q_m\}$$

provided that $\Sigma_m O_{nm}\, O_{lm} = \delta_{nl} = \Sigma_m O_{mn}\, O_{ml}$. In studying both the classical theory and the quantum theory this invariance group will play an important role.*

The model in question is just one example of a class of rotationally symmetric models given, for example, by classical Hamiltonians of the form

$$H(p,q) = \tfrac{1}{2}\Sigma(p_n^2 + m_o^2\, q_n^2) + \Sigma_{j=2}^{J}\lambda_j[\Sigma q_n^2]^j\,, \qquad J < \infty\,,$$

where a sufficient condition on the set of coefficients $\{\lambda_j\}_{j=2}^{J}$ for acceptability is given by $\lambda_j \geq 0$ for all j. When all $\lambda_j = 0$ we again deal with a free theory; in the contrary case the theory has a nonlinear interaction. Although the details are certainly different, the qualitative properties of all interacting theories of the present type are similar. For convenience, therefore, we focus our principal attention on the initial and basic model with a quartic nonlinearity.

Of course, we can express the initial model in rather different ways by introducing *fields* in place of *coordinates*. As we have done several times previously, let $\{h_n(\mathbf{x})\}_{n=1}^{\infty}$ denote a real, complete set of orthonormal functions for some space, say $L^2(\mathbb{R}^s)$ to be specific. Then we introduce the real momentum and coordinate fields

$$f(\mathbf{x}) \equiv \sum_{n=1}^{\infty} p_n\, h_n(\mathbf{x})\,, \qquad g(\mathbf{x}) \equiv \sum_{n=1}^{\infty} q_n\, h_n(\mathbf{x})\,,$$

for time $t = 0$, and more generally, say, for $t \geq 0$,

$$f(t,\mathbf{x}) \equiv \sum_{n=1}^{\infty} p_n(t)\, h_n(\mathbf{x})\,, \qquad g(t,\mathbf{x}) \equiv \sum_{n=1}^{\infty} q_n(t)\, h_n(\mathbf{x})\,.$$

In terms of these functions the Hamiltonian (at time zero) reads

$$H(f,g) = \tfrac{1}{2}\int [f(\mathbf{x})^2 + m_o^2\, g(\mathbf{x})^2]\, d^s x + \lambda\{\int g(\mathbf{x})^2\, d^s x\}^2\,.$$

Evidently this expression is well defined for all $f \in L^2$ and $g \in L^2$, and $H(f,g) \geq 0$ with $H(f,g) = 0$ holding only when $f = g = 0$ (almost everywhere). The field formulation of the model also exhibits a rotational symmetry since the Hamiltonian depends only on the two variables

$$X \equiv (f,f) \equiv \int f(\mathbf{x})^2\, d^s x\,, \qquad Z \equiv (g,g) \equiv \int g(\mathbf{x})^2\, d^s x\,,$$

---

* Our presentation is largely based on [Kl 65].

which are alternative expressions for the length squared of the vectors p and q. Here, transformations of the group $O(\infty)$ may be realized by means of real, distributional, integral kernels $O(\mathbf{x}, \mathbf{y})$ in the form

$$f(\mathbf{x}) \rightarrow f_O(\mathbf{x}) \equiv \int O(\mathbf{x}, \mathbf{y}) f(\mathbf{y}) \, d^s y \,,$$
$$g(\mathbf{x}) \rightarrow g_O(\mathbf{x}) \equiv \int O(\mathbf{x}, \mathbf{y}) g(\mathbf{y}) \, d^s y,$$

provided that

$$\int f_O(\mathbf{x})^2 \, d^s x = \int f(\mathbf{x})^2 \, d^s x = \int f_{\bar{O}}(\mathbf{x})^2 \, d^s x \,,$$
$$\int g_O(\mathbf{x})^2 \, d^s x = \int g(\mathbf{x})^2 \, d^s x = \int g_{\bar{O}}(\mathbf{x})^2 \, d^s x$$

holds for all $f \in L^2$ and all $g \in L^2$, where $f_{\bar{O}}(\mathbf{x}) \equiv \int O(\mathbf{y}, \mathbf{x}) f(\mathbf{y}) \, d^s y$, and likewise for $g_{\bar{O}}$. A few examples of such transformations are given by

$$f(\mathbf{x}) \rightarrow f(\mathbf{x} - \mathbf{a}) \,,$$
$$f(\mathbf{x}) \rightarrow |b|^{s/2} f(b\mathbf{x}) \,, \qquad b \neq 0 \,,$$
$$f(\mathbf{x}) \rightarrow |\det(\partial \mathbf{y}/\partial \mathbf{x})|^{1/2} f(\mathbf{y}(\mathbf{x})) \,,$$

etc.; although not explicitly indicated here each transformation of f is accompanied by exactly the same transformation of g. In the last transformation indicated we observe that a general, nonsingular, invertible, and *nonlinear* coordinate transformation $\mathbf{x} \rightarrow \mathbf{y}(\mathbf{x})$ is even possible as an allowed transformation, and provided that f and g transform as tensor densities of the indicated weight, then the Hamiltonian remains invariant. Of course, the first two examples are just special cases of the last transformation.

As an illustrative example of such transformations, we shall occasionally appeal to the space-translation transformation $f(\mathbf{x}) \rightarrow f(\mathbf{x}-\mathbf{a})$ and $g(\mathbf{x}) \rightarrow g(\mathbf{x}-\mathbf{a})$ for arbitrary $\mathbf{a} \in \mathbb{R}^s$.

Equations of motion

The Hamiltonian equations of motion are readily seen to be

$$\dot{q}_n(t) = \{q_n(t), H\} = p_n(t) \,,$$
$$\dot{p}_n(t) = \{p_n(t), H\} = -m_o^2 \, q_n(t) - 4\lambda \, q_n(t) \left[\Sigma \, q_m(t)^2 \right] \,,$$

or in field form as

$$\dot{g}(t, \mathbf{x}) = f(t, \mathbf{x}) \,,$$
$$\dot{f}(t, \mathbf{x}) = -m_o^2 \, g(t, \mathbf{x}) - 4\lambda \, g(t, \mathbf{x}) \left[\int g(t, \mathbf{y})^2 \, d^s y \right] \,.$$

It matters little whether we follow the coordinate form or the field form of the equations of motion. For convenience, at present, we shall maintain the field notation, and as a further convenience we introduce the three real, time-dependent, auxiliary variables

$$X(t) \equiv \int f(t,\mathbf{x})^2 \, d^s x = (f,f) \,,$$
$$Y(t) \equiv \int f(t,\mathbf{x}) \, g(t,\mathbf{x}) \, d^s x = (f,g) \,,$$
$$Z(t) \equiv \int g(t,\mathbf{x})^2 \, d^s x = (g,g) \,;$$

for notational convenience we suppress the time (t) in the latter, inner-product notation. It follows that $X \geq 0$, $Z \geq 0$, and $XZ \geq Y^2$ as follows from the Schwarz inequality. Based on the equations of motion for the fields we find (choosing $m_o = 1$ for the rest of this section, for simplicity) that

$$\dot{X}(t) = 2(f,\dot{f}) = -2(f,g) - 8\lambda(f,g)(g,g)$$
$$= -2Y(t) - 8\lambda Y(t)\, Z(t) \,,$$
$$\dot{Y}(t) = (f,\dot{g}) + (\dot{f},g) = (f,f) - (g,g) - 4\lambda(g,g)^2$$
$$= X(t) - Z(t) - 4\lambda Z(t)^2 \,,$$
$$\dot{Z}(t) = 2(g,\dot{g}) = 2(g,f)$$
$$= 2Y(t) \,.$$

Observe that the field equations for this problem have led to three closed, coupled, generally nonlinear equations for the variables X, Y, and Z. An integral of the motion is the classical energy

$$E = \tfrac{1}{2}[X(t) + Z(t)] + \lambda Z(t)^2 = \tfrac{1}{2}[X(0) + Z(0)] + \lambda Z(0)^2 \,,$$

where in the last relation we have indicated the form of the energy as expressed in the initial field values $f(\mathbf{x}) = f(0,\mathbf{x})$ and $g(\mathbf{x}) = g(0,\mathbf{x})$ at time $t = 0$. If $f \equiv g \equiv 0$, then $E = 0$ and no classical motion occurs; hereafter we exclude this trivial case. From the equations of motion

$$k = X(t)\, Z(t) - Y(t)^2 = X(0)\, Z(0) - Y(0)^2$$

is also a constant of the motion proportional to the square of the angular momentum (see below). There are two distinct cases, one when $k = 0$, the other when $k > 0$. First, if $k = 0$, it follows that f and g are in essence proportional to each other, i.e., that there exist real constants a and b (not both zero) such that $af(\mathbf{x}) = bg(\mathbf{x})$ holds (almost everywhere) for all $\mathbf{x} \in \mathbb{R}^s$. If $a = 0$ or $b = 0$, then $Y = 0$ and either $X = 0$ or $Z = 0$. If $ab \neq 0$, then X, Y, and Z are all different from zero. We shall argue as if $ab \neq 0$ although the discussion also applies to the case that either a or b is zero. When $k = 0$ it follows, as vectors in the L^2 Hilbert space, that the initial momentum and field are *collinear*, and in reality only one degree of freedom is excited initially. Due to the special nature of these models only that particular single degree of freedom will ever be relevant. Indeed, we could orient the first of the orthonormal functions along the common field direction of the initial momentum and field so that only p_1 and q_1 would ever differ from zero. Consequently, all other momenta and coordinates which started with zero would remain zero for all time: $p_n(t) = q_n(t) = 0$ for all $t \geq 0$ for all $n \geq 2$. In this case the dynamical solution is indistinguishable from that of a *single quartic anharmonic oscillator*.

If $k > 0$, on the other hand, it means that both the initial field g and the initial momentum f are not identically zero, and additionally they are *not* collinear. Instead these two functions span a *two*-dimensional subspace in $L^2(\mathbb{R}^s)$. If we like, we can choose one of these fields (say g) proportional to the first orthonormal function and the second field (f) to be a linear combination of the first and the second orthonormal functions. In other words, we can identify

$$g(\mathbf{x}) = q_1 \, h_1(\mathbf{x}) \, ,$$
$$f(\mathbf{x}) = p_1 \, h_1(\mathbf{x}) + p_2 \, h_2(\mathbf{x}) \, .$$

Hence $X = p_1^2 + p_2^2$, $Y = p_1 q_1$, and $Z = q_1^2$. The two constants of the motion are

$$E = \tfrac{1}{2}(X + Z) + \lambda Z^2 = \tfrac{1}{2}(p_1^2 + p_2^2 + q_1^2) + \lambda q_1^4 \, ,$$
$$k = XZ - Y^2 = q_1^2 \, p_2^2 \, .$$

If $p_2 = 0$, then $k = 0$ and the problem effectively reduces to a single mode. Thus in the most general case the solutions of this field theoretic model are entirely equivalent to those of a one or at most a two degree-of-freedom, rotationally symmetric quartic model! If a direction in L^2 (or equivalently in l^2) is not excited at time zero in either the field or the momentum, then it will remain unexcited for all time in the special class of models under consideration in this chapter. This simplicity has helped us in the classical case and will help us in establishing the quantum theory for such models; of course, most theories of physical interest are not anywhere nearly so simple.

The basic equation of motion

$$\ddot{Z}(t) = 2X(t) - 2Z(t) - 8\lambda Z(t)^2 \, ,$$

in conjunction with the definition of the energy, leads to the relation that

$$\dot{Z}(t)\ddot{Z}(t) = 4E\dot{Z}(t) - 4Z(t)\dot{Z}(t) - 12\lambda Z(t)^2 \dot{Z}(t)$$

which on integration leads to

$$\dot{Z}(t)^2 = 8EZ(t) - 4Z(t)^2 - 8\lambda Z(t)^3 - 4k \, .$$

Note that we have set the constant of integration in this case proportional to k. To verify that choice we observe that $X = 2E - Z - 2\lambda Z^2$ and therefore $\dot{Z}(t)^2 = 4Y(t)^2 = 4X(t)Z(t) - 4k$ as required. Incidentally, this remark confirms that $XZ - Y^2$ is a constant of the motion. With the equation for $\dot{Z}(t)^2$ confirmed, we are led to the implicit solution for $Z(t)$ given by

$$\int_{Z_0}^{Z} \frac{dZ}{\sqrt{8EZ - 4Z^2 - 8\lambda Z^3 - 4k}} = t - t_0 \, ,$$

and this formula applies whenever $E > 0$ and for any compatible Z_0 and $k \geq 0$. Given $Z(t)$, then $X(t)$ follows from the conservation of energy and $Y(t) = \dot{Z}(t)/2$. These expressions may then be put back into the field equations of motion to determine the time dependence of $f(t, \mathbf{x})$ and $g(t, \mathbf{x})$.

Although the foregoing discussion amply illustrates the nontrivial nature of the classical solutions to the quartic coupled rotationally symmetric model, one may also study the equations of motion on a computer. The most general classical solution to the model under discussion may be obtained by examining the two degrees of freedom set of equations of motion

$$\ddot{q}_j(t) = -q_j(t) - 4\lambda q_j(t) \left[q_1(t)^2 + q_2(t)^2\right] , \qquad j = 1, 2 ,$$

subject to the general initial condition $q_j = q_j(0)$ and $p_j = \dot{q}_j(0)$, $j = 1, 2$. The main lesson one draws from such a study is that for $k > 0$ the solutions for q_j, $j = 1, 2$, exhibit a doubly periodic behavior, one period controlled by the energy E, the other period controlled by the size of k. In particular, if $k = 0$ only the former period remains.

Taylor series representation

Before leaving the classical theory we wish to present one additional characterization of the solution. Clearly, the solution is implicitly defined as a function of the time t and the initial data. We can make explicit that dependence with the help of a Taylor series expansion of the time dependence of the solution about $t = 0$. For this formulation we return to the field notation, and for convenience, we continue to maintain $m_o = 1$. In these variables the Hamiltonian (at time zero) reads

$$H = \tfrac{1}{2}[(f, f) + (g, g)] + \lambda(g, g)^2 .$$

Let us introduce the multiple Poisson brackets given by

$$g^{(1)}(0, \mathbf{x}) \equiv \{g(\mathbf{x}), H\} = f(\mathbf{x}) ,$$
$$g^{(2)}(0, \mathbf{x}) \equiv \{\{g(\mathbf{x}), H\}, H\} = -g(\mathbf{x}) - 4\lambda g(\mathbf{x}) (g, g) ,$$
$$g^{(3)}(0, \mathbf{x}) \equiv \{\{\{g(\mathbf{x}), H\}, H\}, H\}$$
$$= -f(\mathbf{x}) - 4\lambda f(\mathbf{x})(g, g) - 8\lambda g(\mathbf{x})(f, g)$$
$$g^{(4)}(0, \mathbf{x}) \equiv \{\{\{\{g(\mathbf{x}), H\}, H\}, H\}, H\}$$
$$= g(\mathbf{x}) + 16\lambda g(\mathbf{x})(g, g) - 8\lambda g(\mathbf{x})(f, f)$$
$$- 16\lambda f(\mathbf{x})(f, g) + 48\lambda^2 g(\mathbf{x})(g, g)^2 ,$$

etc. Since

$$g(t, \mathbf{x}) = \sum_{n=0}^{\infty} \frac{g^{(n)}(0, \mathbf{x}) \, t^n}{n!} ,$$

it follows that

$$g(t, \mathbf{x}) = g(\mathbf{x}) + tf(\mathbf{x}) + t^2[-g(\mathbf{x}) - 4\lambda g(\mathbf{x})(g, g)]/2!$$
$$+ t^3[-f(\mathbf{x}) - 4\lambda f(\mathbf{x})(g, g) - 8\lambda g(\mathbf{x})(f, g)]/3!$$
$$+ t^4[g(\mathbf{x}) + 16\lambda g(\mathbf{x})(g, g) - 8\lambda g(\mathbf{x})(f, f)$$
$$- 16\lambda f(\mathbf{x})(f, g) + 48\lambda^2 g(\mathbf{x})(g, g)^2]/4! + \cdots .$$

The result is a general power series expression for the full time dependence of $g(t, \mathbf{x})$ expressed in terms of the initial values $f(\mathbf{x})$ and $g(\mathbf{x})$. We shall have reason to refer to this formula when we study the quantum theory for these models.

7.2 Conventional quantum formulation

Noninteracting models: $\lambda = 0$

The free theory for the present model is represented by one of the free models extensively studied in Chapter 5, namely that for $\Omega(\mathbf{k}) = m_o$, which is no longer constrained to be unity. The classical Hamiltonian for the free model is given by

$$H_0(f, g) = \tfrac{1}{2}[(f, f) + m_o^2 (g, g)] .$$

In turn, we know that the quantum Hamiltonian is given by

$$\mathcal{H}_0 = m_o \int a^\dagger(\mathbf{k}) a(\mathbf{k}) \, d^s k$$
$$= \tfrac{1}{2} \int : [\pi(\mathbf{x})^2 + m_o^2 \phi(\mathbf{x})^2] : d^s x$$

expressed first in terms of conventional creation and annihilation operators and second in a formal configuration space form. In like manner, the space translation generators are given by

$$\mathcal{P} = \int a^\dagger(\mathbf{k}) \, \mathbf{k} \, a(\mathbf{k}) \, d^s k$$
$$= -\int : \pi(\mathbf{x}) \nabla \phi(\mathbf{x}) : d^s x .$$

We let $|0\rangle$ denote the unique, normalized ground state for the Hamiltonian: $\mathcal{H}|0\rangle = 0$. If $\varphi(f)$ and $\pi(g)$ denote the smeared field and momentum, respectively, then we know that

$$\langle 0| \exp\{i[\varphi(f) - \pi(g)]\} |0\rangle = \exp\{-\tfrac{1}{4}[m_o^{-1}(f, f) + m_o(g, g)]\} .$$

The representation of the operators $\varphi(f)$ and $\pi(g)$, for all square integrable f and g, is irreducible, and is unitarily inequivalent for distinct m_o values; see Exercise 7.2.

Interacting models: $\lambda > 0$

Let us initially approach the interacting models by introducing a cutoff that consists of including only N degrees of freedom, $N < \infty$, and then taking the limit $N \to \infty$ at the end of the calculation. Thus our classical Hamiltonians for the cutoff systems are given by

$$H_N(p, q) \equiv \tfrac{1}{2} \Sigma_{n=1}^N [p_n^2 + m_o^2 q_n^2] + \lambda [\Sigma_{n=1}^N q_n^2]^2 \ .$$

In the conventional approach to such a system, the quantum Hamiltonian is taken, e.g., without normal ordering, to be

$$\mathcal{H}_N \equiv \tfrac{1}{2} \Sigma_{n=1}^N [P_n^2 + m_N^2 Q_n^2] + \lambda_N [\Sigma_{n=1}^N Q_n^2]^2 - C_N \ ,$$

where $[Q_n, P_m] = i\delta_{nm}\mathbf{1}$, $1 \le n, m \le N$, with all other kinematical commutators vanishing. We have taken the liberty of introducing a constant m_N^2 to reflect a possible renormalization of the mass, and also allowed for a possible N-dependence of the coupling constant, i.e., $\lambda \to \lambda_N$.

Let us see what we can say about the properties of the quantum model with infinitely many degrees of freedom, defined as the limit as $N \to \infty$ of a sequence of cutoff models with N degrees of freedom. In particular, for each member of the sequence it is natural to assume that the ground state is unique and is rotationally invariant. We can study the limiting behavior related to this ground state in several different ways. For the present, let us consider the expectation function of the cutoff field operator in a distribution determined by the ground state wave function $\phi_N(x)$, $x \in \mathbb{R}^N$. This expression is given by

$$\langle 0_N | e^{ip \cdot Q} | 0_N \rangle = \int e^{ip \cdot x} |\phi_N(x)|^2 \, d^N x \ ,$$

or taking into account that the ground state is spherically symmetric and depends only on r, where $r^2 \equiv \Sigma_1^N x_n^2$ (and likewise for $p^2 \equiv \Sigma_1^N p_n^2$), we have

$$C_N(p) \equiv \langle 0_N | e^{ip \cdot Q} | 0_N \rangle$$
$$= \int e^{ipr \cos(\theta)} |\psi_N(r)|^2 r^{N-1} \, dr \, \sin^{N-2}(\theta) \, d\theta \, d\Omega_{N-2} \ ,$$

where $d\Omega_{N-2}$ is the differential measure for the remaining angles. For very large N the integration over θ can be carried out by steepest descent leading to the sufficiently accurate form given by

$$C_N(p) = J \int e^{-p^2 r^2 / 2(N-2)} |\psi_N(r)|^2 \, r^{N-1} \, dr + O(1/N) \ ;$$

here we have incorporated the integral over the remaining angles into the constant J, which may also be implicitly determined by the normalization $C_N(0) = 1$ for all N. By a suitable change of variables $[r^2 \to 2(N-2)b, \ b \ge 0]$, it follows that

$$C_N(p) = \int_0^\infty e^{-bp^2} \rho_N(b) \, db + O(1/N) \ ,$$

and in the limit that $N \to \infty$, and assuming convergence, we are led to conclude, for all $p \in l^2$, that

$$C(p) = \lim_{N \to \infty} C_N(p) = \int_0^\infty e^{-bp^2} \, d\mu(b) \, ,$$

for some probability measure μ. When we return to the field language, the result we have just derived asserts that

$$\langle 0|e^{i\varphi(f)}|0\rangle = \int_0^\infty e^{-b\|f\|^2} \, d\mu(b) \, ,$$

for all $f \in L^2$, where

$$\|f\|^2 \equiv (f, f) \, .$$

Let us introduce the states

$$|f\rangle \equiv e^{i\varphi(f)} |0\rangle \, ,$$

and make the assumption that these states *span the Hilbert space*, i.e., the set of such states for all $f \in L^2$ forms a *total set* (see below for the rationale for this assumption). The matrix elements of the unit operator become simply

$$\langle f|f'\rangle = \langle 0|e^{-i[\varphi(f)-\varphi(f')]}|0\rangle = \int e^{-b\|f-f'\|^2} \, d\mu(b) \, .$$

In turn, we can introduce the unitary translation operators $U(\mathbf{a})$ by means of the definition

$$U(\mathbf{a}) |f'\rangle \equiv |f'_{\mathbf{a}}\rangle \, , \qquad f'_{\mathbf{a}}(\mathbf{x}) \equiv f'(\mathbf{x} - \mathbf{a}) \, .$$

Clearly, if $f' = 0$, then $U(\mathbf{a})|0\rangle = |0\rangle$. Furthermore, since $(f, f'_{\mathbf{a}}) \to 0$ as $|\mathbf{a}| \to \infty$ – thanks to the Riemann–Lebesgue Lemma (see, e.g., Exercise 6.2) and the fact that both f and f' are in L^2 – we note that

$$\lim_{|\mathbf{a}| \to \infty} \langle f|U(\mathbf{a})|f'\rangle = \int e^{-b[\|f\|^2 + \|f'\|^2]} \, d\mu(b)$$

$$\equiv \langle f| A |f'\rangle \, ;$$

moreover, it follows that $A|0\rangle = |0\rangle$. In like manner it readily follows that $A = A^\dagger = A^2$, i.e., that A is a projection operator. In addition, since the models in question exhibit time-reversal symmetry, and based on the analysis in Chapter 6, we are immediately able to write down the matrix elements of the Hamiltonian operator as

$$\langle f|\mathcal{H}|f'\rangle = \tfrac{1}{2}(f, f')\langle f|f'\rangle \, ,$$

and in particular that

$$\langle f|\mathcal{H}U(\mathbf{a})|f'\rangle = \tfrac{1}{2}(f, f'_{\mathbf{a}})\langle f|f'_{\mathbf{a}}\rangle \, .$$

A consequence of this relation is that

$$\lim_{|\mathbf{a}| \to \infty} \langle f|\mathcal{H}U(\mathbf{a})|f'\rangle = \langle f|\mathcal{H}A|f'\rangle \equiv 0 \, .$$

Since, by assumption, the set of vectors $\{|f\rangle\}$ span the space, we learn that

$$\mathcal{H}A|f'\rangle = 0$$

for all f'. Because $A \neq 0$ (by construction), and based on the requirement that $|0\rangle$ be the unique ground state for the Hamiltonian $\mathcal{H}$, we deduce that $A|f'\rangle = a|0\rangle$ holds for all f' with a suitable nonzero coefficient $a \in \mathbf{C}$. Thanks to the assumption that vectors of the form $|f'\rangle$ span the space, the only possibility is that $A = |0\rangle\langle 0|$, with $a = \langle 0|f'\rangle$. In other words, insisting on a unique ground state has led us to the conclusion that

$$\lim_{|\mathbf{a}|\to\infty} \langle f|f'_{\mathbf{a}}\rangle = \langle f|0\rangle\langle 0|f'\rangle \,,$$

which can only arise if

$$\langle 0|f\rangle = \int e^{-b\|f\|^2}\,d\mu(b) \equiv e^{-\bar{b}\|f\|^2}$$

holds for some real $\bar{b} > 0$, i.e., that

$$d\mu(b) = \delta(b - \bar{b})\,db\,.$$

Thus the expectation functional for the field operator is necessarily identical to the expectation functional for one of the free models!

In recognition of the fact that the expectation functional coincides with that for a free theory, let us change notation and assert that the result for the expectation value is given by

$$\langle 0|f\rangle \equiv e^{-(1/4m)\|f\|^2}$$

for some as yet undetermined mass parameter m.

Determining the mass parameter

The general argument given above shows that the expectation functional is identical to that for a free theory, but it has not determined the mass parameter m that characterizes such a theory. Conventional "$1/N$" arguments [Ya 82] will enable us to evaluate that parameter. According to the conventional viewpoint, a theory with a symmetry group $O(N)$, such as we are dealing with at present, is accompanied by a coupling constant λ_N that differs from its original value λ by a factor $1/N$, specifically by the replacement $\lambda \to \lambda/N$ in the Hamiltonian. This change leads to the revised cutoff quantum operator

$$\mathcal{H}_N = \tfrac{1}{2}\Sigma\,(P_n^2 + m_o^2\,Q_n^2) + (\lambda/N)[\Sigma\,Q_n^2]^2 - c'_N\,,$$

where $1 \leq n \leq N$. Later we shall offer some comments regarding this major change, but for now let us pursue the consequences of such a modification. As $N \to \infty$, it follows that the factor $L_N \equiv \Sigma\,Q_n^2/N$ tends to a multiple of the unit operator. This fact can be established by the following argument. Since $[Q_k, L_N] = 0$ and $[P_k, L_N] = -2iQ_k/N$, for $N > k$, it follows that

$L \equiv \lim_{N \to \infty} L_N$ commutes with all the Q_k and P_k, $1 \leq k < \infty$, which form an irreducible set of operators. Hence $L = l\mathbf{1}$, and we may evaluate l from the fact that $\langle 0|e^{ipQ_1}|0\rangle = \exp[-p^2/(4m)]$ which implies that $\langle 0|Q_1^2|0\rangle = 1/(2m) = l$; in point of fact, $l = \hbar/(2m)$. In application within the Hamiltonian we observe that the operator $(\Sigma Q_n^2)^2/N = (\hbar/m)\Sigma Q_n^2$ up to an additive multiple of the unit operator and terms that are $O(1/N)$. These facts are a direct consequence of the readily established property that

$$\langle f| \exp(-t\Sigma Q_n^2)|f'\rangle = (1 + t/m)^{-N/2} \exp\{-\|f - f'\|^2/[4(m+t)]\}$$

provided that the support of f and f' is contained within the space spanned by the first N basis functions. When N is very large, therefore, the Hamiltonian effectively becomes

$$\begin{aligned}
\mathcal{H} &= \tfrac{1}{2}\Sigma\,(P_n^2 + m_o^2\,Q_n^2) + (\hbar\lambda/m)\Sigma\,Q_n^2 - c_N \\
&= \tfrac{1}{2}\Sigma\,[P_n^2 + (m_o^2 + 2\hbar\lambda/m)Q_n^2] - c_N \\
&\equiv \tfrac{1}{2}\Sigma\,[P_n^2 + m^2 Q_n^2] - c_N \;.
\end{aligned}$$

As a consequence we are led to a self-consistent determination of the parameter m, namely, the relation

$$m^3 = mm_o^2 + 2\hbar\lambda \;,$$

with interest focused on the solution to this equation that is continuously connected to the value $m = m_o$ when $\lambda = 0$. This solution rises monotonically from $m = m_o$ and becomes asymptotic to $m = (2\hbar\lambda)^{1/3}$ for large $\hbar\lambda$. Incidentally, we also determine that to leading order $c_N = \hbar mN/2$, in order that the ground state energy be zero.

Observe how different scaling factors for λ_N would affect the result. If we were to choose $\lambda_N = \lambda/N^\theta$, then the equation of self consistency would become $m^3 = mm_o^2 + 2\hbar\lambda N^{1-\theta}$. For $\theta > 1$ the effective mass becomes $m = m_o$ as $N \to \infty$, and there would be no influence on the solution from the quartic interaction term. For $\theta < 1$, on the other hand, $m \to \infty$ as $N \to \infty$, and there is no acceptable solution.

Equation of motion

The basic reasoning underlying the need for a rescaling of the quartic coupling constant, i.e., $\lambda \to \lambda/N$, may be seen in a study of the equations of motion. Based on the cutoff Hamiltonian (setting $m_o = 1$ for convenience)

$$\mathcal{H}_N = \tfrac{1}{2}\Sigma_{n=1}^N(P_n^2 + Q_n^2) + \lambda[\Sigma_{n=1}^N Q_n^2]^2 - C_N$$

we learn (adopting the summation convention) that

$$(d/dt)\,Q_j = P_j \;,$$
$$(d/dt)^2\,Q_j = -Q_j - 4\lambda Q_j\,(Q_k Q_k) \;,$$

$$(d/dt)^3 Q_j = -P_j - 2\lambda P_j (Q_k Q_k) - 2\lambda (Q_k Q_k) P_j - 2\lambda Q_j (Q_k P_k)$$
$$- 2\lambda (P_k Q_k) Q_j - 2\lambda Q_j (P_k Q_k) - 2\lambda (Q_k P_k) Q_j \,,$$
$$(d/dt)^4 Q_j = Q_j + 16\lambda Q_j (Q_k Q_k) + 48\lambda^2 Q_j (Q_k Q_k)(Q_l Q_l)$$
$$- 4\lambda P_j (P_k Q_k) - 4\lambda (Q_k P_k) P_j - 4\lambda Q_j (P_k P_k)$$
$$- 4\lambda (P_k P_k) Q_j - 4\lambda (P_k Q_k) P_j - 4\lambda P_j (Q_k P_k) \,,$$

etc. These equations of motion evidently involve expressions of the form $\lambda(Q_k Q_k)$, $\lambda(P_k P_k)$, and $\lambda(Q_k P_k + P_k Q_k)$. Observe how each occurrence of a potentially infinite operator sum is accompanied by a coupling constant λ. In the limit, the first two operators have a c-number divergent coefficient, while the third one does not. Such divergences must be tamed in order for the Taylor series representation of the operator solution $Q(t)$ to make sense. The symmetry of the model enforces that the divergence is proportional to the number of degrees of freedom, and therefore this divergence can be tamed by the cited rescaling of the coupling constant, i.e., $\lambda \to \lambda/N$.

Normal ordering

Although we have consistently subtracted off the ground state energy in our discussion of the Hamiltonian, we have not introduced normal ordering (symbolized by : :) of the Hamiltonian, nor discussed its consequences for the equations of motion. In the cutoff case with N degrees of freedom, and since the field is Gaussian, we have (with $\hbar = 1$)

$$: Q_k Q_k Q_l Q_l : = Q_k Q_k Q_l Q_l - 2Q_k Q_k \langle 0|Q_l Q_l|0\rangle - 4Q_k Q_l \langle 0|Q_k Q_l|0\rangle$$
$$+ \langle Q_k Q_k\rangle \langle Q_l Q_l\rangle + 2\langle Q_k Q_l\rangle \langle Q_k Q_l\rangle$$
$$= Q_k Q_k Q_l Q_l - m^{-1}(N+2) Q_k Q_k + \tfrac{1}{4}m^{-2}(N^2 + 2N) \,;$$

this property is a consequence of the more general relation that

$$: \exp(-t Q_k Q_k) : \equiv (1 - t/m)^{-N/2} \exp\{-[tm/(m-t)]Q_k Q_k\}$$

whenever $0 \le t < m$. Thus, if we introduce a normally ordered cutoff Hamiltonian, we are led to

$$\mathcal{H}_N = \tfrac{1}{2}\Sigma_{n=1}^N : (P_n^2 + m^2 Q_n^2) : + (\lambda/N) : [\Sigma_{n=1}^N Q_n^2]^2 : \,,$$

and we see that this expression is equivalent to

$$\mathcal{H}_N = \tfrac{1}{2}\Sigma_{n=1}^N (P_n^2 + m^2 Q_n^2) + (\lambda/N)[\Sigma_{n=1}^N Q_n^2]^2$$
$$- (\lambda/m)(1 + 2/N)\Sigma_{n=1}^N Q_n^2 - k_N'$$
$$= \tfrac{1}{2}\Sigma_{n=1}^N (P_n^2 + m^2 Q_n^2) - k_N + O(1/N) \,,$$

which means as $N \to \infty$ that in the normally ordered form the mass parameter that appears is already the final mass parameter. Additionally, it follows that the leading behavior of the constant $k_N = mN/2$.

7.3 Alternative derivation of free limit

Although it may seem that we have been beating an essentially dead horse in our discussion of the fact that the quantum theory of the quartic rotationally symmetric model is equivalent to a free theory, with a suitable, self-consistently determined mass parameter, it is useful to analyze the problem one more time as a preparation for the next several sections in which a *nontrivial* solution will be advanced. In this analysis we assume no cutoff and proceed to discuss the case of an infinite number of degrees of freedom directly based on a few assumptions. Additionally, we shall exploit the use of *tags* associated with representations, the concept of which was introduced in Chapter 5.

Let us first introduce a minimum number of postulates relative to the quantum theory of the rotationally symmetric field theories. We assume that there exists a Hilbert space $\mathfrak{H}$ spanned by the normalized states $|f\rangle$ for all real $f \in L^2(\mathbb{R}^s)$. We also assume there exists a unitary representation of the group $O(\infty)$ such that for each element $O \in O(\infty)$, the unitary operator $U(O)$ acts so that $U(O)|f\rangle = |Of\rangle$, which is then extended to all of $\mathfrak{H}$ by linearity and continuity. The invariant state $|0\rangle$ (set $f \equiv 0$) is identified as the nondegenerate ground state of the nonnegative, self-adjoint Hamiltonian operator $\mathcal{H}$ with the energy adjusted so that $\mathcal{H}|0\rangle = 0$. Additionally, there is a local self-adjoint field operator $\varphi(\mathbf{x})$, $\mathbf{x} \in \mathbb{R}^s$, for which the smeared field $\varphi(f) = \int \varphi(\mathbf{x}) f(\mathbf{x})\, d^s x$ is self adjoint, and $|f\rangle \equiv \exp[i\varphi(f)]|0\rangle$, for all real $f \in L^2(\mathbb{R}^s)$. The momentum operator may be defined by

$$\pi(g) \equiv i[\mathcal{H}, \varphi(g)]$$

and the commutation relations are assumed to be satisfied in the form

$$[\varphi(f), \pi(g)] = i(f, g)\mathbf{1} \ .$$

We further assume that $\mathcal{H}$ is time-reversal invariant [heuristically, $\mathcal{H} = \frac{1}{2}\int \pi(\mathbf{x})^2\, d^s x + \mathcal{V}(\varphi)$], and therefore the Araki relations $\langle f|\pi(\mathbf{x})|f'\rangle = \frac{1}{2}[f(\mathbf{x}) + f'(\mathbf{x})]\langle f|f'\rangle$ and $\langle f|\mathcal{H}|f'\rangle = \frac{1}{2}(f, f')\langle f|f'\rangle$ hold for all $f, f' \in L^2$ (see Section 6.9). In virtue of the symmetry of the rotationally symmetric model and the assumed $O(\infty)$ rotational invariance of the ground state $|0\rangle$, it follows that

$$E(X) \equiv \langle 0|f\rangle \ , \qquad X \equiv (f, f) \equiv \|f\|^2 \geq 0 \ .$$

Observe that E is a function of just *one*, nonnegative variable, and that moreover E is necessarily a real, *continuous function*; these properties follow from (i) $E(X)^* = \langle f|0\rangle = \langle 0| - f\rangle = E(X)$, and (ii) the fact that for any fixed $f \in L^2$, $U(t) \equiv \exp[it\varphi(f)]$, regarded as a unitary one-parameter group, has arbitrary matrix elements which are continuous in the real variable t, $t \in \mathbb{R}$.

Next, consider the family of operators

$$U[r_\mathbf{a}] \equiv \exp[i\varphi(r_\mathbf{a})] = \exp[i\int \varphi(\mathbf{x})\, r(\mathbf{x} - \mathbf{a})\, d^s x]$$

for an arbitrary function $r \in L^2(\mathbb{R}^s)$ and an arbitrary $\mathbf{a} \in \mathbb{R}^s$. The matrix elements of this operator in the states $|f\rangle$, for general $f \in L^2$, are given by

$$\langle f|U[r_{\mathbf{a}}]|f'\rangle = \langle 0|f' - f + r_{\mathbf{a}}\rangle$$
$$= E(\|f' - f + r_{\mathbf{a}}\|^2)$$
$$= E(\|f' - f\|^2 + 2(f' - f, r_{\mathbf{a}}) + \|r\|^2) ,$$

where in the last line we have made use of the fact that $\|r_{\mathbf{a}}\| = \|r\|$ for all $r \in L^2$ and all $\mathbf{a} \in \mathbb{R}^s$. Now consider the limit $|\mathbf{a}| \to \infty$ from which we learn that

$$\lim_{|\mathbf{a}| \to \infty} \langle f|U[r_{\mathbf{a}}]|f'\rangle = E(\|f' - f\|^2 + \|r\|^2)$$
$$\equiv E(\|f' - f\|^2 + x) , \qquad x \equiv (r, r) \equiv \|r\|^2 ,$$
$$\equiv \langle f|C(x)|f'\rangle ,$$

which holds because the function E is continuous and the middle term, $(f' - f, r_{\mathbf{a}})$, vanishes by the Riemann–Lebesgue Lemma. Observe that in this fashion we have *defined* a set of operators $C(x)$, for all $x \geq 0$, by means of a total set of matrix elements. It is straightforward to see that the operator $C(x)^\dagger = C(x)$, that $C(x)C(x') = C(x + x')$, and finally, for any two states $|\psi\rangle$ and $|\tau\rangle$, that $\langle \psi|C(x)|\tau\rangle$ is continuous in x simply because E is already continuous. It follows that there exists a self-adjoint operator B such that $C(x) = \exp(-Bx)$. By construction $C(x) \leq 1$ and therefore $B \geq 0$. Finally, we can give an integral representation for the function E when we diagonalize B leading to

$$E(X) = \langle 0|C(X)|0\rangle = \int_0^\infty e^{-bX} \, d\mu(b)$$

for some probability measure μ concentrated on the interval $0 \leq b < \infty$. In other words, we have again been led to the conclusion that

$$\langle 0|e^{i\varphi(f)}|0\rangle = \int_0^\infty e^{-b(f,f)} \, d\mu(b) .$$

Thereafter the argument proceeds just as before. In particular, we deduce that

$$\langle f|\mathcal{H}A|f'\rangle \equiv \lim_{|\mathbf{a}| \to \infty} \langle f|\mathcal{H}U(\mathbf{a})|f'\rangle$$
$$= \lim_{|\mathbf{a}| \to \infty} \tfrac{1}{2}(f, f'_{\mathbf{a}})\langle f|f'_{\mathbf{a}}\rangle$$
$$= 0 ,$$

which implies that $A|f'\rangle = a|0\rangle$ for all f', and since $A = A^\dagger = A^2$, we find that $A = |0\rangle\langle 0|$ and $a = \langle 0|f'\rangle$. As a consequence, uniqueness of the ground state forces $\langle f|f'_{\mathbf{a}}\rangle \to \langle f|0\rangle\langle 0|f'\rangle$ as $|\mathbf{x}| \to \infty$ and thus that $d\mu(b) = \delta(b - \bar{b}) \, db$ for some $0 < \bar{b} < \infty$, leading to the equivalence with some particular free theory with mass parameter $m = 1/4\bar{b}$.

Is a free theory reasonable?

Although we have inevitably been led to a free theory for the quantum theory of a classical, quartic (or any other!) rotationally symmetric model, we can properly ask whether this result is satisfactory. *Our answer is decidedly no!* One of the standard tenets of quantum theory is that under suitable conditions, such as appropriately large quantum numbers, the quantum theory should limit to the classical theory. Although somewhat vague, this quantum-to-classical limit is inherent in the conventional correspondence principle [Di 76]. Nevertheless, how can a *trivial* ($\equiv$ free) quantum theory limit to a *nontrivial* ($\equiv$ nonfree) classical theory? More specifically, the quantum equation of motion is *linear* in the fields and therefore, for general $\mathbf{x}$, $\mathbf{y}$, and $\mathbf{z}$,

$$[\pi(\mathbf{x}), [\pi(\mathbf{y}), \ddot{\varphi}(\mathbf{z})]] \equiv 0 \, ,$$

while it is the definite *non*vanishing of the corresponding Poisson bracket,

$$\{f(\mathbf{x}), \{f(\mathbf{y}), \ddot{g}(\mathbf{z})\}\} \not\equiv 0 \, ,$$

that expresses the nontriviality and nonlinearity of the classical theory.

We trace this unsatisfactory situation to the rescaling of the coupling constant $\lambda \to \lambda/N$ which is a *totally classical redefinition of the original theory* having nothing to do with the quantum parameter $\hbar$. With such a rescaling the classical theory becomes free in the limit $N \to \infty$, and thus it is no surprise that the quantum theory is free as well. However, it must be emphasized that the original classical model under consideration was *not* free. In essence, the quantum theory we have arrived at following conventional procedures cannot be the quantum theory of the original nonlinear classical field theory!

7.4 An alternative formulation

In the present section we shall briefly sketch the principal results derived in detail in the following sections of this chapter dealing with an alternative formulation of the quantum theory of rotationally symmetric models. We do so to prepare the reader for the similarities – as well as the profound differences – in the alternative approach as compared to the conventional approach discussed above. In the section following this one we explain the alternative quantization in very simple terms including a natural cutoff to make the visualization of what is going on as clear as possible.

To begin with we replace the assumption that the set of states $\{|f\rangle\}$, for all $f \in L^2(\mathbb{R}^s)$ is total (spans the Hilbert space) by the assumption that the set of states of the form

$$|f, g\rangle \equiv \exp\{i[\varphi(f) - \pi(g)]\} |0\rangle \, , \qquad f, g \in L^2(\mathbb{R}^s) \, ,$$

is total in $\mathfrak{H}$. Under a conservative and reasonable choice of assumptions, the set of all representations of the Weyl operators for this problem are found, and they are such that only the functions given by

$$\langle 0|f,g\rangle = \exp\{-\tfrac{1}{4}[\xi m^{-1}(f,f) + m(g,g)]\}\,, \qquad m > 0\,, \quad \xi \geq 1$$

are allowed. The parameter m, as usual, is mainly a scale factor needed to account for dimensions. If $\xi = 1$, the representation of the Weyl operators is *irreducible*, while if $\xi > 1$, the representation of the Weyl operators is *reducible* (see below for a proof). The irreducible representation accommodates the free Hamiltonian and describes the free rotationally symmetric model and no other. All interacting models are accommodated within the reducible representations. Hereafter, we restrict our remarks to the reducible representations where $\xi > 1$. In particular, the smeared momentum operator $\pi(e)$ is defined by

$$\langle f,g|\pi(e)|f',g'\rangle = \tfrac{1}{2}[(e,f+f') + im(e,g-g')]\,\langle f,g|f',g'\rangle\,,$$

while the space translation generators $\mathcal{P}$ are defined by the matrix elements

$$\langle f,g|\mathcal{P}|f',g'\rangle$$
$$= -\tfrac{1}{2}[(f,\nabla g') - (g,\nabla f') + i\xi m^{-1}(f,\nabla f') + im(g,\nabla g')]\,\langle f,g|f',g'\rangle\,.$$

It is evident, but nevertheless worth emphasizing, that these operators are completely determined from the properties of the Weyl operators and the given form for the overlap $\langle f,g|f',g'\rangle$. On general grounds we will show that the Hamiltonian operator $\mathcal{H}$ is defined by the matrix elements

$$\langle f,g|\mathcal{H}|f',g'\rangle$$
$$= \{\tfrac{1}{2}[(f+img,f'-img') + m^2\zeta^2(g,g')] + Vm^4\zeta^4(g,g')^2\}\langle f,g|f',g'\rangle\,,$$

where we have introduced the parameters $V \geq 0$ and

$$\zeta^2 \equiv 1 - \xi^{-1} < 1\,.$$

The association of these operators with the classical motivating theory is given through the weak correspondence principle, specifically by the diagonal matrix elements

$$\langle f,g|\mathcal{P}|f,g\rangle \equiv \mathbf{P}(f,g) = -(f,\nabla g)\,,$$
$$\langle f,g|\mathcal{H}|f,g\rangle \equiv H(f,g) = \tfrac{1}{2}[(f,f) + m_o^2(g,g)] + \lambda(g,g)^2\,,$$

where $m_o^2 \equiv m^2(1+\zeta^2)$ and $\lambda \equiv Vm^4\zeta^4$. Observe, in particular, that there is no rescaling of the coupling constant λ (such as $\lambda \to \lambda/N$) involved; in fact, all expressions are given for the full field theory without cutoffs.

Despite the apparently rather unremarkable nature of the expectation functional and the matrix elements of the operators involved there is something quite extraordinary that is hidden within the operators themselves. Whenever one deals with a reducible representation of the fundamental field and momentum operators – as is here the case when $\xi > 1$ – it follows that basic operators such as the space translation generators or the Hamiltonian *cannot be expressed solely in terms of the field and momentum operators.* This conclusion holds whenever we require that the ground state be the only state up to a phase factor that is invariant under space translations, or whenever the ground state is a unique eigenstate of the Hamiltonian with eigenvalue zero. The proof is straightforward. Let

$$\varphi(f) = \varphi_1(f) \oplus \varphi_2(f) ,$$
$$\pi(g) = \pi_1(g) \oplus \pi_2(g)$$

denote a reducible set of field and momentum operators, and assume that the generator $\mathcal{G}$ is a function solely of φ and π. In that case

$$\mathcal{G} = \mathcal{G}_1 \oplus \mathcal{G}_2$$

holds as well. Now in the same decomposition let $|0\rangle = \alpha_1|0\rangle_1 \oplus \alpha_2|0\rangle_2$, where $|\alpha_1|^2 + |\alpha_2|^2 = 1$, satisfy $\mathcal{G}|0\rangle = 0$. This equality requires that $\mathcal{G}_j|0\rangle_j = 0$ for $j = 1, 2$. But then it follows that the state $|0'\rangle \equiv \beta_1|0\rangle_1 \oplus \beta_2|0\rangle_2$, where $|\beta_1|^2 + |\beta_2|^2 = 1$, satisfies $\mathcal{G}|0'\rangle = 0$ for all choices of β_j, $j = 1, 2$. Now, if the operator $\mathcal{G}$ is such that it should have a *unique invariant state* $|0\rangle$, then it follows that $\mathcal{G}$ *cannot* be expressed solely as a function of the (reducible) field and momentum! In fact, if the reducible Weyl algebra is augmented by just *one* such operator $\mathcal{G}$, the resultant algebra becomes *irreducible*. For the proof, assume that C commutes with the Weyl operators and with $\mathcal{G}$. Then $\mathcal{G}C|0\rangle = C\mathcal{G}|0\rangle = 0$ which due to the uniqueness of the invariant state asserts that $C|0\rangle = c|0\rangle$ for some c-number c. Then $CU[f, g]|0\rangle = U[f, g]C|0\rangle = cU[f, g]|0\rangle$ holds for all f, g, which being a total set implies that $C = c\mathbb{1}$ as needed to claim irreducibility of the operators involved. Of course, other families of operators may, when added to the Weyl operators, make an irreducible set, and they may be more natural and easier to deal with; we will see an example of that situation shortly.

All this rather unconventional behavior applies to the rotationally symmetric models whenever one seeks a nontrivial solution that requires a reducible representation of the Weyl operators. The following section discusses a particularly transparent formulation of the Hamiltonian of the quartic interaction rotationally symmetric model. Rotationally symmetric models with higher interaction powers may be treated in an analogous manner [Kl 65].

7.5 Elementary formulation of the dynamics

We return to the case of finitely many degrees of freedom, and besides the family of N canonical sets of operators $\{Q_n\}_{n=1}^N$ and $\{P_n\}_{n=1}^N$, we introduce a *second, independent set* of N canonical operators, $\{S_n\}_{n=1}^N$ and $\{R_n\}_{n=1}^N$, all of which satisfy (for $1 \leq j, k \leq N$)

$$[Q_j, P_k] = i\delta_{jk}\mathbf{1}\,, \qquad\qquad [S_j, R_k] = i\delta_{jk}\mathbf{1}\,,$$
$$[Q_j, Q_k] = [P_j, P_k] = [S_j, S_k] = [R_j, R_k] = 0\,,$$
$$[R_j, Q_k] = [R_j, P_k] = [S_j, Q_k] = [S_j, P_k] = 0\,.$$

The cutoff Hamiltonian – implied by the matrix elements given in the preceding section and established in the following sections of this chapter – then takes the form

$$\mathcal{H}_N = \tfrac{1}{2}\Sigma_{n=1}^N [: P_n^2 + m^2(Q_n + \zeta S_n)^2 : + : R_n^2 + m^2(S_n + \zeta Q_n)^2 :]$$
$$+ V : \{\Sigma_{n=1}^N [R_n^2 + m^2(S_n + \zeta Q_n)^2]\}^2 : \,.$$

This expression seems unusual and out of place, but we claim that it is the unique operator associated with the (cutoff) Hamiltonian matrix elements given in the preceding section. Observe that the cutoff Hamiltonian is in fact *not* a function solely of the canonical operators, $\{Q_n\}$ and $\{P_n\}$, but involves other operators as well, here not simply represented by a single additional operator but by another set of canonical operators. It is as if the problem was really a *two*-field problem rather than a one-field problem.

The advantages of and the role played by the additional operators as $N \to \infty$ can readily be made clear. Simply put, the operator $\Sigma_{n=1}^N : Q_n Q_n :$, that appears in the conventional formulation, ceases to be an operator as $N \to \infty$ (with or without normal ordering), while the operator $\Sigma_{n=1}^N : [R_n^2 + m^2(S_n + \zeta Q_n)^2] :$, that appears in the unconventional treatment, does remain an operator as $N \to \infty$. The auxiliary operators see to it!

Let us examine the weak correspondence principle in detail in the present language. If we set

$$|p, q\rangle = \exp[i\Sigma_{n=1}^N (p_n Q_n - q_n P_n)]\,|0\rangle \equiv \exp[i(p \cdot Q - q \cdot P)]\,|0\rangle\,,$$

using a natural vector notation appropriate to a finite number of degrees of freedom, then the weak correspondence principle asserts that

$$H(p, q) = \langle p, q| \tfrac{1}{2} : [P^2 + m^2(Q + \zeta S)^2] : + \tfrac{1}{2} : [R^2 + m^2(S + \zeta Q)^2] :$$
$$+ V : [R^2 + m^2(S + \zeta Q)^2]^2 : |p, q\rangle$$
$$= \langle 0| \tfrac{1}{2} : [(P + p)^2 + m^2(Q + q + \zeta S)^2] :$$
$$+ \tfrac{1}{2} : [R^2 + m^2(S + \zeta(Q + q))^2] :$$
$$+ V : [R^2 + m^2(S + \zeta(Q + q))^2]^2 : |0\rangle$$
$$= \tfrac{1}{2}[p^2 + m^2(1 + \zeta^2)q^2] + Vm^4\zeta^4(q^2)^2$$
$$= \tfrac{1}{2}[p^2 + m_o^2 q^2] + \lambda (q^2)^2\,,$$

where we identify $m_o^2 = m^2(1 + \zeta^2)$ and $\lambda = Vm^4\zeta^4$. Hence, despite the need for and the consequent appearance of a second set of canonical variables, the weak correspondence principle automatically leads to the proper expression for the classical Hamiltonian.

The formula for the cutoff Hamiltonian operator $\mathcal{H}_N$ also permits us to study the operator equations of motion. In particular we learn (retaining the vector notation for inner products) that

$$(d/dt)\, Q_j = P_j\,,$$
$$(d/dt)^2\, Q_j = -m^2(1+\zeta^2)Q_j - \zeta^2 m^2 S_j$$
$$- 2V\zeta m^2 : (S_j + \zeta Q_j)[R^2 + m^2(S + \zeta Q)^2] :$$
$$- 2V\zeta m^2 : [R^2 + m^2(S + \zeta Q)^2](S_j + \zeta Q_j) : \,,$$

etc. These equations of motion evidently involve expressions of the form $\lambda :$ $[R^2 + m^2(S + \zeta Q)^2] :$ and $\lambda : [P^2 + m^2(Q + \zeta S)^2] :$, which, unlike the analogous situation before, *do* remain operators as $N \to \infty$.

The weak correspondence principle extends to the equations of motion simply be taking the diagonal expectation value in terms of the states $|p, q\rangle$. It follows that $(d/dt)\, q_j \equiv \langle p, q|(d/dt)\, Q_j|p, q\rangle = \langle p, q|P_j|p, q\rangle = p_j$ and

$$(d/dt)^2\, q_j \equiv \langle p, q|(d/dt)^2\, Q_j|p, q\rangle$$
$$= -m^2(1+\zeta^2)q_j - 4V\zeta^4 m^4 q_j(q, q)$$
$$= -m_o^2\, q_j - 4\lambda\, q_j(q, q)\,,$$

etc., which, on comparison with the first few terms of the Taylor expansion of the classical solution, yield exactly what is desired!

An even simpler formulation of the Hamiltonian can be given in terms of creation and annihilation operators related to the formulation just presented. Let $\{a_n^\dagger\}_{n=1}^\infty$, $\{a_n\}_{n=1}^\infty$ and $\{b_n^\dagger\}_{n=1}^\infty$, $\{b_n\}_{n=1}^\infty$ be two sets of independent creation and annihilation operators that satisfy the familiar commutation relations

$$[a_n, a_m^\dagger] = \delta_{nm}\mathbf{1} = [b_n, b_m^\dagger]$$

with all other commutators vanishing. We choose variables so that

$$m\Sigma\, b_n^\dagger b_n = \tfrac{1}{2}\Sigma : [R_n^2 + m^2(S_n + \zeta Q_n)^2] :\,,$$

and if we let $\sin(\alpha) \equiv \zeta$ and $\cos(\alpha) \equiv \sqrt{1 - \zeta^2}$, then it follows that

$$m\Sigma\,[\cos(\alpha)a_n^\dagger + \sin(\alpha)b_n^\dagger][\cos(\alpha)a_n + \sin(\alpha)b_n]$$
$$= \tfrac{1}{2}\Sigma : [P_n^2 + m^2(Q_n + \zeta S_n)^2] :\,.$$

Consequently, the Hamiltonian for the quartic interacting rotationally symmetric model may be given by

$$\mathcal{H} = m\Sigma\,[\cos(\alpha)a_n^\dagger + \sin(\alpha)b_n^\dagger][\cos(\alpha)a_n + \sin(\alpha)b_n]$$
$$+ m\,\Sigma\, b_n^\dagger b_n + 4Vm^2\,\Sigma\, b_n^\dagger b_k^\dagger b_k b_n\,.$$

What could be more simple!

In this latter form, the existence of the Hamiltonian for the interacting models is self evident. On the other hand, what is less transparent is whether this form for the Hamiltonian is really the one that should apply to the original models. The doubts of the reader can readily be appreciated. However, all we ask is that the reader withhold final judgment until he or she has read the next two sections.

7.6 Beyond conventional quantization

If we are to find an alternative quantization for rotationally symmetric models, then we must start with a reexamination of the postulates introduced earlier. The *physical assumptions* – such as the existence of a nonnegative, self-adjoint (hence observable!), time-reversal invariant Hamiltonian operator $\mathcal{H}$; the uniqueness, time-reversal invariance and full $O(\infty)$ rotational invariance of the ground state $|0\rangle$, $\mathcal{H}|0\rangle = 0$; the existence of self-adjoint smeared field operators $\varphi(f)$ – all seem inevitable and will be retained in the new formulation. The existence of a canonical momentum operator $\pi(g) = i[\mathcal{H}, \varphi(g)]$ may be questioned, but continuing in a conservative fashion, we will retain it along with the canonical commutation relations $[\varphi(f), \pi(g)] = i(f, g)\mathbf{1}$. That only leaves the *technical assumption* that the states $|f\rangle \equiv \exp[i\varphi(f)]|0\rangle$ form a *total set*, i.e., that these states span the full Hilbert space $\mathfrak{H}$. This is the assumption we shall give up, but before doing so it is useful to recall why we assumed it in the first place.

The Schrödinger representation for a system with one variable is described by a wave function $\psi(x)$, $x \in \mathbb{R}$. If this state is the ground state of a reasonable Hamiltonian operator [e.g., $\frac{1}{2}P^2 + V(Q)$ with a piece-wise continuous potential V – decidedly not an infinite square-well potential], then $\psi(x)$ is either nowhere vanishing, or else it vanishes on a set of (Lebesgue) measure zero. Consequently, the set of functions $\psi_p(x) \equiv e^{ipx}\psi(x)$, for all $p \in \mathbb{R}$, is a total set for $L^2(\mathbb{R})$. This fact follows from the observation that if $\chi \in L^2$, then $\int \psi_p(x)^*\chi(x)\,dx = 0$ for all p implies that $\psi(x)^*\chi(x) = 0$ for almost all x, and hence that $\chi(x) = 0$ for almost all x. The same kind of argument extends to any finite number of degrees of freedom, which therefore becomes the basis for our assumption of this condition for a field theory. But this postulate must be recognized for what it is, namely, an assumption.

What conditions exist in the case of a field theory – and correspondingly do not exist for finitely many degrees of freedom – that could prohibit the states $|f\rangle$, for all $f \in L^2$, from spanning the Hilbert space? The central difference lies in the fact that for *finitely* many degrees of freedom there exists only *one* unitarily inequivalent, irreducible representation of the Weyl form of the commutation relations, and thus all representations of the subalgebra generated by the coordinate operators are unitarily equivalent. In the case of fields, on the other hand, there exist *uncountably* many unitarily inequivalent, irreducible representations of the Weyl form of the commutation relations. In consequence, the subalgebras formed by the field operators alone may (and do) exist in uncountably many inequivalent representations; this is already the case for the field operators characterized by $\langle 0| \exp[i\varphi(f)]|0\rangle = \exp[-(f,f)/4m]$, which correspond to unitarily inequivalent field algebras for each m, $0 < m < \infty$.

We first assume that the smeared momentum operator $\pi(g)$ is self adjoint for each $g \in L^2$, and that the Weyl operators

$$U[f,g] \equiv \exp\{i[\varphi(f) - \pi(g)]\}$$

form a unitary family of operators for all $f, g \in L^2$ (or a dense set thereof), which satisfy the Weyl form of the commutation relations, i.e.,

$$U[f,g]\,U[f',g'] = \exp\{(i/2)[(f,g') - (g,f')]\}\,U[f+f',g+g']\,.$$

We next assume that the states

$$|f,g\rangle \equiv U[f,g]\,|0\rangle = \exp\{i[\varphi(f) - \pi(g)]\}\,|0\rangle$$

defined for all f, g are a total set and span the Hilbert space $\mathfrak{H}$ of interest. Observe that this assumption does not preclude that the subset of states $|f\rangle \equiv |f,0\rangle$ for all f form a total set. Rather it includes that situation and allows for wider possibilities as well. No condition of irreducibility on the representation of the Weyl operators is assumed, but it is not excluded; rather we adopt only the weaker condition that the states $|f,g\rangle$ for all f and g span the Hilbert space. There is one further assumption that will be introduced later.

Even in the case that the states $|f\rangle = |f,0\rangle$ for all f do not span the entire Hilbert space they nevertheless span a *subspace* in that Hilbert space. Everything that has been concluded previously about matrix elements involving the states $|f\rangle$ holds true just as before *within* the subspace such states do span. In particular, in that subspace we still have relations such as $\langle f|\pi(\mathbf{x})|f'\rangle = \frac{1}{2}[f(\mathbf{x}) + f'(\mathbf{x})]\langle f|f'\rangle$ and $\langle f|\mathcal{H}|f'\rangle = \frac{1}{2}(f,f')\langle f|f'\rangle$ since they required only general properties for their validity. In the present circumstances, what is implied is that these matrix elements generally do not uniquely determine either the momentum or the Hamiltonian operator whenever the states $|f\rangle$ by themselves do not span the whole Hilbert space.

Our next goal is to determine the class of acceptable representations of the Weyl operators.

Determination of the fundamental operator representations

We focus our attention on the expectation functional

$$\langle 0| \exp\{i[\varphi(f) - \pi(g)]\}|0\rangle = \langle 0|f,g\rangle\,.$$

We assume the existence of a unitary representation of the group $O(\infty)$ in the form $U(O)$ for each $O \in O(\infty)$ with a self-evident multiplication rule. We assume that $U(O)|f,g\rangle = |Of,Og\rangle$ for all arguments, and as a consequence that

$U(O)|0\rangle = |0\rangle$ for every $O \in O(\infty)$. It follows that $\langle 0|Of, Og\rangle = \langle 0|f, g\rangle$ for all f, g, and therefore we necessarily have the functional form

$$E(X, Y, Z) \equiv \langle 0| \exp\{i[\varphi(f) - \pi(g)]\}|0\rangle$$

which depends on only *three variables*,

$$X \equiv (f, f), \qquad Y \equiv (f, g), \qquad Z \equiv (g, g)$$

defined for all $f, g \in L^2(\mathbb{R}^s)$ and which clearly satisfy $X \geq 0$, $Z \geq 0$, and $XZ \geq Y^2$. The function E is jointly continuous in the three variables in the appropriate domain as follows from the continuity of the various one-parameter unitary groups. Moreover, $E(X, Y, Z)^* = \langle f, g|0\rangle = \langle 0| - f, -g\rangle = E(X, Y, Z)$, while time-reversal invariance of the ground state implies that $E(X, Y, Z)^* = E(X, -Y, Z)$. Thus, $E(X, Y, Z)^* = E(X, Y, Z) = E(X, -Y, Z)$. We next proceed to determine acceptable functional forms for E.

Consider the expression

$$\langle 0|U[r_{\mathbf{a}}, s_{\mathbf{a}}]|f, g\rangle = e^{i[(r_{\mathbf{a}}, g) - (s_{\mathbf{a}}, f)]/2} \langle 0|f + r_{\mathbf{a}}, g + s_{\mathbf{a}}\rangle,$$

where $r_{\mathbf{a}}(\mathbf{x}) = r(\mathbf{x} - \mathbf{a})$ and $s_{\mathbf{a}}(\mathbf{x}) = s(\mathbf{x} - \mathbf{a})$. Now it follows that

$$\lim_{|\mathbf{a}| \to \infty} \langle 0|U[r_{\mathbf{a}}, s_{\mathbf{a}}]|f, g\rangle = E(X + x, Y + y, Z + z),$$

where X, Y, and Z are as previously given, and

$$x \equiv (r, r), \qquad y \equiv (r, s), \qquad z \equiv (s, s),$$

with the obvious requirements that $x \geq 0$, $z \geq 0$, and $xz \geq y^2$. A comparable calculation shows that

$$\lim_{|\mathbf{a}| \to \infty} \langle f, g|U[r_{\mathbf{a}}, s_{\mathbf{a}}]|f', g'\rangle = e^{i[(g, f') - (f, g')]/2}$$

$$\times E(\|f' - f\|^2 + x, (f' - f, g' - g) + y, \|g' - g\|^2 + z)$$

$$\equiv \langle f, g|C(x, y, z)|f', g'\rangle.$$

Observe that the last relation introduces the operator $C(x, y, z)$, which, under the assumption that the states $|f, g\rangle$ form a total set, is uniquely defined. Since $U[r_{\mathbf{a}}, s_{\mathbf{a}}]^\dagger = U[-r_{\mathbf{a}}, -s_{\mathbf{a}}]$, it follows that $C(x, y, z)^\dagger = C(x, y, z)$. Additionally, by essentially repeating the calculation just given, we may learn that

$$\langle f, g|C(x, y, z) C(x', y', z')|f', g'\rangle = \langle f, g|C(x + x', y + y', z + z')|f', g'\rangle$$

holds for all arguments. Thus we are led to a family of self-adjoint operators $C(x, y, z)$ that satisfy the multiplication rule

$$C(x, y, z) C(x', y', z') = C(x + x', y + y', z + z')$$

on the indicated domain of allowed values for the arguments. Furthermore, since E is continuous it follows that arbitrary matrix elements of $C(x, y, z)$ are jointly

continuous in the three arguments. The most general solution of these conditions
leads to the representation

$$C(x, y, z) = \exp(-Ax - By - Cz)$$

in terms of three, mutually commuting, self-adjoint operators A, B, and C.
Using the relation

$$E(X, Y, Z) = \langle 0|C(X, Y, Z)|0\rangle \,,$$

we can find a representation for E by the spectral representation for A, B, and
C, namely,

$$E(X, Y, Z) = \int_{\mathcal{D}} e^{-aX - bY - cZ} \, d\mu(a, b, c) \,,$$

where the domain of integration $\mathcal{D}$ is symmetric in b so as to satisfy the condition
$E(X, Y, Z) = E(X, -Y, Z)$. To actually correspond to a representation of the
Weyl algebra, it is necessary that some restriction be placed on the domain of
allowed values of (a, b, c). Since $\|C(X, Y, Z)\| \leq 1$ holds as an operator bound,
it follows that one requirement on the acceptable domain of integration comes
from the condition that $(aX \pm bY + cZ) \geq 0$ for all possible (X, Y, Z). For
example, it follows that $a \geq 0$ as well as $c \geq 0$, as well as $4ac \geq b^2$. Furthermore,
the Heisenberg uncertainty relation forces $ac \geq 1/16$. In summary, the final
conditions on the domain of integration $\mathcal{D}$ are given by $a \geq 0$, $c \geq 0$, and
$ac \geq (1 + 4b^2)/16$, with a symmetric inclusion of $-b$ for every $b > 0$; see Exercise
7.1. From an argument based on tags (analogous to the argument of Exercise
7.2) it follows that each primitive element in the integrand, namely $\exp(-aX - bY - cZ)$, corresponds to an inequivalent representation of the Weyl operators
for each distinct set of acceptable values for (a, b, c). Therefore, we have found
a preliminary set of acceptable representations of the Weyl algebra.

Our next step follows one that we took in the previous sections. In particular,
we study

$$\lim_{|\mathbf{a}| \to \infty} \langle f, g|U(\mathbf{a})|f', g'\rangle = \lim_{|\mathbf{a}| \to \infty} e^{i[(g, f'_\mathbf{a}) - (f, g'_\mathbf{a})]/2} \langle 0|f'_\mathbf{a} - f, g'_\mathbf{a} - g\rangle$$

$$= E(X + X', Y + Y', Z + Z')$$

$$\equiv \langle f, g|A|f', g'\rangle \,,$$

where $X' \equiv (f', f')$, $Y' \equiv (f', g')$, and $Z' \equiv (g', g')$. This expression determines
the bounded operator A, and it readily follows that $A = A^\dagger = A^2$, i.e., that A is a
projection operator. By construction $U(O)A = A = AU(O)$ for any $O \in O(\infty)$,
and therefore A is a projection operator onto the subspace of fully rotationally
symmetric vectors, and in particular, it follows that $A|0\rangle = |0\rangle$. In the case that
the set of states $\{|f\rangle\}$ spanned the Hilbert space, we could show that $A = |0\rangle\langle 0|$.
In the present case, where we assume only that the set of states $\{|f, g\rangle\}$ spans
the Hilbert space, we shall need to assume that $A = |0\rangle\langle 0|$, which is our final
assumption to build the models. If this assumption were to lead to undesirable

results, then we would reject it immediately; as it will turn out, however, we will find a perfectly acceptable solution to our model problems based on this assumption.

If, as we now assume, $A = |0\rangle\langle 0|$, then we deduce the functional equation for E given by

$$E(X, Y, Z)\, E(X', Y', Z') = E(X + X', Y + Y', Z + Z')\,,$$

which, as we have now seen many times, has, for a continuous function E, the unique solution $E(X, Y, Z) = \exp(-aX - bY - cZ)$ for some fixed real constants a, b, and c. Since $E(X, -Y, Z) = E(X, Y, Z)$, it follows that $b = 0$. Moreover $ac \geq 1/16$. Thus we are led to the final form for the expectation functional given by

$$\langle 0|\exp\{i[\varphi(f) - \pi(g)]\}|0\rangle = \exp\{-\tfrac{1}{4}[\xi m^{-1}(f, f) + m(g, g)]\}\,, \qquad \xi \geq 1\,.$$

The range of ξ is chosen in order to satisfy the Heisenberg uncertainty inequality. We note in addition that the case $\xi = 1$ corresponds to an *irreducible* representation of the Weyl operators, while the cases $\xi > 1$ all correspond to *reducible* representations of the Weyl operators (proved in the next paragraph). The states $\{|f\rangle\}$ form a total set and span the Hilbert space only in the case $\xi = 1$ corresponding to the irreducible representation. On the other hand, the set of states $\{|f, g\rangle\}$ – and *not* the smaller set $\{|f\rangle = |f, 0\rangle\}$ – span the Hilbert space whenever $\xi > 1$. Thus, in essence, all that we have bought by giving up the assumption that the set of states $\{|f\rangle\}$ span the space is a single new parameter ξ and a collection of reducible representations of the Weyl operators! Fortunately, this little bit extra will prove adequate to our needs.

Let us first demonstrate the reducibility of the Weyl operators whenever $\xi > 1$. We first observe that the representation characterized by the expectation functional

$$E_\chi(f, g) \equiv \exp\{-\tfrac{1}{4}[m^{-1}(f, f) + m(g, g)] + i(f, \chi)\}$$

is irreducible for every $\chi \in L^2(\mathbb{R}^s)$. Indeed, this representation is well defined and irreducible for every $\chi \in \mathcal{S}'$, the space of tempered distributions, provided we restrict f to lie in the space $\mathcal{S}$ of test functions for tempered distributions. To prove this remark we revert to a coordinate formulation for which

$$E_\chi(f, g) = \Pi_{n=1}^\infty \exp\{-\tfrac{1}{4}[m^{-1}p_n^2 + mq_n^2] + ip_n c_n\}\,,$$

where $f = \Sigma\, p_n h_n$, $g = \Sigma\, q_n h_n$, and $\chi = \Sigma\, c_n h_n$, and $\{h_n = h_n(\mathbf{x})\}_{n=1}^\infty$ is a set of real orthonormal functions each in $\mathcal{S}(\mathbb{R}^s)$. Since

$$\exp\{-\tfrac{1}{4}[m^{-1}p^2 + mq^2] + ipc\} = \int \phi(x + q/2)^* e^{ipx} \phi(x - q/2)\, dx\,,$$

for

$$\phi(x) \equiv (m/\pi)^{1/4} e^{-m(x-c)^2/2}\,,$$

it follows that this representation is an irreducible product representation for every set $\{c_n\}$. In particular, this implies that the representation for $\chi \equiv 0$ (our case $\xi = 1$) is irreducible. Next, for $\xi > 1$, we note that

$$
\exp\{-\tfrac{1}{4}[\xi m^{-1}(f,f) + m(g,g)]\}
$$
$$
= \mathcal{N} \int \exp\{-\tfrac{1}{4}[m^{-1}(f,f) + m(g,g)] + i(f,\chi)\}
$$
$$
\times \exp\{-m(\chi,\chi)/(\xi-1)\}\,\delta\chi \,,
$$

written as a formal functional integral, or more precisely as

$$
\exp\{-\tfrac{1}{4}[\xi m^{-1}(f,f) + m(g,g)]\}
$$
$$
= \int \exp\{-\tfrac{1}{4}[m^{-1}(f,f) + m(g,g)] + i(f,\chi)\}\,dG_\xi(\chi)
$$

in terms of a Gaussian probability measure G_ξ. This formula illustrates a decomposition of the Weyl operator representation into irreducible components, and thereby demonstrates the reducibility of the representation.*

There is another useful way to illustrate the reducibility of the Weyl operator representation when $\xi > 1$. Let us first recall the real parameter ζ, $0 < \zeta < 1$, defined by $\zeta = \sqrt{1 - 1/\xi}$. Since

$$
\xi m^{-1}(f,f) + m(g,g)
$$
$$
\equiv \tfrac{1}{2}[m(1+\zeta)]^{-1}(f,f) + \tfrac{1}{2}[m(1+\zeta)](g,g)
$$
$$
+ \tfrac{1}{2}[m(1-\zeta)]^{-1}(f,f) + \tfrac{1}{2}[m(1-\zeta)](g,g) \,,
$$

it follows that

$$
\langle 0| \exp\{i[\varphi(f) - \pi(g)]\}|0\rangle
$$
$$
= \exp(-\tfrac{1}{4}\{\rho_+[(m_+)^{-1}(f,f) + m_+(g,g)]
$$
$$
+ \rho_-[(m_-)^{-1}(f,f) + m_-(g,g)]\}) \,,
$$

where $\rho_\pm \equiv 1/2$ and $m_\pm \equiv m(1 \pm \zeta)$. We also note that

$$
\tfrac{1}{2}(m_+ + m_-) \equiv m \,, \qquad \tfrac{1}{2}(m_+^{-1} + m_-^{-1}) \equiv \xi m^{-1} \,.
$$

In turn, this Weyl operator expectation functional resembles the expectation functional for a generalized free field introduced in the exercises for Chapter 5.

---

* It is interesting to note that this simple example may also be decomposed into alternative families of irreducible representations no member of which is unitarily equivalent to any of those in any other decomposition! To illustrate one such alternative, inequivalent decomposition, we offer the formal Gaussian integral

$$
\mathcal{N} \int \exp\{-\tfrac{1}{4}[\xi m^{-1}(f,f) + \xi^{-1}m(g,g)] + i(g,\chi)\} \exp\{-\xi(\chi,\chi)/[m(\xi-1)]\}\,\delta\chi \,.
$$

We further observe that the reducible operator representation involved here is a so-called *factor representation*, and specifically a *factor of type III* [Em 72].

In particular, it suggests that we introduce *two independent fields*, φ_+, φ_-, and *their associated momenta*, π_+, π_-, in the fashion

$$\langle 0| \exp\{i[\varphi_+(f_+) - \pi_+(g_+)] + i[\varphi_-(f_-) - \pi_-(g_-)]\}|0\rangle$$
$$= \exp(-\tfrac{1}{4}\{[(m_+)^{-1}(f_+, f_+) + m_+(g_+, g_+)]$$
$$+ [(m_-)^{-1}(f_-, f_-) + m_-(g_-, g_-)]\}) .$$

Observe in this case that we deal with two independent fields which taken together form an irreducible representation of the *two*-field Weyl algebra. The field commutation relations in this case read

$$[\varphi_\pm(f), \pi_\pm(g)] = i(f, g)\mathbf{1} , \qquad [\varphi_\pm(f), \pi_\mp(g)] = 0 .$$

Now we introduce a "new" field φ and its associated momentum π by insisting that $f_+ = f_- \equiv f/\sqrt{2}$ and $g_+ = g_- \equiv g/\sqrt{2}$. In this case,

$$\langle 0| \exp\{i[\varphi_+(f_+) - \pi_+(g_+)] + i[\varphi_-(f_-) - \pi_-(g_-)]\}|0\rangle$$
$$\equiv \langle 0| \exp\{i[\varphi(f) - \pi(g)]\}|0\rangle ,$$

where

$$\varphi(f) \equiv [\varphi_+(f) + \varphi_-(f)]/\sqrt{2} ,$$
$$\pi(g) \equiv [\pi_+(g) + \pi_-(g)]/\sqrt{2} ,$$

a prescription which has the virtue that

$$[\varphi(f), \pi(g)] = \tfrac{1}{2}[\varphi_+(f) + \varphi_-(f), \pi_+(g) + \pi_-(g)] = i(f, g)\mathbf{1}$$

as desired. This construction confirms the realization of the Weyl operator representation as arising from an irreducible two-field construction as stated in earlier sections.

Incidentally, it follows from this representation that the space translation generators $\mathcal{P}$ are given by

$$\mathcal{P} = -\int : [\pi_+(\mathbf{x})\boldsymbol{\nabla}\varphi_+(\mathbf{x}) + \pi_-(\mathbf{x})\boldsymbol{\nabla}\varphi_-(\mathbf{x})] : d^s x$$

and in no way (if $\xi > 1$) can it be considered that $\mathcal{P} = -\int : \pi(\mathbf{x})\boldsymbol{\nabla}\varphi(\mathbf{x}) : d^s x$. Here again is a reflection of the fact that in a reducible field operator representation, an operator with a unique invariant state cannot be constructed solely out of the fundamental field operators.

7.7 Determination of the dynamics

To analyze the dynamics of rotationally symmetric models it is convenient if we employ a coherent state representation of the Hilbert space. We first deal with the case of an irreducible Weyl operator representation, in which case the states $|f\rangle$ for all $f \in L^2$ serve as our coherent states. In particular, we note that

$$\langle f|f'\rangle = \exp[-(1/4m)\|f - f'\|^2] \equiv NN' \exp[(u^*, u')],$$

where the inner product remains real, $u(\mathbf{x}) \equiv \sqrt{1/2m}\, f(\mathbf{x})$ and $N = \exp[-\frac{1}{2}(u^*, u)]$ represents a normalization factor. Let us next introduce the elements $U(O)$ of the unitary representation of the symmetry group $O(\infty)$ according to the prescription

$$\langle f|U(O)|f'\rangle = \langle f|Of'\rangle = NN' \exp[(u^*, Ou')].$$

Recall that the Hamiltonian commutes with all the operators $U(O)$ as does the unitary time evolution operator $U(t) \equiv \exp(-i\mathcal{H}t)$, and since the representations of $O(\infty)$ are mutually disjoint on each multiparticle subspace (see Exercise 7.3), it follows that $U(t)$ has matrix elements given by

$$\langle f|U(t)|f'\rangle = NN' \exp[(u^*, e^{-imt} u')]$$

for some parameter m. In turn the matrix elements of the Hamiltonian are given by expanding the previous expression to first order in t leading to

$$\langle f|\mathcal{H}|f'\rangle = m(u^*, u')NN' \exp[(u^*, u')] = \tfrac{1}{2}(f, f')\langle f|f'\rangle$$

as expected. This concludes the discussion of the dynamics for the irreducible representation case ($\xi = 1$).

We follow a similar line of argument in the case of the reducible representations ($\xi > 1$). First, we present a coherent state representation of the basic states. To that end we note that

$$\langle f, g|f', g'\rangle = NN' \exp[(u^*, u') + (v^*, v')]$$

where, in the present case, $u(\mathbf{x}) \equiv \sqrt{1/2m}\,[m\,g(\mathbf{x}) + if(\mathbf{x})]$, $v(\mathbf{x}) \equiv \sqrt{(\xi - 1)/2m}\,if(\mathbf{x})$, and $N = \langle f, g|0\rangle = \exp\{-\frac{1}{2}[(u^*, u) + (v^*, v)]\}$. This formula clearly relates to two fields, but in this formulation it is also clear that the states $|f, g\rangle$ span the two-field Hilbert space inasmuch as f effectively spans the second field space while g spans the first field space. The elements of the group $O(\infty)$ are represented by

$$\langle f, g|U(O)|f', g'\rangle = \langle f, g|Of', Og'\rangle = NN' \exp[(u^*, Ou') + (v^*, Ov')].$$

We also offer a second coherent-state representation of the basic states given by

$$\langle f, g|f', g'\rangle = \mathsf{N}\mathsf{N}' \exp[(\mathsf{u}^*, \mathsf{u}') + (\mathsf{v}^*, \mathsf{v}')]$$

where $\mathsf{u}(\mathbf{x}) \equiv \sqrt{(1 - \zeta^2)/2m}\,[m\,g(\mathbf{x}) + i\xi f(\mathbf{x})]$, $\mathsf{v}(\mathbf{x}) \equiv \sqrt{m/2}\,\zeta\,g(\mathbf{x})$, and $\mathsf{N} = \langle f, g|0\rangle = \exp\{-\frac{1}{2}[(\mathsf{u}^*, \mathsf{u}) + (\mathsf{v}^*, \mathsf{v})]\}$. Again this relates to two fields, but nevertheless, just as in the former case, the space is spanned by the states $|f, g\rangle$. The elements of the group $O(\infty)$ are represented, just as before, by

$$\langle f,g|U(O)|f',g'\rangle = \langle f,g|Of',Og'\rangle = NN' \exp[(u^*,Ou') + (v^*,Ov')]\,.$$

We shall have occasion to refer to these two different decompositions below. For the present we use their existence to infer the general form of matrix elements for operators – most notably the Hamiltonian – which are invariant under the symmetry group $O(\infty)$.

Consider the Hamiltonian operator $\mathcal{H}$ which commutes with all the unitary operators $U(O)$. It follows from the coherent state representation of the Hilbert space, that the matrix elements of such an operator necessarily have the form

$$\langle f,g|\mathcal{H}|f',g'\rangle = G[(f,f'),(f,g'),(g,f'),(g,g')]\,\langle f,g|f',g'\rangle$$

for some function G of the four variables indicated. It readily follows that

$$\langle f,g|e^{-i\tau\varphi(e)}\mathcal{H}e^{i\tau\varphi(e)}|f',g'\rangle$$
$$= G[(f+\tau e,f'+\tau e),(f+\tau e,g'),(g,f'+\tau e),(g,g')]\,\langle f,g|f',g'\rangle\,.$$

Next we observe that

$$\langle f,g|\pi(e)|f',g'\rangle = \tfrac{1}{2}[(e,f+f') + im(e,g-g')]\,\langle f,g|f',g'\rangle$$
$$= \frac{d}{d\tau}\langle f,g|e^{-i\tau\varphi(e)}\mathcal{H}e^{i\tau\varphi(e)}|f',g'\rangle\Big|_{\tau=0}$$
$$\equiv \{(e,f+f')\frac{\partial}{\partial a} + (e,g')\frac{\partial}{\partial b} + (e,g)\frac{\partial}{\partial c}\} G[a,b,c,d]\,\langle f,g|f',g'\rangle\,,$$

where $a \equiv (f,f')$, $b \equiv (f,g')$, $c \equiv (g,f')$, and $d \equiv (g,g')$. Since e is an arbitrary function, the general solution to this differential equation is

$$G[a,b,c,d] = \tfrac{1}{2}[a - im(b - c) + m^2 d] + F[d]$$
$$= \tfrac{1}{2}(f + img, f' - img') + F[(g,g')]\,,$$

where F is an arbitrary function. We fix F by means of the weak correspondence principle, i.e.,

$$H(f,g) = \langle f,g|\mathcal{H}|f,g\rangle = \tfrac{1}{2}[(f,f) + m^2(g,g)] + F[(g,g)]$$
$$= \tfrac{1}{2}[(f,f) + m_o^2(g,g)] + \lambda(g,g)^2\,,$$

which means that

$$F[(g,g)] = \tfrac{1}{2}(m_o^2 - m^2)(g,g) + \lambda(g,g)^2 = \tfrac{1}{2}m^2\zeta^2(g,g) + Vm^4\zeta^4(g,g)^2\,.$$

Based on this analysis, we build the Hamiltonian out of two operators $\mathcal{H}_o$ and $\mathcal{W}$ which in turn are defined by

$$\langle f,g|\mathcal{H}_o|f',g'\rangle = \tfrac{1}{2}(f + img, f' - img')\langle f,g|f',g'\rangle$$
$$\langle f,g|\mathcal{W}|f',g'\rangle = \tfrac{1}{2}m^2\zeta^2(g,g')\langle f,g|f',g'\rangle\,.$$

As we shall see, the interaction can be made out of normally ordered powers of $\mathcal{W}$. Let us recall the two distinct coherent state decompositions of Hilbert space given previously. Expressed in terms of the first of those decompositions, we observe that the operator $\mathcal{H}_o$ is uniquely defined as the generator of the action

$$\langle f, g| e^{-i\mathcal{H}_o t} |f', g'\rangle = NN' \exp[(u^*, e^{-imt} u') + (v^*, v')] .$$

In like manner,

$$\langle f, g| e^{-i\mathcal{W}t} |f', g'\rangle = \mathsf{N}\mathsf{N}' \exp[(u^*, u') + (v^*, e^{-imt} v')] .$$

Differentiation with respect to t confirms the desired matrix elements, and moreover that

$$\langle f, g| \mathcal{W}^2 - m\mathcal{W} |f', g'\rangle = m^2 (v^*, v')^2 \langle f, g|f', g'\rangle$$
$$= \tfrac{1}{4} m^4 \zeta^4 (g, g')^2 \langle f, g|f', g'\rangle$$

implying that the full Hamiltonian is given by

$$\mathcal{H} = \mathcal{H}_o + \mathcal{W} + 4V[\mathcal{W}^2 - m\mathcal{W}] .$$

We observe that the time-dependent, two-point function is determined completely by the first two terms of the Hamiltonian and takes the form

$$\langle 0| \varphi(\mathbf{x}) e^{-i\mathcal{H}t} \varphi(\mathbf{y}) |0\rangle = \tfrac{1}{2} [\rho_+ m_+^{-1} e^{-im_+ t} + \rho_- m_-^{-1} e^{-im_- t}] ,$$

where $\rho_\pm = 1/2$ and $m_\pm = m(1 \pm \zeta)$, as noted previously. Some properties of the four-point function have been discussed elsewhere [AbKT 66].

In relation to the simple description given in Section 7.5 we note that the operator $\mathcal{H}_o = \tfrac{1}{2} \Sigma : [P_n^2 + m^2 (Q_n + \zeta S_n)^2] :$ and likewise $\mathcal{W} = \tfrac{1}{2} \Sigma : [R_n^2 + m^2 (S_n + \zeta Q_n)^2] :$. This identification makes it clear that $\mathcal{W}^2 - m\mathcal{W} =: \mathcal{W}^2 :$ as claimed. The reader may ask why is it necessary to include the term linear in $\mathcal{W}$ without a coefficient λ. The reason for this is to ensure that the ground state of the Hamiltonian is nondegenerate. For example, consider the alternative Hamiltonian

$$\mathcal{H} = m\Sigma [\cos(\alpha) a_n^\dagger + \sin(\alpha) b_n^\dagger][\cos(\alpha) a_n + \sin(\alpha) b_n]$$
$$+ Wm \Sigma b_n^\dagger b_n + 4Vm^2 \Sigma b_n^\dagger b_k^\dagger b_k b_n$$

where we have introduced another parameter, W, which up to this point has been chosen as unity. If instead $W = 0$, then besides the ground state $|0\rangle$, the states $[\sin(\alpha) a_j^\dagger - \cos(\alpha) b_j^\dagger]|0\rangle$ for all j would be eigenstates of the Hamiltonian with zero energy. To avoid this situation we must choose $W > 0$, but that does not require that $W = 1$ as has been the case so far. For each choice of W we set $m_o^2 = m^2 (1 + W\zeta^2)$ and $\lambda = 4Vm^4 \zeta^4$. Each choice of $W > 0$ leads to an alternative quantization with an alternative spectrum, but one that differs from the one implicitly given by our choice of Hamiltonian only by terms of order $\hbar$. This kind of change reflects the ever-present $\hbar$ ambiguity that always exists when quantizing any system, a point already stressed in Chapter 4. What is important is that all of these quantizations reduce to the correct classical theory in the limit of large quantum numbers, a fact implicit in the property that the Hamiltonian operators of all of these quantizations have identical (leading order in $\hbar$) classical Hamiltonians according to the weak correspondence principle. This favorable fact could in no way be obtained by the conventional quantization procedure!

7.8 Path integrals for rotationally symmetric models

In the light of the discussion of previous sections, it is instructive to look briefly at the rotationally symmetric model from a path integral point of view. For the free theory the problem is straightforward, being just the infinite sum of separate and equal harmonic oscillators. Interest therefore centers on the nonfree cases such as the quartic rotationally symmetric model. For this example, one is naively led to initially consider the expression given by

$$\mathcal{M} \int \exp(i\int \{(f,\dot{g}) - \tfrac{1}{2}[(f,f) + m_o^2(g,g)] - \lambda(g,g)^2\}\,dt)\,\mathcal{D}f\,\mathcal{D}g \,.$$

In order for this expression to make sense a cutoff is required such that the number of degrees of freedom is limited, and the most natural way to do so is to interpret $(f,f) = \Sigma_1^N p_n^2$, etc., with a limit $N \to \infty$ reserved to the final step after the functional integration has been carried out. In order for a meaningful limit to emerge, it is necessary to rescale the coupling constant in the manner discussed in Section 7.2, specifically $\lambda \to \lambda/N$. The outcome for the functional integral – just as was the case in the operator study – is a free theory with a resultant mass term that generally involves the quartic coupling constant as derived earlier. In other words, there is no "miracle" which the functional integral can perform that leads to a nontrivial, i.e., a non-Gaussian, result. After all, such functional integrals are designed to reproduce operator results, and so it should be no surprise that trivial results again emerge. Does that mean that there is no way to obtain the nontrivial operator results found with the use of reducible representations of the field and momentum operators? No, it does not mean that, but the price of obtaining those results in a functional integral formulation is rather high as we now explain.

Instead of the previously given functional integral expression we now consider

$$\mathcal{M} \int \exp(i\int \{(f,\dot{g}) + (r,\dot{s})$$
$$- \tfrac{1}{2}[(f,f) + m^2(g + \zeta s, g + \zeta s) + (r,r) + m^2(s + \zeta g, s + \zeta g)]$$
$$- V[(r,r) + m^2(s + \zeta g, s + \zeta g)]^2\}\,dt)\,\mathcal{D}f\,\mathcal{D}g\,\mathcal{D}r\,\mathcal{D}s \,,$$

which involves *two* sets of fields (f,g) and (r,s). The full propagator, as usual, is given by integrating over all fields subject to the boundary conditions $g(t'') = g''$, $s(t'') = s''$ and $g(t') = g'$, $s(t') = s'$. To obtain the results for the rotationally symmetric models it is only necessary to take matrix elements in the states $\{|f,g\rangle\}$. Although this two field formulation is correct it is not a very practical way to characterize the quantization of such fields *ab initio*; if such auxiliary fields would be required to quantize other models with less symmetry, then it would generally be difficult, if not impossible, to guess how the additional fields should enter into a properly extended model.

Thus although there is indeed a functional integral formulation for the rotationally symmetric models, it is definitely *not* the conventional functional integral formulation. As we have seen, to obtain an appropriate procedure it has been necessary to go beyond the conventional formulation.

This concludes our discussion of interacting rotationally symmetric quantum field models.

7.9* Cell models

The kind of model studied in this chapter may be extended to a wide variety of qualitatively similar models known as *cell models* [Gr 70]. Consider the class of Hamiltonians given by

$$H(f,g) = \tfrac{1}{2}(f,f) + V[(g_{\mathbf{j}}, g_{\mathbf{k}})] ,$$

where, for example,

$$(g_{\mathbf{j}}, g_{\mathbf{k}}) \equiv \int_{\mathbf{x} \in \Delta} g(\mathbf{x} + \mathbf{j})\, g(\mathbf{x} + \mathbf{k})\, d^s x .$$

Here Δ equals a (hyper)cubic cell of side length unity centered at the origin, and $\mathbf{j}$ and $\mathbf{k}$ denote integer-valued vectors, e.g., $\mathbf{j} = (j^1, \ldots, j^s)$, $j^l \in \{0, \pm 1, \pm 2, \ldots\}$ for all l, and likewise for $\mathbf{k}$. Among such models is one where

$$V[(g_{\mathbf{j}}, g_{\mathbf{k}})] = \tfrac{1}{2} C \Sigma_{<\mathbf{j}, \mathbf{k}>} (g_{\mathbf{j}} - g_{\mathbf{k}}, g_{\mathbf{j}} - g_{\mathbf{k}}) + \tfrac{1}{2} m_o^2 (g, g) + \lambda \Sigma_{\mathbf{k}} (g_{\mathbf{k}}, g_{\mathbf{k}})^2 ,$$

where C is a constant and the notation $<\mathbf{j}, \mathbf{k}>$ means that the sum is restricted to nearest neighbors. This model does not enjoy the full rotational symmetry of the original models, but does enjoy a smaller rotational symmetry in which each cell undergoes an *identical* rotational transformation, $f_{\mathbf{k}} \to O f_{\mathbf{k}}$ and $g_{\mathbf{k}} \to O g_{\mathbf{k}}$, where the transformation $O \in O(\infty)$ is independent of the cell label. Confining attention to such models, or to ones with a comparable symmetry, Grimmer has shown that the representation of the field and momentum operators is determined by the expectation functional

$$E(f,g) = \exp\{-\tfrac{1}{4} \int [\xi(\mathbf{k})\, m(\mathbf{k})^{-1} |\tilde{f}(\mathbf{k})|^2 + m(\mathbf{k}) |\tilde{g}(\mathbf{k})|^2]\, d^s k\} ,$$

where the functions $\xi(\mathbf{k})$ and $m(\mathbf{k})$ are independently defined within the first Brillouin zone, i.e., for $-\pi < k^j \leq \pi$, $1 \leq j \leq s$, and are defined to be periodic

---

* This section lies somewhat outside the main line of discussion and may be omitted on a first reading.

outside that interval; in addition, $1 \leq \xi(\mathbf{k})$ and $0 < m(\mathbf{k}) < \infty$. If $\xi(\mathbf{k}) = 1$ at some point $\mathbf{k}$ then the representation of the canonical fields is irreducible there, otherwise it is reducible.

As was the case in the original models, a nontrivial interaction requires that the representation be reducible. A nontrivial dynamics can be introduced and analyzed in much the same way for the cell models as was the case for the original models. We shall not pursue the dynamics any further here; a rather complete discussion is available in the literature [Gr 70].

Exercises

7.1 For a single degree of freedom, let

$$\int \psi(x + q/2, y)^* \, e^{ipx} \, \psi(x - q/2, y) \, dx \, dy = \int e^{-ap^2 - bpq - cq^2} \, d\rho(a, b, c) \,,$$

where ρ denotes a probability measure with equal weight for $(a, -b, c)$ as for (a, b, c). Find the functions $\psi(x, y)$ that admit this representation, and thereby determine that the support of ρ is confined to $a \geq 0$, $c \geq 0$, and $ac \geq (1 + 4b^2)/16$.

7.2 Let $U[f, g] \equiv \exp\{i[\varphi(f) - \pi(g)]\}$ and

$$\langle 0|U[f, g]|0 \rangle = \exp\{-\tfrac{1}{4}[m_o^{-1}(f, f) + m_o(g, g)]\}$$

characterize an irreducible representation of the field and momentum operators. Let $V(\mathbf{a}) = U[r_{\mathbf{a}}, s_{\mathbf{a}}]$, where $r_{\mathbf{a}}(\mathbf{x}) = r(\mathbf{x} - \mathbf{a})$ and $s_{\mathbf{a}}(\mathbf{x}) = s(\mathbf{x} - \mathbf{a})$, with $r, s \in L^2$. Show that $V \equiv \lim_{|\mathbf{a}| \to \infty} V(\mathbf{a})$ exists weakly and commutes with $U[f, g]$, $VU[f, g] = U[f, g]V$, for all $f, g \in L^2$. Since $\{U[f, g]\}$ is irreducible, $V = v\mathbf{1}$. Show that $v = \langle 0|U[r, s]|0 \rangle$, and as a consequence, show that the representations of the field and momentum for two different values of m_o are unitarily inequivalent.

7.3 Let a sequence of orthogonal matrices $O^{(r)} = \{O_{mn}^{(r)}\}$ be defined by the condition

$$\begin{aligned}
O_{mn}^{(r)} &= \cos(\beta) \, \delta_{m,n} \,, & 1 \leq m, n \leq r \,, \\
&= -\sin(\beta) \, \delta_{m+r,n} \,, & 1 \leq m \leq r \,, \;\; r+1 \leq n \leq 2r \,, \\
&= \sin(\beta) \, \delta_{m,n+r} \,, & r+1 \leq m \leq 2r \,, \;\; 1 \leq n \leq r \,, \\
&= \cos(\beta) \, \delta_{m,n} \,, & r+1 \leq m, n \leq 2r \,, \\
&= \delta_{m,n} \,, & 2r+1 \leq m \text{ or } n \,.
\end{aligned}$$

Show that this sequence of orthogonal transformations satisfies

$$\lim_{r \to \infty} (u, O^{(r)}v) = \cos(\beta) \, (u, v)$$

for all $u, v \in l^2$.

Let $U_N(g)$ and $U_M(g)$, $g \in G$, denote two unitary representations of a group G. If there exists a nonzero operator V for which $VU_N(g) = U_M(g)V$ holds for all g, then V *intertwines* the two representations and establishes the unitary equivalence of a subrepresentation of U_N with a subrepresentation of U_M. On the other hand, if $V \equiv 0$, then the two representations are mutually disjoint. Let us apply these notions to the problem at hand. In particular, show how the limiting behavior of $O^{(r)}$ can be used to establish that no subrepresentation of the unitary representation of $O(\infty)$ acting on the N-particle subspace, $N < \infty$, is unitarily equivalent to any subrepresentation of the unitary representation of $O(\infty)$ acting on the M-particle subspace, $M < \infty$, $N \neq M$.

7.4 In Section 7.5 it was argued that the Hamiltonian for a quartic rotationally symmetric quantum field theory is given by

$$\mathcal{H} = m\Sigma \left[\cos(\alpha)a_n^\dagger + \sin(\alpha)b_n^\dagger \right]\left[\cos(\alpha)a_n + \sin(\alpha)b_n\right]$$
$$+ m \Sigma b_n^\dagger b_n + 4Vm^2 \Sigma b_n^\dagger b_k^\dagger b_k b_n ,$$

where $\sin(\alpha) = \zeta$ and $\cos(\alpha) = \sqrt{1 - \zeta^2}$, and where $\{a_n\}_{n=1}^\infty$, $\{a_n^\dagger\}_{n=1}^\infty$ and $\{b_n\}_{n=1}^\infty$, $\{b_n^\dagger\}_{n=1}^\infty$ are two sets of independent annihilation and creation operators. In this expression the nonlinear interaction enters with a coefficient V. If this is the case show that a meaningful perturbation theory in the nonlinear interaction does in fact exist for the rotationally symmetric models, without any divergences, in seeming contradiction to the wisdom suggested by Haag's Theorem (see, e.g., Section 6.5).

7.5 So far our attention has been confined to models with full rotational symmetry. Now consider a model with the classical Hamiltonian

$$H(p,q) = \tfrac{1}{2}\Sigma \left(p_n^2 + q_n^2 \right) + (\Sigma q_n^2)^2 + \lambda q_1^6 , \qquad \lambda \geq 0 .$$

Show that the quantum formulation of such a model can be accommodated within the reducible representations appropriate to the model with $\lambda = 0$.

7.6 In the spirit of the preceding problem, examine the quantum theory associated with the classical Hamiltonian

$$H(p,q) = \tfrac{1}{2}\Sigma \left(p_n^2 + q_n^2 \right) + (\Sigma q_n^2)^2 + \lambda (\Sigma e^{-an} q_n^2)^3 , \qquad \lambda \geq 0 .$$

Show that such models can be accommodated within the reducible representations appropriate to the model with $\lambda = 0$. Show that as $a \to 0$ the resultant quantum theory enjoys full rotational invariance.

8

Continuous and Discontinuous Perturbations

WHAT TO LOOK FOR

In the course of the next several sections we shall illustrate some systems with a rather strange property. The common property shared by these systems may be readily described. In brief, if $\mathcal{H}_0$ denotes a free Hamiltonian and $\mathcal{V}$ an interaction, both of which may be assumed bounded below, then it seems self evident that the family of Hamiltonians defined by

$$\mathcal{H} \equiv \mathcal{H}_0 + \lambda \mathcal{V}$$

for all $\lambda \geq 0$ has the property, as $\lambda \to 0^+$, that $\mathcal{H} \to \mathcal{H}_0$. Indeed, there are a great many cases when this natural property holds true, and for evident reasons, in this case, we say that we are dealing with a *continuous perturbation*. However, there are counterexamples to this natural limiting behavior in which case $\mathcal{H} \to \mathcal{H}_0' \not\equiv \mathcal{H}_0$ in the sense that, in general, the eigenvectors and eigenvalues are different in the two cases. When this is the case, we say that we are dealing with a *discontinuous perturbation*. After a general discussion, we show that discontinuous perturbations can already exist for simple one-dimensional quantum mechanical examples. Finally, we argue that singular problems in quantum field theory may possibly find an interpretation as discontinuous perturbations as well. Throughout this chapter we choose units such that $\hbar = 1$.

8.1 General qualitative characterization

When discussing the quantization of a given system, one may adopt any one of several different, but equivalent, viewpoints. To illustrate the general principles behind both continuous and discontinuous perturbations, it is somewhat easier initially to employ an imaginary-time path integral formulation of quantization. (Although we shall speak here of *paths*, the following discussion is meant to apply to the case of *fields* as well.) Thus, in highly symbolic language, we consider the path integral expression for general free and interacting imaginary-time propagators schematically given, respectively, by

$$P_0 \equiv \sum e^{-W_0} \,,$$

$$P \equiv \sum e^{-(W_0 + \lambda W_1)} \,,$$

where W_0 and W_1 denote the contributions to the classical (Euclidean) action represented by $\mathcal{H}_0$ and $\mathcal{V}$. In the first line we have symbolically illustrated the path summation for the free system; in the second line is the expression appropriate to the interacting theory. Here the "sum" stands for a summation (more precisely an integration) over those paths consistent with the pregiven boundary conditions. In the case of P_0, we naively assume that the criterion for the admissibility of paths in the first formula is $W_0 < \infty$ on the grounds that if $W_0 = \infty$ then the contribution of such a path would be infinitely suppressed. (This argument is *not* entirely correct, but let us nevertheless follow its consequences to the end. We will do better later.) Thus the criterion $W_0 < \infty$ delineates those paths which are included. In the case of the interacting theory, with $\lambda > 0$, the "sum" runs over those paths for which $W_0 + \lambda W_1 < \infty$; paths for which $W_0 + \lambda W_1 = \infty$ are excluded by the same reasoning that they were excluded for the free propagator. Let us further assume that both W_0 and W_1 are nonnegative (or bounded below), so that with $\lambda > 0$ it follows that the requirement for the second path summation expression may be stated as $W_0 < \infty$ *and* $W_1 < \infty$.

We can now imagine two different relevant scenarios. In the first situation, let us assume that if $W_0 < \infty$ then it follows that $W_1 < \infty$. If this is the case, then the set of paths in the two expressions are the same, and as a consequence we expect that

$$\lim_{\lambda \to 0^+} \sum e^{-(W_0 + \lambda W_1)} = \sum e^{-W_0} = P_0 \,;$$

in other words, as the coupling constant goes to zero, the interacting theory passes continuously to the free theory. We shall refer to this situation as a *continuous perturbation*. This fact may seem quite obvious, but it nonetheless bears repeating.

Now consider a different situation, namely that $W_0 < \infty$ does *not* imply that $W_1 < \infty$. This means that some of the paths which are allowed for the free theory because $W_0 < \infty$ are forbidden by the interaction term because $W_1 = \infty$. Therefore the interacting theory may also be written in the form

$$P = \sum e^{-(W_0 + \lambda W_1)} \equiv \sum{}' e^{-(W_0 + \lambda W_1)} \,,$$

where the prime on the second sum acknowledges that certain of the paths allowed by the free action are omitted or projected out by the interaction term for any value of $\lambda > 0$. Thus it follows that

$$\lim_{\lambda \to 0^+} \sum{}' e^{-(W_0 + \lambda W_1)} = \sum{}' e^{-W_0} \equiv P_0' \neq P_0 \,.$$

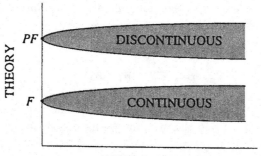

Fig. 8.1. Qualitative representation of Theory vs. Coupling Constant. The vertical axis schematically characterizes the theory by means, say, of the collection of eigenfunctions and eigenvalues, or perhaps the collection of correlation functions, etc. The horizontal axis represents the magnitude of the coupling constant λ, $\lambda \geq 0$. For $\lambda = 0$ there are two points, one labelled F for free, the other labelled PF for pseudofree. If the coupling is turned on for a continuous perturbation, the theory follows a curve in the lower branch marked continuous. If instead the coupling is turned on for a discontinuous perturbation, then the theory jumps instantly from the point F to a point just inside the upper branch marked discontinuous. It then follows a curve in the upper branch. If the coupling constant is turned off, the theory returns by continuity to the point PF rather than F. If the interaction is turned on again, then the theory follows the same curve as before but now as a continuous perturbation starting from the point PF.

In other words, in the second case, the limit of the interacting theory as the coupling constant goes to zero does *not* lead back to the starting point of the free theory. Instead, it leads to what may be called a *pseudofree* theory. This second case corresponds to what we shall call a *discontinuous perturbation*. So strong is such a perturbation that the interacting theory is not even continuously connected to the free theory! Once turned on, *such an interaction leaves an indelible imprint on the system!* It is as if the interaction has behaved as a *hard core* in path space projecting out certain paths and their contribution that would otherwise have been included had the interaction term been entirely absent. On the other hand, if one were to *re*introduce the interaction term again, then the behavior is that of a continuous perturbation, not about the free theory but rather about the pseudofree theory which already has the affected paths already projected out. The two categories of continuous and discontinuous perturbations may be given a qualitative graphical representation in Fig. 8.1.

8.2 Hard-core interactions in quantum mechanics

As our first set of problems with continuous and discontinuous perturbations, we focus on some elementary examples related to one-dimensional quantum mechanics. As our choice for the free Hamiltonian we adopt the harmonic oscillator with unit frequency, namely,

$$\mathcal{H}_0 = \tfrac{1}{2}(P^2 + Q^2) \,,$$

while for the interaction potential we focus on

$$\mathcal{V} = |Q|^{-\alpha}$$

where α denotes a positive real parameter. We shall therefore study the λ and α dependence of the quantum system

$$\mathcal{H} = \tfrac{1}{2}(P^2 + Q^2) + \lambda|Q|^{-\alpha} \,, \qquad \lambda \geq 0 \,, \qquad \alpha > 0 \,.$$

Such problems have long been studied for cases where $\alpha < 0$, and suffice it to say that these potentials give rise to continuous perturbations of $\mathcal{H}_0$ in every such case. Our interest centers instead on the case where $\alpha > 0$ leading to a singular behavior at the coordinate origin, and therefore, our focus is on the influence of the singular behavior when $Q = 0$. In that case, the behavior at infinity is not at issue, and we have chosen a familiar and convenient free Hamiltonian (harmonic oscillator) so as to control the behavior at infinity in a natural and very simple way; a qualitatively similar set of results holds with – or even without – any other confining potential. On the other hand, there is some utility in maintaining a free Hamiltonian and an additional potential with an even parity in the coordinate, i.e., $\mathcal{H}_0(P, -Q) = \mathcal{H}_0(P, Q)$ and $\mathcal{V}(-Q) = \mathcal{V}(Q)$, so that we can exploit this particular symmetry.

Path integral representation

Following the procedure sketched in the preceding section, let us first examine this problem by means of an imaginary-time path integral. The propagator in that case is formally given by the expression

$$P(x'', T; x', 0) = \mathcal{N} \int \exp(-\int \{\tfrac{1}{2}[\dot{x}(t)^2 + x(t)^2] + \lambda|x(t)|^{-\alpha}\}\, dt)\, \mathcal{D}x$$

integrated over continuous paths $x(t)$, $0 \leq t \leq T$, that connect the point $x' \equiv x(0)$ and $x'' \equiv x(T)$. Intuitively, those paths that stay away from $x = 0$ are contributing and do so in the limit $\lambda \to 0$ just as in the harmonic oscillator case. However, those paths that reach or cross $x = 0$, may well do so in such a manner that they give rise to an infinite value for the integral in the exponent leading, when $\lambda > 0$, therefore, to an infinite suppression of that path. As we shall shortly see, whenever $\alpha < 1$ no divergences arise from the paths that reach or cross $x = 0$, while on the other hand, when $\alpha > 2$ every path that reaches or crosses $x = 0$ leads to a divergent integral and its contribution is totally suppressed in the integral for any $\lambda > 0$ no matter how small. Staying with the case $\alpha > 2$ for the moment, it would follow that as $\lambda \to 0$ the omitted paths would never reappear and so the limiting theory is *not* that of the free harmonic oscillator, but rather corresponds to a pseudofree harmonic oscillator in which only paths that are *always positive*, $x(t) > 0$, $0 \leq t \leq T$, or *always negative*, $x(t) < 0$, $0 \leq t \leq T$, are incorporated. As a consequence, the propagator P_0' for the pseudofree harmonic oscillator *vanishes* whenever $x''x' < 0$ and can be nonvanishing only when $x''x' > 0$. If we denote by $P_0(x'', T; x', 0)$ the imaginary-time propagator for the harmonic oscillator, then it is not difficult to argue that the propagator for the pseudofree harmonic oscillator is given by

$$P_0'(x'', T; x', 0) \equiv \theta(x''x')[P_0(x'', T; x', 0) - P_0(x'', T; -x', 0)] \,,$$

where $\theta(y) \equiv 1$, $y > 0$, and $\theta(y) \equiv 0$, $y < 0$. The simplest way to argue for such a formula is to make a sketch that indicates several paths and to observe, with the given combination of terms, that the contribution from any path starting at x' that reaches or crosses the axis $x = 0$ is canceled out by the contribution of a suitably matched path starting at $-x'$. We encourage the reader to make such a sketch! See Exercise 8.1.

The propagator can also be decomposed in terms of the eigenfunctions and eigenvalues of the quantum system which leads, in the case of the harmonic oscillator, to the decomposition given by

$$P_0(x'', T; x', 0) = \sum_{n=0}^{\infty} h_n(x'')e^{-E_n T} h_n(x') \,,$$

where the set of functions $\{h_n\}_{n=0}^{\infty}$ refers to the Hermite functions introduced in Chapter 3, and the values $E_n = n + 1/2$ are just the eigenvalues of the (unit frequency) harmonic oscillator. The Hermite functions have a parity directly related to their index, namely, $h_n(-x) = (-1)^n h_n(x)$, a fact which will strongly influence the expression for the pseudofree propagator. In particular,

$$P_0'(x'', T; x', 0) = \theta(x''x')\{\sum_{n=0}^{\infty} h_n(x'')e^{-(n+1/2)T}[h_n(x') - h_n(-x')]\}$$

$$\equiv \theta(x''x') \sum_{n=0}^{\infty} h_n'(x'')e^{-E_n' T} h_n'(x') \,,$$

where $E_{2n}' \equiv E_{2n+1}' \equiv E_{2n+1} = 2n + 3/2$, and

$$h_{2n}'(x) \equiv \text{sign}(x)h_{2n+1}(x) \,,$$
$$h_{2n+1}'(x) \equiv h_{2n+1}(x) \,.$$

In other words, for the pseudofree harmonic oscillator, every odd-parity eigenfunction of the harmonic oscillator and its associated eigenvalue is retained, while every even-parity eigenfunction of the harmonic oscillator is dropped and replaced with the next order odd-parity harmonic oscillator eigenfunction continued so as to be an *even* parity eigenfunction. This leads to a double degeneracy of the eigenvalues. As stated earlier, the introduction and subsequent removal of a perturbation of the form $|x|^{-\alpha}$, $\alpha > 2$, necessarily results in the pseudofree propagator as $\lambda \rightarrow 0^+$. For $\alpha < 1$ the limit as $\lambda \rightarrow 0^+$ is always the harmonic oscillator. For $1 \leq \alpha \leq 2$ the situation is more complicated and far more interesting, and it will be discussed below.

Brownian motion paths have the property of being continuous but nowhere differentiable. The latter property readily follows from the fact, established in Chapter 5, that for a mean zero Gaussian variable X the condition $\langle X^2 \rangle < \infty$ implies that $X^2 < \infty$ almost everywhere, while the condition $\langle X^2 \rangle = \infty$ implies that $X^2 = \infty$ almost everywhere. Let us consider the variable $X \equiv [x(t) - x(s)]/|t - s|^\beta$. Since $\langle [x(t) - x(s)]^2 \rangle = |t - s|$ for a standard Brownian motion (see Section 6.8), it follow that $\langle X^2 \rangle = |t - s|^{1-2\beta}$. As $t \rightarrow s$ this expression goes to zero if $\beta < 1/2$ and diverges if $\beta > 1/2$. This criterion implies that with probability one there exists a stochastic variable C such that

$$|x(t) - x(s)| < C\,|t - s|^{\frac{1}{2}-\epsilon}\,, \qquad \epsilon > 0\,, \qquad |t - s| \leq 1\,,$$

while with probability one such a stochastic variable does *not* exist if $\epsilon < 0$. In other words, almost all Brownian paths satisfy the indicated inequality for any $\epsilon > 0$, while almost none of them satisfy it for $\epsilon < 0$. This condition is especially relevant for the case $\epsilon = -1/2$ (i.e., $\beta = 1$) needed to form the derivative $\dot{x}(s) = \lim_{t \rightarrow s^+}[x(t) - x(s)]/|t - s|$; in fact, it follows therefore that the derivative of a Brownian motion path is everywhere divergent with probability one.

It is interesting to observe that the previous inequality on paths leads to a criterion for including or excluding paths that is not identical to the one of finite action that we naively assumed in the previous section, but in some sense this criterion is very close to it. The Schwarz inequality for classical functions of finite action implies that

$$|x(t) - x(s)| = |\textstyle\int_s^t \dot{x}(u)\,du| \leq \int_s^t |\dot{x}(u)|\,du \leq \sqrt{|t - s|}\sqrt{\int \dot{x}(u)^2\,du}\,.$$

Thus, for classical functions of finite action, it follows that the paths satisfy $|x(t) - x(s)| < C'\,|t - s|^{1/2}$ while true Brownian motion paths satisfy $|x(t) - x(s)| < C\,|t - s|^{1/2-\epsilon}$ for any $\epsilon > 0$. Although the two descriptions of the paths are close to each other in the sense indicated, they differ widely when it comes to the action integral. In particular, since the derivative of a Brownian path diverges with probability one, it follows that $\frac{1}{2}\int \dot{x}(u)^2\,du = \infty$ for almost all such paths, whereas the classical paths with finite action are just that, paths with finite classical action! Here we encounter the first difference from our heuristic description in Section 8.1, specifically, if the free action was that of a free particle, then instead of the paths entering for which the action integral is finite, almost all paths that enter the path integral are such that the action integral diverges.

We can use the characteristic behavior of Brownian motion paths to establish that those cases where $\alpha < 1$ lie in the class of continuous perturbations. For that purpose we need only establish that for almost all Brownian paths the interaction action is finite. Based on the single time distribution function for a general Wiener process given in Section 6.8, we observe that

$$\langle J \rangle \equiv \langle \int_0^t |x(s)|^{-\alpha} \, ds \rangle = \int_{-\infty}^{\infty} |y|^{-\alpha} \int_0^t p(y, s; x'', x') \, ds \, dy \, ,$$

$$p(y, s; x'', x') \equiv \frac{1}{\sqrt{2\pi s(1 - s/t)}} \exp\{-\frac{[y - x'(1 - s/t) - x''s/t]^2}{2s(1 - s/t)}\} \, .$$

To show the convergence of $\langle J \rangle$ when $\alpha < 1$, consider that

$$\langle J \rangle \leq \int_{-\infty}^{\infty} |y|^{-\alpha} \int_0^t p(y, s; x'', x') \, ds \, dy + \int_{-1}^1 \int_0^t p(y, s; x'', x') \, ds \, dy$$

$$\leq \int_{-\infty}^{\infty} \int_0^t p(y, s; x'', x') \, ds \, dy + \int_{-1}^1 |y|^{-\alpha} \int_0^t p(y, s; x'', x') \, ds \, dy$$

$$\leq t + \int_{-1}^1 |y|^{-\alpha} \int_0^t \frac{1}{\sqrt{2\pi s(1 - s/t)}} \, ds \, dy = t + \frac{2}{(1 - \alpha)} \sqrt{\frac{\pi t}{2}} < \infty \, ,$$

as required. For such α, $\langle J \rangle < \infty$ implies that $J < \infty$ for almost all paths (for in the contrary case $\langle J \rangle$ would diverge!). Thus we draw the important conclusion that perturbations for which $\alpha < 1$ are continuous perturbations, i.e., they lie among those perturbations whose influence vanishes as the coupling constant goes to zero [ShEK 74].

It is important to add that with singular but harmless perturbations such as $|x|^{-\alpha}$, $\alpha < 1$, we learn that it is *not* the criterion of *finite classical energy* – for which it would be necessary that $|x(t)|^{-\alpha} < \infty$ and hence $\alpha \leq 0$ – which decides harmless contributions, but rather whether $\int |x(t)|^{-\alpha} \, dt < \infty$ with probability one.

Harmless terms

Suppose we are assured that some particular potential $\mathcal{X}(Q)$ is a continuous perturbation for a particular $\mathcal{H}_0$ such that $\lim_{\lambda \to 0^+} [\mathcal{H}_0 + \lambda \mathcal{X}(Q)] = \mathcal{H}_0$, then it follows that both operators

$$\mathcal{H}_0 + \lambda \mathcal{V}(Q) \, ,$$

$$\mathcal{H}_0 + \lambda \mathcal{V}(Q) + O(\lambda) \mathcal{X}(Q)$$

converge* as $\lambda \to 0$ either to $\mathcal{H}_0$ or to $\mathcal{H}_0'$ – that is, $\mathcal{V}$ is either a continuous or a discontinuous perturbation of $\mathcal{H}_0$ – quite independently of whether $O(\lambda) \mathcal{X}(Q)$ is present or not. In short, the addition of terms that are guaranteed to disappear in the limit that the coupling constant vanishes does not change the qualitative

---

* For example, convergence may be considered in the sense of the unitaries generated by these self-adjoint operators.

conclusion of whether a given perturbation will be a continuous or a discontinuous perturbation. We call such contributions to the potential *harmless terms*. Since we are unable to solve analytically the problem of interest to us we will appeal to related problems that we can deal with analytically and which differ from the problem of interest only by harmless terms.

Differential equation viewpoint

In a Schrödinger representation for this problem we are asked to consider the differential operator eigenvalue equation

$$-\frac{1}{2}\frac{d^2}{dx^2}\psi(x) + \frac{1}{2}x^2\psi(x) + \frac{\lambda}{|x|^\alpha}\,\psi(x) = E\psi(x)\,.$$

We note, in this language, that harmless terms include potentials of the form $O(\lambda)|x|^{-\gamma}$, where $\gamma < 1$. In addition, we can also include the potential $O(\lambda)\delta(x)$ among the harmless terms as well. To arrive at this conclusion, recall that a $\lambda\delta$ potential at the origin is incorporated by a sudden change in slope of the eigenfunction given by $\psi'(0^+) - \psi'(0^-) = 2\lambda\psi(0)$. As $\lambda \to 0$ this change of slope smoothly disappears leaving no trace of the $\lambda\delta$ potential.

Since the important part of the eigenfunction equation occurs near $x = 0$, we are justified in studying the approximate equation

$$\frac{1}{2}\frac{d^2}{dx^2}\psi(x) = \frac{\lambda}{|x|^\alpha}\psi(x)$$

that holds very close to $x = 0$. We focus much of our attention on this reduced equation which is valid near the origin.

As a next step, let us regularize the singularity of the potential $|x|^{-\alpha}$ by introducing a set of potentials $V_\epsilon(x)$, $\epsilon > 0$, characterized by the property that

$$\lim_{\epsilon \to 0} V_\epsilon(x) = |x|^{-\alpha}\,, \qquad x \neq 0\,;$$

in other words, the regularization is chosen so that it yields $|x|^{-\alpha}$ for all x *except possibly at $x = 0$*. As a first and natural example, we may consider letting $V_\epsilon(x) = (|x| + \epsilon)^{-\alpha}$, which is everywhere positive. The result in this case is that for all $\alpha \geq 1$ the perturbation is discontinuous. To reach this conclusion we consider the form domain [Si 73] of $\mathcal{H}_0 + V_\epsilon$ given by

$$\frac{1}{2}\int \phi'^*(x)\psi'(x)\,dx + \int \phi^*(x)\left[\frac{1}{2}x^2 + \lambda(|x| + \epsilon)^{-\alpha}\right]\psi(x)\,dx$$

defined for those $\phi, \psi \in L^2(\mathbb{R})$ for which the indicated integrals are well defined and remain meaningful in the limit $\epsilon \to 0$. For this limit to be well defined and preserve linearity, when $\alpha \geq 1$, it is necessary that both $\phi(0) = 0$ and $\psi(0) = 0$ for all $\lambda > 0$. Thanks to the presence of the gradient terms, it is not possible as $\lambda \to 0$ for a function that is nonvanishing at the origin to arise, leading to the conclusion that we are dealing with a discontinuous perturbation.

Alternative regularizations

To escape this conclusion when α satisfies $1 \leq \alpha \leq 2$ it is necessary to choose *alternative regularizations* $V_\epsilon(x)$ of the potential $|x|^{-\alpha}$. Rather than choose a regularization initially, let us *derive* the regularization, up to harmless terms, by first choosing an analytic form for the eigenfunction $\psi(x)$. In order that the perturbation behaves as a continuous perturbation, it is sufficient to show that as $\lambda \to 0$ the assumed form of $\psi(x)$ continuously passes to an acceptable behavior of the unperturbed system at $x = 0$. Since the odd parity case is not at issue we concentrate on the even parity case, i.e., to show that we may choose ψ so that as $\lambda \to 0$, $\psi(x) \to 1$ in the vicinity of $x = 0$ [Kl 73b, EzKS 74, EzKS 75].

As a first example let us assume that $\alpha < 2$ and consider the proposal

$$\psi(x) = \exp[-k_1(2-\alpha)^{-1}\lambda(|x|+\epsilon)^{2-\alpha}]\,,$$

where k_1 is a (λ-independent) constant to be determined. Evidently this function approaches unity as $\lambda \to 0$. In taking the first and second derivatives of this and related functions, we shall encounter $(d/dx)(|x|+\epsilon) \equiv \epsilon(x)$ and $(d/dx)\epsilon(x) = 2\delta(x)$. Here we have introduced the standard notation for the function $\epsilon(x) \equiv 1$ for $x > 0$ and $\epsilon(x) \equiv -1$ for $x < 0$. Observe that we therefore have two distinct uses for the symbol ϵ; one is a small positive regularization parameter, while the other is a function. We shall consistently use the functional notation in the second case, and there should be no cause for confusion.

As we already have noted, near $x = 0$ the singularity and the second-order derivatives are the most important part of the time-independent Schrödinger equation. In this first example let us proceed slowly and observe first that

$$\frac{1}{\psi(x)}\frac{d\psi(x)}{dx} = -k_1\lambda(|x|+\epsilon)^{1-\alpha}\epsilon(x)\,.$$

Consequently it follows that the expression

$$\frac{1}{2}\frac{1}{\psi(x)}\frac{d^2\psi(x)}{dx^2} = \frac{1}{2}\{-k_1(1-\alpha)\lambda(|x|+\epsilon)^{-\alpha} - 2k_1\lambda\epsilon^{1-\alpha}\,\delta(x)$$
$$+ k_1^2\lambda^2(|x|+\epsilon)^{2-2\alpha}\}\,.$$

In deriving this expression we have used the facts that $\epsilon(x)^2 \equiv 1$ (except possibly at the point $x = 0$) and $(|x|+\epsilon)^{1-\alpha}\delta(x) = \epsilon^{1-\alpha}\delta(x)$. There are three terms that make up the potential here. As $\epsilon \to 0$, the first term becomes proportional to $|x|^{-\alpha}$, namely to the singular potential we want. As $\epsilon \to 0$, the third term becomes proportional to $|x|^{2-2\alpha}$, which is a harmless term provided that $2 - 2\alpha > -1$, namely when $\alpha < 3/2$. Finally the second term, proportional to $\delta(x)$ with a diverging coefficient for $1 < \alpha < 3/2$, provides the necessary alternative regularization which leads to a continuous perturbation. To match coefficients with the coupling constant it is necessary that $k_1 \equiv 2/(\alpha - 1)$. In summary, we have found that the regularization

$$V_\epsilon(x) = \frac{1}{(|x| + \epsilon)^\alpha} - \frac{2\epsilon^{1-\alpha}}{(\alpha - 1)} \delta(x)$$

leads, as $\epsilon \to 0$, to a continuous perturbation of the harmonic oscillator in the range $1 < \alpha < 3/2$.

In going from $\alpha < 1$ to $1 < \alpha < 3/2$ in the preceding discussion we have skipped over the point $\alpha = 1$, and so let us focus on this case. The proposed form of ψ in this case is given by

$$\psi(x) = \exp[-\overline{k}_1 \lambda(|x| + \epsilon) \ln(|x| + \epsilon)] ,$$

for which $\psi'(x)/\psi(x) = -\overline{k}_1 \lambda \ln(|x| + \epsilon)\epsilon(x) - \overline{k}_1 \lambda \epsilon(x)$, and

$$\begin{aligned}
\tfrac{1}{2}\psi''(x)/\psi(x) = &-\tfrac{1}{2}\overline{k}_1 \lambda(|x| + \epsilon)^{-1} - \overline{k}_1 \lambda[\ln(\epsilon) + 1]\,\delta(x) \\
&+ \tfrac{1}{2}\overline{k}_1^2 \lambda^2[\ln(|x| + \epsilon) + 1]^2 .
\end{aligned}$$

It follows that the first term becomes the desired perturbation provided $\overline{k}_1 = -2$. The third term is harmless, while the second term provides the necessary regularization to ensure a continuous perturbation. In particular, when $\alpha = 1$ we learn that the regularization

$$V_\epsilon(x) = \frac{1}{(|x| + \epsilon)} - 2|\ln(\epsilon)|\,\delta(x)$$

leads to a continuous perturbation of the harmonic oscillator. Observe that the correct logarithmic expression arises when we augment our previous result for $1 < \alpha < 3/2$ by a harmless term so that

$$V_\epsilon(x) = \frac{1}{(|x| + \epsilon)^\alpha} - \frac{2[\epsilon^{1-\alpha} - 1]}{(\alpha - 1)} \delta(x) ,$$

subsequently followed by a limit in which $\alpha \to 1$.

Brownian motion for $1 \le \alpha < 3/2$

Before proceeding to higher values of α, it is interesting to examine the range $1 \le \alpha < 3/2$ from another perspective, namely from the point of view of Brownian motion. Heuristically, the Brownian motion paths zigzag back and forth, and in so doing they tend to *sample an average of the potential*. Let us first consider the integral of the regularized potential that we have found appropriate for the case $1 \le \alpha < 3/2$. In particular, incorporating a suitable harmless term and integrating over the region of the singularity,

$$\int_{-1}^{1} \left[\frac{1}{(|x| + \epsilon)^\alpha} - \frac{2[\epsilon^{1-\alpha} - 1]}{(\alpha - 1)} \delta(x) \right] dx = \frac{2[1 - (1 + \epsilon)^{1-\alpha}]}{(\alpha - 1)}$$

which remains finite as $\epsilon \to 0$. The case $\alpha = 1$ may be obtained through a limit of this expression as $\alpha \to 1$. Thus we see that the proper regularization in the interval $1 \leq \alpha < 3/2$ is such that the integral over the regularized potential is uniformly bounded in the regularization parameter ϵ.

The implication of this heuristic discussion can be confirmed by appealing to the concept of *local time* [ItM 65]. Local time is a stochastic variable that, roughly speaking, is a measure of the "time" a Brownian path spends at a certain level, say $y \in \mathbb{R}$. Local time is defined by

$$t^*(y) \equiv \int_0^T \delta(x(u) - y)\, du \; ;$$

here "*" has nothing to do with complex conjugation! It may be shown (see Exercise 8.2) that

$$\int [t^*(y) - t^*(z)]^2 \, d\mu(x) \leq c|y - z|$$

for some finite positive constant c when $|y - z| \leq 1$. Hence it follows that there exists a stochastic variable C_β such that

$$|t^*(y) - t^*(z)| \leq C_\beta |y - z|^\beta$$

holds with probability one whenever $\beta < 1/2$ and $|y - z| \leq 1$. The stochastic integral of the regularized potential reads

$$\int_0^T V_\epsilon(x(t))\, dt = \int V_\epsilon(y) t^*(y)\, dy$$

$$= \int_{-1}^1 V_\epsilon(y) t^*(y)\, dy + \int_{|y|>1} V_\epsilon(y) t^*(y)\, dy \; .$$

The first term in the last line can be rewritten as

$$\int_{-1}^1 V_\epsilon(y)[t^*(y) - t^*(0)]\, dy + t^*(0) \int_{-1}^1 V_\epsilon(y)\, dy \; .$$

In view of the bound $V_\epsilon(y)|t^*(y) - t^*(0)| \leq C_\beta |y|^{\beta-\alpha}$, $y \neq 0$, which is integrable in y around the origin provided $\beta - \alpha > -1$. For any $\beta < 1/2$, it follows that the first term is well defined for all α such that $\alpha < 3/2 \, (= 2 - 1/2)$. Because of the regularizing δ-function in V_ϵ, the second term, proportional to $t^*(0)$, has been arranged to converge as $\epsilon \to 0$, and therefore we learn that the interaction term remains well defined in the limit and thus the perturbation in question is a continuous perturbation. This concludes our examination of this problem from the point of view of Brownian motion; for further details see [ShEK 74, EzKS 74].

Regularization for $3/2 \leq \alpha < 5/3$

For $3/2 \leq \alpha$ the nature of the problem changes dramatically. The term that is proportional to $(|x| + \epsilon)^{2-2\alpha}$, which was harmless so long as $\alpha < 3/2$, is no longer harmless for larger α values. Thus we must introduce an explicit counterterm in our ansatz for $\psi(x)$ to cancel this particular harmful term. Let us first concentrate on $3/2 < \alpha$. To that end we next consider the expression

$$\psi(x) = \exp[-k_1(2-\alpha)^{-1}\lambda(|x|+\epsilon)^{2-\alpha} - k_2(4-2\alpha)^{-1}\lambda^2(|x|+\epsilon)^{4-2\alpha}] \,.$$

An easy calculation shows that

$$\begin{aligned}
\tfrac{1}{2}\psi''(x)/\psi(x) = \tfrac{1}{2}\{ &- k_1(1-\alpha)\lambda(|x|+\epsilon)^{-\alpha} - k_2\lambda^2(3-2\alpha)(|x|+\epsilon)^{2-2\alpha} \\
&- 2[k_1\lambda\epsilon^{1-\alpha} + k_2\lambda^2\epsilon^{3-2\alpha}]\,\delta(x) \\
&+ [k_1\lambda(|x|+\epsilon)^{1-\alpha} + k_2\lambda^2(|x|+\epsilon)^{3-2\alpha}]^2\} \,.
\end{aligned}$$

With $k_1 = 2/(\alpha-1)$, the first term represents the desired perturbation, just as before. The term $k_1^2\lambda^2(|x|+\epsilon)^{2-2\alpha}$, arising from the square in the last factor, is the harmless-turned-harmful term that we must cancel. To do so, we choose $k_2(3-2\alpha) = k_1^2$, i.e., we set

$$k_2 = \frac{k_1^2}{(3-2\alpha)} = \frac{4}{(\alpha-1)^2(3-2\alpha)} \,.$$

With the offending term canceled, we must examine the remaining terms and test them for potential harm. The next leading order term is proportional to $(|x|+\epsilon)^{4-3\alpha}$, and this term remains harmless so long as $4 - 3\alpha > -1$, namely, for $\alpha < 5/3 \ (= 2 - 1/3)$. Accepting this restriction, we conclude, for the range $3/2 < \alpha < 5/3$, that the regularization

$$V_\epsilon(x) = (|x|+\epsilon)^{-\alpha} - [k_1\epsilon^{1-\alpha} + k_2\lambda\epsilon^{3-2\alpha}]\,\delta(x) \,,$$

$$k_1 = \frac{2}{(\alpha-1)} \,, \qquad k_2 = \frac{k_1^2}{(3-2\alpha)}$$

will lead to a continuous perturbation of the harmonic oscillator.

An analysis of the case $\alpha = 3/2$ shows that a logarithmic term arises again, and in particular, it is necessary to choose

$$V_\epsilon(x) = (|x|+\epsilon)^{-3/2} - [k_1\epsilon^{-1/2} + \overline{k}_2\lambda|\ln(\epsilon)|]\,\delta(x) \,,$$

$$k_1 = 4 \,, \qquad \overline{k}_2 = -k_1^2 = -16 \,.$$

As was the case for $\alpha = 1$ the appropriate coefficient of the logarithm may be obtained by a suitable limit in which a harmless term is introduced, specifically, by considering the expression

$$V_\epsilon(x) = (|x|+\epsilon)^{-\alpha} - [k_1\epsilon^{1-\alpha} + k_2\lambda(\epsilon^{3-2\alpha} - 1)]\,\delta(x)$$

followed by the limit that $\alpha \to 3/2$. Evidently, the correct coefficient of the logarithm may be found from the expression $[k_2(3-2\alpha)]\{[\epsilon^{3-2\alpha} - 1]/(3-2\alpha)\} \to \overline{k}_2|\ln(\epsilon)|$.

The regularization depends on the coupling constant!

It is extremely important to appreciate the price we have had to pay to obtain a continuous perturbation in this case; in particular, the price has been that *the regularizing potential must depend on the strength of the coupling constant* λ. There is no escaping this fact. In order for the effects of the potential to disappear as $\lambda \to 0$, it is necessary that the regularized perturbation depends *nonlinearly* on the coupling constant. This fact will be important below when we make analogies to regularization and renormalization as they appear in quantum field theory.

General regularization strategy for all $\alpha < 2$

The preceding discussion has shown how the use of suitable counterterms enables us to ensure a continuous perturbation in those cases where $\alpha < 5/3$. We next present the general solution for all $\alpha < 2$, focusing initially on cases where $(2 - \alpha)^{-1}$ is *not* an integer (saving the logarithmic cases until later). Guided by the cases already treated, let us introduce the ansatz

$$\psi(x) = \exp\{-\Sigma_{j=1}^{J} k_j [j(2 - \alpha)]^{-1} \lambda^j (|x| + \epsilon)^{j(2-\alpha)}\},$$

where the upper limit to the sum, J, is to be determined. It follows that

$$\begin{aligned}
\tfrac{1}{2}\psi''(x)/\psi(x) = &-\tfrac{1}{2}\Sigma_{j=1}^{J} k_j \left[j(2 - \alpha) - 1\right] \lambda^j (|x| + \epsilon)^{j(2-\alpha)-2} \\
&- \left[\Sigma_{j=1}^{J} k_j \lambda^j \epsilon^{j(2-\alpha)-1}\right] \delta(x) \\
&+ \tfrac{1}{2}\Sigma_{p,q=1}^{J} k_p k_q \lambda^{p+q} (|x| + \epsilon)^{(p+q)(2-\alpha)-2} .
\end{aligned}$$

With $k_1 = 2/(\alpha - 1)$, it follows that

$$\begin{aligned}
\tfrac{1}{2}\psi''(x)/\psi(x) = &\lambda(|x| + \epsilon)^{-\alpha} - \left[\Sigma_{j=1}^{J} k_j \lambda^j \epsilon^{j(2-\alpha)-1}\right] \delta(x) \\
&- \tfrac{1}{2}\Sigma_{j=2}^{J} k_j \left[j(2 - \alpha) - 1\right] \lambda^j (|x| + \epsilon)^{j(2-\alpha)-2} \\
&+ \tfrac{1}{2}\Sigma_{p,q=1}^{J} k_p k_q \lambda^{p+q} (|x| + \epsilon)^{(p+q)(2-\alpha)-2} .
\end{aligned}$$

In this version the first line contains the desired perturbation as well as the necessary regularization to secure a continuous perturbation, while the last two lines represent unwanted contributions. The goal is to choose the coefficients k_j, $2 \le j \le J$, so that all harmful terms in the last two lines cancel leaving only harmless terms. Cancellation may be achieved if, with $k_1 = 2/(\alpha - 1)$, we determine higher order k_j from the recursion relation

$$k_j \equiv -\frac{1}{[1 - j(2 - \alpha)]} \sum_{q=1}^{j-1} k_{j-q} k_q , \qquad 2 \le j \le J .$$

With this choice for the coefficients, all terms in the last two lines with powers λ^j, $2 \le j \le J$, cancel, leaving behind terms proportional to λ^{J+r}, $r \ge 1$. Finally we choose J so that all the remaining terms are harmless. The first and most singular term *not* included in the sum is proportional to $(|x| + \epsilon)^{(J+1)(2-\alpha)-2}$ which is harmless whenever $(J+1)(2-\alpha) - 2 > -1$, i.e., when $J + 1 > 1/(2-\alpha)$. Since we assume for the present that $1/(2-\alpha)$ is not an integer, it follows that

$$J = [(2-\alpha)^{-1}] \, ;$$

here the bracket notation $[Y]$ denotes the *integer part* of Y. For example, if $\alpha = 17/10$ and $Y \equiv 1/(2 - \alpha) = 10/3 = 3.333\ldots$, then $[Y] = 3$.

In summary, for all α, $1 < \alpha < 2$, for which $1/(2 - \alpha)$ is not an integer, the perturbation $|x|^{-\alpha}$ can be made into a continuous perturbation of the harmonic oscillator by choosing the regularized potential

$$V_\epsilon(x) = \frac{1}{(|x| + \epsilon)^\alpha} - [\sum_{j=1}^{[(2-\alpha)^{-1}]} k_j \lambda^{j-1} \epsilon^{j(2-\alpha)-1}] \, \delta(x) \, ,$$

where $k_1 = 2/(\alpha - 1)$ and the remaining factors k_j, $2 \leq j \leq J$, satisfy the previously given recursion relation.

We may include the presently excluded points where $1/(2-\alpha)$ equals an integer J in just the same way we did for $\alpha = 1$ and $\alpha = 3/2$. When $J = 1/(2 - \alpha) \geq 2$ is an integer the regularized potential reads

$$V_\epsilon(x) = \frac{1}{(|x| + \epsilon)^\alpha} - [\sum_{j=1}^{J-1} k_j \lambda^{j-1} \epsilon^{j(2-\alpha)-1} + \overline{k}_J \lambda^{J-1} |\ln(\epsilon)|] \, \delta(x) \, ,$$

where $\overline{k}_J \equiv - \sum_{q=1}^{J-1} k_{J-q} k_q$.

The results presented above extend the analysis of continuous perturbations of the harmonic oscillator by the perturbation $|x|^{-\alpha}$ to any value of $\alpha < 2$. *Observe, as $\alpha \to 2$, that the required regularization [proportional to $\delta(x)$] involves a polynomial of ever increasing order in the coupling constant λ, each term of which has a distinct and characteristic divergent behavior as the regularization parameter $\epsilon \to 0$.*

Regularization for $\alpha = 2$

We shall look at this case in two different ways. First, let us simply extend the previous analysis directly to the case $\alpha = 2$ as a limit from below. In this case the regularizing potential is formally given by

$$V_\epsilon(x) = \frac{1}{(|x| + \epsilon)^2} - [\sum_{j=1}^{\infty} c_j \lambda^{j-1}] \epsilon^{-1} \, \delta(x) \, ,$$

where in the present case $c_1 = 2$ and $c_j = -\sum_{q=1}^{j-1} c_{j-q} c_q$. In this case it is to be noted that every term in the series has an *identical* divergence in the regularization parameter, namely ϵ^{-1}, and the polynomial in the coupling constant has turned into a power series in the coupling constant λ. The functional form of that power series is not difficult to determine, but we choose to defer that calculation until we discuss the second approach to the present case.

For our second look at the case $\alpha = 2$ let us return to the basic differential equation valid near the point of singularity, i.e., to the equation

$$\frac{1}{2}\frac{d^2\psi(x)}{dx^2} = \frac{\lambda}{(|x| + \epsilon)^2}\,\psi(x)\,, \qquad x \neq 0\,.$$

In this case the form of the solution is given simply by

$$\psi(x) = (|x| + \epsilon)^\gamma\,,$$

where $2\lambda = \gamma(\gamma - 1)$. This quadratic equation for γ has two solutions, namely,

$$\gamma = -\frac{4\lambda}{(1 + \sqrt{1 + 8\lambda})}\,, \qquad \gamma = \tfrac{1}{2}(1 + \sqrt{1 + 8\lambda})\,.$$

For $\lambda > 0$, we observe that the first of these solutions satisfies $\gamma < 0$, while the second solution satisfies $\gamma > 1$. As $\lambda \to 0$ the first solution passes to $\psi(x) = 1$, while the second solution becomes $\psi(x) = |x|$. For $x > 0$ these are representatives of the even and odd solutions, respectively, and so we focus on the first solution. [We could have explicitly obtained the odd solution x as $\lambda \to 0$ if we had instead started with the ansatz that $\psi(x) = x(|x| + \epsilon)^{\gamma-1}$.] Finally, if we choose $\psi(x) = (|x| + \epsilon)^\gamma$ for the solution, then it follows that $\frac{1}{2}\psi''(x)/\psi(x) = \lambda/(|x| + \epsilon)^2 + \gamma\epsilon^{-1}\,\delta(x)$, or in particular,

$$V_\epsilon(x) = \frac{1}{(|x| + \epsilon)^2} - \frac{4\epsilon^{-1}}{(1 + \sqrt{1 + 8\lambda})}\,\delta(x)\,.$$

In comparing this result with the one obtained by means of the power series, it is reassuring, provided $\lambda < 1/8$ (so as to avoid the singularity), that

$$\frac{4}{(1 + \sqrt{1 + 8\lambda})} = \sum_{j=1}^\infty c_j\lambda^{j-1}\,,$$

where the coefficients c_j are exactly those defined by the recursion relation given above. In other words, the direct solution has yielded the same solution that we had found by means of a perturbative power series within its radius of convergence.

Satisfactory as this all may seem, there is one point worthy of further discussion. Given the form of the solution as $(|x| + \epsilon)^\gamma$, we may (and should!) ask whether the given solution is square integrable in the neighborhood of $x = 0$, and remains so as $\epsilon \to 0$, as must be the case for it to represent a possible Hilbert space vector. For that to be the case it is necessary that $2\gamma > -1$, or that $8\lambda < 1 + \sqrt{1 + 8\lambda}$, a condition that requires $\lambda < 3/8$. When $\lambda \geq 3/8$, on the other hand, it follows that the chosen solution (with $\gamma < 0$) is *not* square integrable, and thus is unacceptable. When $\lambda \geq 3/8$ there is only one solution which, as $\epsilon \to 0$, remains square integrable at $x = 0$, and that solution is of the form

$$(|x| + \epsilon)^{(1+\sqrt{1+8\lambda})/2} .$$

If this solution were *analytically continued* back to $\lambda = 0$, it would *not* correspond to a continuous perturbation of the harmonic oscillator. Of course, no one *forces* us to analytically continue the only solution valid for $\lambda \geq 3/8$ back to $\lambda = 0$; rather, we are perfectly free to exploit the existence of the two square-integrable solutions that hold when $\lambda < 3/8$. And since we are in the business of trying to extend the family of continuous perturbations to as many cases as possible we definitely choose to arrange matters so that as $\lambda \to 0$ we retain only the first solution with $\gamma < 0$, and thereby ensure that x^{-2} constitutes a continuous perturbation of the harmonic oscillator.

Situation for $\alpha > 2$

For $\alpha > 2$ and for any $\lambda > 0$, the essential portion of the differential equation of interest near $x = 0$ reads

$$\tfrac{1}{2}\psi''(x) = \frac{\lambda}{(|x| + \epsilon)^\alpha}\psi(x) ,$$

which may be solved in terms of Bessel functions [Sz 59]. In particular, let $Z_\nu(y)$ denote a generic solution of the imaginary argument Bessel's equation, i.e.,

$$\frac{d^2 Z_\nu(y)}{dy^2} + \frac{1}{y}\frac{dZ_\nu(y)}{dy} - (1 + \frac{\nu^2}{y^2})Z_\nu(y) = 0 .$$

Then it follows that $w(x) \equiv x^{1/2}Z_\nu(\beta x^{-\gamma})$ – with a new meaning for γ – satisfies the differential equation

$$\frac{1}{2}\frac{d^2 w(x)}{dx^2} = \tfrac{1}{2}\beta^2\gamma^2 x^{-2(1+\gamma)}w(x)$$

provided that $\nu\gamma = \pm 1/2$. We let $\lambda = \tfrac{1}{2}\beta^2\gamma^2$, $\alpha = 2(1+\gamma)$, $\gamma > 0$, (since $\alpha > 2$), and identify $\psi(x)$ with $w(x)$. We further choose $\beta > 0$. The behavior of these solutions near $x = 0$ may be obtained from the asymptotic behavior of the Bessel function for large argument. In particular, for $x > 0$, it follows that

$$\psi(x) \propto x^{(1+\gamma)/2} \exp(\pm\beta x^{-\gamma}) .$$

Only one of the two solutions (the one with the "−") is square integrable at the origin, and it evidently has the behavior that $\psi(x) \to 0$ as $x \to 0$ for any $\lambda > 0$. If the solution for $\lambda = 0$ is approached through this family of solutions – for which we need $\beta x^{-\gamma} \gg 1$ – then the resultant solution vanishes at the origin. This result establishes that the perturbation $|x|^{-\alpha}$ is a discontinuous perturbation of the harmonic oscillator for any $\alpha > 2$, as was to be shown.

Dimensionality and its relation to renormalization

We have emphasized the importance of formulating the Schrödinger equation near $x = 0$ in the form given by

$$\mathcal{H}\psi(x) = -\frac{1}{2}\frac{d^2\psi(x)}{dx^2} + \frac{\lambda}{(|x|+\epsilon)^\alpha}\,\psi(x) - C_\alpha(\lambda,\epsilon)\,\delta(x)\,\psi(x)\,,$$

and we have gone to great lengths detailing the coefficient $C_\alpha(\lambda,\epsilon)$ that multiplies the operator of renormalization $\delta(x)$. Dimensionality can be put to use for us to say something about the form of C_α. In the present case, let the symbol $[Z]$ represent the units of Z. The basic variable x has the units of length, and we write $[x] = 1$, where "1" means length to the first power. For ϵ we also have $[\epsilon] = 1$, while for the Hamiltonian operator $\mathcal{H}$, $[\mathcal{H}] = -2$. It follows that $[\lambda] = (\alpha - 2)$. Since $[\delta(x)] = -1$ we must have $[C_\alpha(\lambda,\epsilon)] = -1$. These facts express themselves in the form taken by the factor of regularization C_α. In particular, if we suppose that C_α admits a power series representation such as

$$C_\alpha(\lambda,\epsilon) = \sum_{j=1}^{J} k_j \lambda^j \epsilon^{f(j)}$$

in terms of pure numbers $\{k_j\}$, then it follows simply on dimensional grounds that $f(j) = j(2 - \alpha) - 1$ in order that each term in the sum has a dimension of -1. Of course, dimensional arguments cannot by themselves determine the numerical coefficients $\{k_j\}$. As before, only enough terms need to be taken to include every relevant contribution; terms for which the exponent $f(j) > 0$ need not be included as they represent harmless terms. If $\alpha < 2$, the coefficients of successive powers of the coupling constant have decreasing singular behavior and such a series eventually terminates. However, if $\alpha > 2$, successive powers of the coupling constant are accompanied by factors of ever *increasing* divergence, and it is clear that such a series has little if anything to do with the discontinuous nature of the perturbation that applies in this case – save perhaps to signal that something dramatic is taking place.

Discussion of the results

It should be noted that the very character of the solution changes dramatically between the cases $\alpha < 2$ and $\alpha > 2$. On the one hand, for $\alpha < 2$ the solution possesses a power series representation in λ; for $\alpha > 2$ that is no longer the case since the solution depends on $\beta \propto \sqrt{\lambda}$. This significant change in behavior reflects the fact that for $\alpha < 2$ the dominant term in the potential arose from the second derivative of the logarithm of the proposed form of $\psi(x)$; for $\alpha > 2$, on the other hand, the dominant term in the potential arose from the square of the first derivative of the logarithm of the proposed form of $\psi(x)$. It is only for the case $\alpha = 2$ that the contribution from these two different terms are of the same order.

This change in character of the solution is accompanied by a change in the possibility of defining a regularized form of the perturbation that leads to a continuous perturbation of the harmonic oscillator, as holds true for $\alpha \leq 2$. On the other hand, for $\alpha > 2$, there is no way of regularization that leads to a continuous perturbation, and consequently a discontinuous perturbation is inevitable. If one recalls Fig. 8.1 then one sees, for the problem at hand, that the branch labelled "continuous" may be identified with (properly regularized forms of) perturbations of the form $|x|^{-\alpha}$ for $\alpha \leq 2$, while the branch marked "discontinuous" may be identified with perturbations of the form $|x|^{-\alpha}$ for $\alpha > 2$ (despite any attempt made to achieve a continuous perturbation through regularization).

Analogies to quantum field theory

Finally, for this quantum mechanical problem, we wish to stress the close connection of our results on continuous and discontinuous perturbations of the harmonic oscillator with rather analogous behavior found in Chapter 5 for scalar quantum field theories. In particular, consider first the interval $1 \leq \alpha < 3/2$ for which a λ-independent regularization was required and for which the average of the regularized potential $V_\epsilon(x)$ over the interval $[-1, 1]$ was uniformly bounded in ϵ. This regime is analogous to the regime in quantum field theory for which *normal ordering* suffices to define the interaction term, namely for a ϕ_n^p theory, for the regime $p < n/(n-2)$. Next consider the interval $3/2 \leq \alpha < 2$ for which the proper regularization $V_\epsilon(x)$ involves a polynomial in the coupling constant λ the degree of that polynomial being given by the integer part of $1/(2-\alpha)$, which therefore increases without limit as $\alpha \to 2$. This regime is analogous to the regime in quantum field theory covered by the phrase *superrenormalizable* for which it is sufficient to consider renormalization counterterms that are polynomials in the coupling constant. For scalar fields this regime lies in the interval $n/(n-2) \leq p < 2n/(n-2)$. The case $\alpha = 2$ for which the regularized potential $V_\epsilon(x)$ entails a function of the coupling constant, not given simply by a polynomial, is analogous to the *strictly renormalizable* case in quantum field theory, i.e., $p = 2n/(n-2)$, which also involves renormalizations that are non-polynomial functions of the coupling constant. Finally, the case $\alpha > 2$, which for the harmonic oscillator results in a discontinuous perturbation, is seen to correspond, by carrying the analogy further, to the case of a *nonrenormalizable quantum field theory*, namely for $p > 2n/(n-2)$. Although what happens in the field theory case in such singular cases is poorly known at best, what happens for the perturbed harmonic oscillator is clear: the perturbation in this case is so strong that it leaves an indelible imprint on the system, so much so that the interacting theories are not continuously connected to the free theory, but rather, as the coupling constant vanishes, converge to an alternative theory that we have termed a pseudofree theory.

It is our avowed purpose to plant in the reader's mind the idea that perhaps covariant nonrenormalizable quantum field theories find their explanation as discontinuous perturbations – and indeed the next two sections of this chapter are devoted to making this idea as plausible as possible; see [Kl 78] for a pedagogical discussion. Moreover, the following two chapters will illustrate soluble, noncovariant, model field theories, each of which is more singular in some sense than a corresponding covariant model, and that fully exhibit the behavior of discontinuous perturbations.

8.3 Hard-core interactions in quantum field theory

In the first section of the present chapter we have presented an heuristic overview of continuous and discontinuous perturbations from a path integral point of view. In the second section we have illustrated this kind of phenomenon with the aid of a simple, one-dimensional harmonic oscillator perturbed by an inverse power potential. In that discussion we were able to examine our heuristic assumptions regarding the onset of discontinuous behavior carefully and found that, although not perfect, they were nevertheless qualitatively correct. This result gives us the courage to apply the heuristic principles of hard-core interactions to far more complicated examples for which, unfortunately, we do not have at the present time any more sophisticated techniques available for our use.

Our goal is to examine quantum field theory models and decide – if such a decision is warranted – whether the individual models in question belong to the class of continuous perturbations or to the class of discontinuous perturbations. We shall appeal to a formal functional integral for our analysis, and base our conclusions on the admittedly naive criterion of whether paths are allowed or forbidden by the free and interaction actions, based simply on the finiteness or divergence of the appropriate action for the path in question. In the case of particle mechanics discussed in the previous section we found that the naive criterion gave essentially the correct answers. Hopefully, that is the case for our analysis in the case of model field theories.

Thus our strategy is exactly that presented in Section 8.1 save in the present case W_0 refers to the free action for a model field theory while W_1 refers to the interaction action for a model field theory.

We start with a simple set of examples which will form the topic of Chapter 9. In particular,

$$W_0 = \tfrac{1}{2} \int \phi(x)^2 \, d^n x \,,$$
$$W_1 = \int |\phi(x)|^p \, d^n x \,,$$

where $p > 2$. There are no derivatives in either of these action expressions. For any $n \geq 1$ it is self evident that there are fields $\phi(x)$ for which $W_0 < \infty$ and $W_1 = \infty$. For example, if

$$\phi(x) = |x|^{-\sigma} e^{-x^2} , \qquad n/p \leq \sigma < n/2 ,$$

we have an example of such a field. Thus, a model quantum field theory based on this choice of free and interaction actions, for $n \geq 1$ and $p > 2$, should by all rights exhibit a hard-core behavior. In point of fact, this is just what is shown in Chapter 9 regarding such unphysical, idealized models.

A next step in considering examples of this sort is given by

$$W_0 = \tfrac{1}{2} \int [\dot{\phi}(x)^2 + \phi(x)^2] \, d^n x ,$$
$$W_1 = \int |\phi(x)|^p \, d^n x ,$$

where once again $p > 2$. Here $\dot{\phi}(x)$ means a derivative with respect to just one of the n variables, say x^0, the (Euclidean) time; there are no derivatives with respect to any of the other independent variables. For $n = 1$ this action refers to Euclidean quantum mechanics, and the discussion in Section 8.2 has already shown that if W_0 is finite so too is W_1. However, this result no longer holds when $n \geq 2$. In particular, it suffices to choose

$$\phi(x) = |\mathbf{x}|^{-\sigma} e^{-x^2} , \qquad (n-1)/p \leq \sigma < (n-1)/2 ,$$

in which case $W_0 < \infty$ while $W_1 = \infty$. Thus a model quantum field theory based on this choice of free and interaction actions, for $n \geq 2$ and $p > 2$, should also exhibit a hard-core behavior. In Chapter 10 we will carefully and fully confirm this heuristic expectation.

It is useful to observe that the finiteness of $A \equiv \int [\dot{\phi}(x)^2 + \phi(x)^2] \, d^n x$ implies the finiteness of $B \equiv \int \phi(x)^2 \, d^n x$, but not *vice versa*. With respect to the use of these quadratic expressions as Euclidean actions within functional integrals, the qualitatively different criterion for the respective finiteness of the Euclidean actions leaves an imprint in the qualitatively different character of the support of the measure in the respective functional integrals. In particular, the action A is finite for fewer functions ϕ than is the case for the action B, and as a consequence, the Gaussian measure, to which the action A gives rise, is supported on qualitatively smoother distributions than is the Gaussian measure that arises from the action B. This kind of qualitative difference is very evident if we discuss an action with a sufficiently high order of derivatives, such as in the case

$$W_0 \equiv \tfrac{1}{2} \int \{ \Sigma_j [\partial_j^k \phi(x)]^2 + \phi(x)^2 \} \, d^n x .$$

Provided that $2k > n$ in this expression, the resultant fields in the support of the Gaussian measure are point-wise defined, continuous functions, which are not singular distributions as in the previous two cases. This smoothing of the support properties of the Gaussian functional integral can qualitatively be read out of the fact that an increased smoothness of the fields for which the Euclidean action is finite is needed; see Exercise 8.3. Indeed, in the case

$$W_0 \equiv \tfrac{1}{2}\int\{\Sigma_j[\partial_j^k\phi(x)]^2 + \phi(x)^2\}\,d^n x\;,$$
$$W_1 \equiv \int|\phi(x)|^p\,d^n x\;,$$

and with $2k > n$, it follows that finiteness of W_0 always implies finiteness of W_1 for any $p > 2$. Thus, with so much smoothness in the free Euclidean action, a polynomial interaction is always a continuous perturbation of the free theory.

We conclude the present discussion with the observation that the form of W_0 introduced above is, for any $k \geq 1$, more restrictive than either the W_0 with no derivatives or the form with just one (Euclidean time) derivative. Thus the support of the Gaussian measure to which the Euclidean action W_0 leads is, for any $k \geq 1$, concentrated on smoother distributions than if the gradients were all removed, or all removed but one.

Covariant theories

Of course, the most interesting case from a physical point of view is that related to a relativistically covariant theory. By this we mean that

$$W_0 \equiv \tfrac{1}{2}\int\{[\nabla\phi(x)]^2 + \phi(x)^2\}\,d^n x\;,$$
$$W_1 \equiv \int|\phi(x)|^p\,d^n x\;,$$

for any $p > 2$; this is just the case discussed above when $k = 1$. The analysis of this particular case is far more involved than any of the comparatively trivial examples given above. For the present, let us simply state the result and defer the proof until Section 8.4. Here, then, is the essential Sobolev inequality [LaSU 68]:

Let ϕ be a suitable function on $\mathbb{R}^n$, $n \geq 3$. Then, provided that

$$p \leq \frac{2n}{n-2}\;,$$

there exists a finite constant C_n, independent of ϕ and depending only on n, and a mass parameter M, $0 < M < \infty$, which may be chosen as small as one likes, such that

$$[\int|\phi(x)|^p\,d^n x]^{2/p} \leq C_n\int\{[\nabla\phi(x)]^2 + M^2\phi(x)^2\}\,d^n x\;,$$

holds for all functions for which the right-hand side is well defined. It is useful to note for $M = 1$ that $C_n \leq 4/3$ holds for any $n \geq 3$.

Generally, such inequalities are first proved for smooth functions with compact support, such as functions in the space of test functions C_0^∞, i.e., functions that have compact support and continuous derivatives of arbitrary order. Once such an inequality is established for such a special and limited class of functions, it is generally acceptable to extend the domain of applicability to include all functions for which both sides make sense. Before we take up the proof of this inequality, let us discuss some of its implications.

Application to functional integrals

As presented in Chapter 5, the Euclidean space quantization of a ϕ_n^p, with p even, scalar field is formally characterized by the expression

$$\mathcal{N} \int \exp[\int (h(x)\phi(x) - \tfrac{1}{2}\{[\nabla\phi(x)]^2 + m^2\phi(x)^2\} - g\phi(x)^p)\, d^n x]\, \mathcal{D}\phi\,.$$

Although formal in nature this expression is rich with potential. This potential is traditionally realized by giving some meaning to this expression by the introduction of a *regularization*. In Chapter 5 we have already discussed a lattice regularization as one possible technique in which the exponent in the integrand is replaced by a Riemann sum approximation on a (hyper)cubic lattice, and the result is obtained as a limit as the lattice spacing goes to zero and the lattice volume goes to infinity. This is implicitly the procedure, or one equivalent to it, that is conventionally used in studying the formal functional integral appearing above. This is a fine procedure and works well – when it works well. However, when it leads to results that are unsatisfactory, then this procedure should, if possible, be replaced with another one that leads to more satisfactory answers.

Let us discuss the formal functional integral given above from an alternative point of view in this section. In particular, let us examine the consequences of the Sobolev inequality given above. It is clear that we may consider the right-hand side of that inequality as representing the free action while the left-hand side involves the interaction action. As noted above, the inequality holds for all functions for which both sides make sense. On the right side this leads to a space of functions where both the function ϕ and its gradient $\nabla\phi$ are square integrable over $\mathbb{R}^n$. The inequality, when satisfied, then asserts that such functions are not only integrable when raised to the pth power, but that the norm squared of that integral is bounded by a constant multiple of the Sobolev norm squared of the function. So much for classical functions, but what about the kind of functions that enter functional integrals?

As is well known, already in the case of simple Wiener integrals, the support of a functional integral is generally concentrated on *distributions* rather than classical functions. This fact would seem to make the Sobolev inequality irrelevant in any proposed application to functional integrals. On the other hand, it is also a fact that any distribution may be approached as a suitable limit of smooth functions, a property already familiar in the use of delta-sequences to characterize a delta distribution as a limit. For smooth, not-identically-zero functions for which the numerator and denominator separately make sense, it follows that

$$\frac{[\int |\phi(x)|^p \, d^n x]^{2/p}}{\int \{[\nabla \phi(x)]^2 + M^2 \phi(x)^2\} \, d^n x} \leq C_n \,,$$

a relation which holds whenever $p \leq 2n/(n-2)$. Since the right-hand side is independent of the nonzero function ϕ, it follows that the bound on this quotient remains valid in a limit in which a sequence of smooth functions converges (weakly) to a generalized function (distribution). In general, the numerator and the denominator both diverge in such a limit, but the given quotient remains uniformly bounded. Since this bound holds for all distributions it holds almost everywhere throughout the support of the measure in the functional integration. As a consequence we learn from this fact that although both the free and interaction actions are infinite, the given quotient, defined by the limiting operation described, remains uniformly bounded on the space of distributions. We infer from this remark that the free action controls the interaction action in the sense that paths – distributional paths to be sure – that enter into the functional integrals are qualitatively the same with or without the interaction, i.e., whether $g > 0$ or $g = 0$. We have introduced the qualifier "qualitatively" to cover the cases where the paths indeed do change but do so in a suitably continuous and harmless way. For example, consider the case where the interaction action deals with a mass perturbation for which $p = 2$. In this case the nature of the paths both before and after the introduction of the perturbation can be readily analyzed, and one learns that the support depends on the value of the mass in such a way that the support regions for any two different masses are completely orthogonal to each other, or as one says they are *mutually orthogonal*. By itself, this situation is not necessarily the source of any trouble. After all, the supports of the two delta functions $\delta(x-1)$ and $\delta(x-c)$ are mutually orthogonal if $c \neq 1$, however, the Fourier transform of these distributions, $\exp(ip)$ and $\exp(ipc)$, respectively, have the property that as $c \to 1$, they coincide. A continuous convergence of characteristic functions of this kind leads to a kind of convergence of the measures which is called *weak convergence;* it is the kind of convergence we have in mind when we speak about continuous perturbations. Returning to the case at hand, we note that since the support of the measure depends, strictly speaking, on the mass value, changing the value of that mass changes the support of the functional integral. Nevertheless, this change is a continuous one since, as the perturbation vanishes, the paths return to those of the free theory, as one may explicitly verify. On the other hand, for a discontinuous perturbation, we have in mind a *significant and qualitative change of the support of the paths*, one that leads to a discontinuity in the result of the functional integral between the value when $g = 0$ and the value when $g > 0$ no matter how small.

8.4* Multiplicative inequalities

In this section we derive some classical multiplicative inequalities that lead to certain Sobolev inequalities, some of which were used in the preceding section in discussing local interactions in functional integrals. Let us begin with some basics.

Tools and notation

Young's inequality states for $a \geq 0$, $b \geq 0$, $\epsilon > 0$, and for $m > 1$, that

$$ab \leq \frac{1}{m}\epsilon^m a^m + \frac{(m-1)}{m}\epsilon^{-m/(m-1)}b^{m/(m-1)} .$$

This relation may be proved by considering the right side to be a function of b and $a \equiv c/b$, and minimizing with respect to b holding c fixed; the resultant minimum is just $c = ab$. With $a_j \geq 0$ for all j, one consequence of this relation is that

$$a_1[a_2 \cdots a_N] \leq \frac{1}{N}a_1^N + \frac{(N-1)}{N}[a_2 \cdots a_N]^{N/(N-1)} ,$$

which leads upon further iteration to the inequality

$$a_1 a_2 \cdots a_N \leq \frac{1}{N}(a_1^N + a_2^N + \cdots + a_N^N) .$$

Holder's inequality is a generalization of Schwarz's inequality and reads for nonnegative functions v and w as

$$\int v^\alpha w^\beta \, d\sigma \leq (\textstyle\int v \, d\sigma)^\alpha (\int w \, d\sigma)^\beta ,$$

where $\alpha \geq 0$, $\beta \geq 0$, and $\alpha + \beta = 1$ (as may be inferred to ensure dimensionality with respect to σ). This relation may be proved by first specializing to piece-wise constant functions over uniform cells so that Holder's inequality becomes

$$\Sigma \, v_j^\alpha w_j^\beta \leq (\Sigma v_j)^\alpha (\Sigma w_j)^\beta .$$

If we set $v_j^\alpha \equiv c_j/w_j^\beta$ and extremize the right-hand side over each w_j holding all c_j fixed, the minimum is just the left-hand side as shown. Limiting behavior from piece-wise constant functions proves the rest. If $v_k \geq 0$, $1 \leq k \leq n$, and we apply Holder's inequality to

$$\int v_1^{\alpha_1}[v_2^{\alpha_2} \cdots v_n^{\alpha_n}] \, d\sigma$$

---

* This section lies somewhat outside the main line of discussion and may be omitted on a first reading.

where $\alpha_j \geq 0$, $\alpha_1 + \alpha_2 + \cdots + \alpha_n = 1$, it follows that

$$\int v_1^{\alpha_1} [v_2^{\alpha_2} \cdots v_n^{\alpha_n}] \, d\sigma$$

$$\leq (\textstyle\int v_1 \, d\sigma)^{\alpha_1} \, (\int [v_2^{\alpha_2} \cdots v_n^{\alpha_n}]^{(\alpha_2 + \cdots + \alpha_n)^{-1}} \, d\sigma)^{(\alpha_2 + \cdots + \alpha_n)}$$

$$\leq (\textstyle\int v_1 \, d\sigma)^{\alpha_1} \, (\int v_2 \, d\sigma)^{\alpha_2}$$

$$\times \, (\textstyle\int [v_3^{\alpha_3} \cdots v_n^{\alpha_n}]^{(\alpha_3 + \cdots + \alpha_n)^{-1}} \, d\sigma)^{(\alpha_3 + \cdots + \alpha_n)}$$

$$\cdots$$

$$\leq (\textstyle\int v_1 \, d\sigma)^{\alpha_1} \, (\int v_2 \, d\sigma)^{\alpha_2} \cdots (\int v_n \, d\sigma)^{\alpha_n} \, .$$

In summary, we learn for nonnegative v_k that

$$\int \Pi v_k^{\alpha_k} \, d\sigma \leq \Pi (\textstyle\int v_k \, d\sigma)^{\alpha_k}$$

for $\alpha_k \geq 0$ and $\Sigma \alpha_k = 1$. This is an important relation in what follows.

For a $C_0^\infty(\mathbb{R}^n)$ function ϕ, we introduce standard norm notation in the form

$$\|\phi\|_p \equiv (\textstyle\int |\phi(x)|^p \, d^n x)^{1/p}$$

for all p, $1 \leq p < \infty$.

We divide our derivation of the multiplicative inequalities into three steps.

Step 1: *To show* $\|\phi\|_p \leq \|\phi\|_s^\alpha \|\phi\|_r^{(1-\alpha)}$

Our first step in obtaining the desired form of inequality proceeds according to the chain

$$\int |\phi|^p \, d^n x = \int |\phi|^{\alpha p} \, |\phi|^{(1-\alpha)p} \, d^n x$$

$$= \int (|\phi|^{\alpha p/\beta})^\beta \, (|\phi|^{(1-\alpha)p/(1-\beta)})^{(1-\beta)} \, d^n x$$

$$\leq (\textstyle\int |\phi|^{\alpha p/\beta} \, d^n x)^\beta \, (\int |\phi|^{(1-\alpha)p/(1-\beta)} \, d^n x)^{(1-\beta)} \, .$$

Consequently,

$$\|\phi\|_p \equiv (\textstyle\int |\phi|^p \, d^n x)^{1/p}$$

$$\leq (\textstyle\int |\phi|^s \, d^n x)^{\alpha/s} \, (\int |\phi|^r \, d^n x)^{(1-\alpha)/r}$$

$$\equiv \|\phi\|_s^\alpha \|\phi\|_r^{(1-\alpha)}$$

where, without loss of generality, we choose $\alpha \geq \beta$, and

$$s \equiv \alpha p/\beta \, , \qquad r \equiv (1-\alpha)p/(1-\beta) \, .$$

Hence $r \leq p \leq s$, and if we eliminate α and β in favor of r and s we find

$$(1-\beta)r = p - \alpha p$$

$$= p - \beta s \, .$$

Thus

$$\beta = \frac{p-r}{s-r} \, ,$$

$$\alpha = \frac{s\,(p-r)}{p\,(s-r)} = \frac{\frac{1}{r} - \frac{1}{p}}{\frac{1}{r} - \frac{1}{s}} \, .$$

Thus we find the relation

$$\|\phi\|_p \le \|\phi\|_s^\alpha \|\phi\|_r^{(1-\alpha)} \ ,$$

where

$$0 \le \alpha \equiv \frac{\frac{1}{r} - \frac{1}{p}}{\frac{1}{r} - \frac{1}{s}} \le 1 \ .$$

Step 2: *To show* $\|u\|_{n/(n-1)} \le 1/(2\sqrt{n})\|\nabla u\|_1$

We next turn our attention to the following argument. We note that

$$\int |2u|^{n/(n-1)} \, d^n x = \int |2u|^{1/(n-1)} \cdots |2u|^{1/(n-1)} \, dx_1 \cdots dx_n \ .$$

For smooth functions $u \in C_0^\infty$ it follows that

$$u(x) = \int_{-\infty}^{x_1} u_1 \, dx_1$$

or

$$u(x) = -\int_{x_1}^{\infty} u_1 \, dx_1$$

and thus

$$2u(x) = \int_{-\infty}^{x_1} u_1 \, dx_1 - \int_{x_1}^{\infty} u_1 \, dx_1$$

where $u_1 \equiv \partial u/\partial x_1$, or more generally,

$$u_i \equiv \frac{\partial u}{\partial x_i} \ .$$

Hence,

$$|2u(x)| \le \int |u_1| \, dx_1 \equiv I_1 \ .$$

By a similar argument

$$|2u(x)| \le \int |u_i| \, dx_i \equiv I_i \ .$$

Note that the integral I_i involves a derivative with respect to x_i and depends on all the x variables *except* for x_i which at this point has been integrated out. Using this fact we learn that

$$\int |2u|^{n/(n-1)} \, dx_1 \le \int I_1^{1/(n-1)} I_2^{1/(n-1)} \cdots I_n^{1/(n-1)} \, dx_1$$
$$\le I_1^{1/(n-1)} (I_{21} \cdots I_{n1})^{1/(n-1)}$$

where we have used Holder's inequality and recognized that I_1 is independent of x_1. The notation is defined by

$$I_{ij} \equiv \int |u_i| \, dx_i \, dx_j \ , \qquad i \ne j$$

and thus the second subscript denotes a new variable of integration – *and so on for each newly added subscript.*

Next we integrate with respect to x_2 and note that

$$\int |2u|^{n/(n-1)}\, dx_1\, dx_2 \leq \int (I_1 I_{21} \cdots I_{n1})^{1/(n-1)}\, dx_2$$
$$\leq I_{21}^{1/(n-1)} (I_{12} I_{312} \cdots I_{n12})^{1/(n-1)} .$$

A further application yields

$$\int |2u|^{n/(n-1)}\, dx_1\, dx_2\, dx_3 \leq \int (I_{21} I_{12} I_{312} \cdots I_{n12})^{1/(n-1)}\, dx_3$$
$$\leq I_{312}^{1/(n-1)} (I_{213} I_{123} I_{4123} \cdots I_{n123})^{1/(n-1)} ,$$

etc. In the case $n = 3$ this chain stops and leads to

$$\int |2u|^{3/2}\, d^3x \leq I_{312}^{1/2} I_{213}^{1/2} I_{123}^{1/2} ,$$

while for $n = 4$ it leads to

$$\int |2u|^{4/3}\, d^4x \leq I_{4123}^{1/3} I_{3124}^{1/3} I_{2134}^{1/3} I_{1234}^{1/3} .$$

This rule evidently generalizes to yield

$$\int |2u|^{n/(n-1)}\, d^n x \leq \Pi_{i=1}^n \left(\int |u_i|\, d^n x \right)^{1/(n-1)} .$$

On the basis of the generalization of Young's inequality it follows that

$$\Pi_i \left(\int |u_i|\, d^n x \right)^{1/n} \leq \frac{1}{n} \int \Sigma_i |u_i|\, d^n x .$$

Since

$$\Sigma_i (|u_i| - \Sigma_j |u_j|/n)^2 \geq 0$$

it follows that

$$\Sigma |u_i|^2 \geq \frac{1}{n} (\Sigma |u_i|)^2 ,$$

and as a consequence

$$\Pi_i \left(\int |u_i|\, d^n x \right)^{1/n} \leq \frac{1}{\sqrt{n}} \int \sqrt{\Sigma |u_i|^2}\, d^n x$$
$$\equiv \frac{1}{\sqrt{n}} \|\nabla u\|_1 ,$$

following the notation of [LaSU 68]. Thus

$$\left(\int |2u|^{n/(n-1)}\, d^n x \right)^{(n-1)/n} \leq \frac{1}{\sqrt{n}} \|\nabla u\|_1 ,$$

or stated otherwise

$$\|u\|_{n/(n-1)} \leq \frac{1}{2\sqrt{n}} \|\nabla u\|_1 .$$

Step 3: *To show* $\|\phi\|_p \leq [\frac{1}{2\sqrt{n}} \frac{(n-1)m}{(n-m)}]^\alpha \|\nabla \phi\|_m^\alpha \|\phi\|_r^{(1-\alpha)}$

We now substitute in the last derived relation the identity

$$u \equiv \phi^\sigma , \qquad \sigma \equiv \frac{(n-1)m}{n-m} ,$$

where $1 \le m < n$ is necessary. Then it follows that

$$\left(\int |\phi|^{nm/(n-m)} \, d^n x \right)^{(n-1)/n} \le \frac{1}{2\sqrt{n}} \frac{(n-1)m}{(n-m)} \int |\phi|^{\frac{(n-1)m}{n-m}-1} \sqrt{(\nabla\phi)^2} \, d^n x$$

$$= \frac{1}{2\sqrt{n}} \frac{(n-1)m}{(n-m)} \int |\phi|^{\frac{n(m-1)}{n-m}} |\nabla\phi|^{\frac{m}{m}} \, d^n x ,$$

which by Holder's inequality becomes

$$\left(\int |\phi|^{nm/(n-m)} \, d^n x \right)^{(n-1)/n} \le \frac{1}{2\sqrt{n}} \frac{(n-1)m}{(n-m)} \|\phi\|^{\frac{nm}{(n-m)}(1-1/m)}_{\frac{nm}{(n-m)}} \|\nabla\phi\|_m .$$

Since

$$\frac{nm}{(n-m)} \left(1 - \frac{1}{n} - 1 + \frac{1}{m} \right) = 1 ,$$

it follows that we have obtained

$$\|\phi\|_{\frac{nm}{(n-m)}} \le \frac{1}{2\sqrt{n}} \frac{(n-1)m}{(n-m)} \|\nabla\phi\|_m .$$

We now combine this last relation with the one derived in Step 1 with the identification

$$s \equiv \frac{nm}{n-m} ,$$

or restated in alternative form as

$$\frac{1}{s} = \frac{1}{m} - \frac{1}{n} .$$

Our final multiplicative inequality then is

$$\|\phi\|_p \le \{ [1/(2\sqrt{n})][(n-1)m/(n-m)] \}^\alpha \|\nabla\phi\|^\alpha_m \|\phi\|^{(1-\alpha)}_r$$

where

$$0 \le \alpha \equiv \frac{\frac{1}{r} - \frac{1}{p}}{\frac{1}{r} + \frac{1}{n} - \frac{1}{m}} \le 1 .$$

This relation holds under the conditions that $1 \le m < n$ and

$$p \le \frac{nm}{n-m} .$$

Note that if $r = p$ and thus $\alpha = 0$ the inequality contains no real information. Hence, we may restrict attention to the case $r < p$ and $\alpha > 0$.

While the preceding proof has assumed that the functions involved were C_0^∞, no great use has been made of that property. Consequently, the indicated multiplicative inequality may be extended (by limits of suitable Cauchy sequences) to any function for which both sides make sense; see Exercise 8.4.

Optimality of the multiplicative inequality

The basic multiplicative inequality derived above *cannot* be extended to higher p values than indicated. Consider the case of a function ψ which, for simplicity, we choose as

$$\psi(x) = (|x| + \epsilon)^{-\gamma}, \qquad \epsilon > 0, \qquad |x| \leq 1,$$

while for $|x| > 1$ we assume that ψ is joined smoothly to any C_0^∞ function you like. Recall, in this context, that

$$|x| \equiv \sqrt{\Sigma x_i^2}.$$

For this function it follows that

$$|\nabla\psi|^2 = \frac{\gamma^2}{(|x| + \epsilon)^{2(\gamma+1)}}, \qquad |x| \leq 1,$$

and thus

$$\int |\nabla\psi|^m \, d^n x = \gamma^m \int_{|x|<1} (|x| + \epsilon)^{-m(\gamma+1)} \, d^n x + \text{(finite terms)}.$$

Provided $m(\gamma + 1) < n$, or stated otherwise

$$\gamma < \frac{n - m}{m},$$

the integral above is convergent and *uniformly* bounded as $\epsilon \to 0$. For the integral of $|\psi|^p$ we see that

$$\int |\psi|^p \, d^n x = \int_{|x|<1} (|x| + \epsilon)^{-p\gamma} \, d^n x + \text{(finite terms)}$$

which is *not* uniformly bounded in ϵ when $p\gamma \geq n$, in particular having the behavior

$$\|\psi\|_p = O(\epsilon^{-\gamma+n/p}), \qquad p\gamma > n,$$
$$= O([-\ln(\epsilon)]^{1/p}), \qquad p\gamma = n$$

for small ϵ. Finally we turn our attention to the third ingredient, the integral of ψ^r. As we have already noted, we can confine our attention to the case $r < p$. Consequently, we see that

$$\|\psi\|_r = O(\epsilon^{-\gamma+n/r}), \qquad r\gamma > n,$$
$$= O([-\ln(\epsilon)]^{1/r}), \qquad r\gamma = n,$$
$$= O(1), \qquad r\gamma < n.$$

Is the qualitative behavior determined above compatible with a multiplicative inequality of the general form found previously, i.e.

$$\|\psi\|_p \leq K' \|\nabla\psi\|_m^\alpha \|\psi\|_r^{(1-\alpha)}$$

for any $0 < \alpha \leq 1$ and $0 < K' < \infty$? For this to be true it would require that

$$O(\epsilon^{-\gamma+n/p}) \leq K'' O(\epsilon^{-\gamma+n/r})$$

holds whenever $r\gamma > n$ for some finite nonzero K'' as $\epsilon \to 0$. However, since $p > r$ this inequality *cannot* hold true because it will always be violated for sufficiently small ϵ. It is easy to see that a similar conclusion holds in the other cases (i.e., $r\gamma = n$ and $r\gamma < n$) as well. Hence, the original inequality as it stands is optimal with regard to the maximum allowed power p being $p = nm/(n-m)$. Notice that this maximum depends only on the parameters in the gradient term and does not depend on the power r in the other factor. So too for the coefficient in front, save for the dependence of α on r.

Application to covariant scalar fields

The Euclidean-space (imaginary-time) formulation of the free action for a relativistic massive scalar field takes the form

$$W_0 = \tfrac{1}{2} \int [(\nabla \phi)^2 + M^2 \phi^2] \, d^n x \,,$$

where $M > 0$ denotes the mass. The typical interaction action associated with this free action is given by

$$W_1 = \lambda \int \phi^p \, d^n x \,, \qquad \lambda \geq 0$$

where p is an even integer. In applying the multiplicative inequality to study this problem we take the variables $r = m = 2$, and learn, for $n > 2$, that

$$\|\phi\|_p \leq \{(1/\sqrt{n})[(n-1)/(n-2)]\}^\alpha \, \|\nabla \phi\|_2^\alpha \, \|\phi\|_2^{(1-\alpha)} \,,$$

where

$$0 \leq \alpha \equiv \frac{n(p-2)}{2p} \leq 1 \,,$$

which implies that

$$p \leq \frac{2n}{(n-2)} \,.$$

It follows that the relation pair

$$p < \frac{2n}{(n-2)} \,, \qquad p = \frac{2n}{(n-2)} \,,$$

is associated with the pair

$$\alpha < 1 \,, \qquad \alpha = 1 \,,$$

respectively.

Suppose first that $0 < \alpha < 1$, and thus that $p < 2n/(n-2)$. Then we can appeal to Young's inequality to show that

$$\|\nabla \phi\|_2^{2\alpha} \, \|\phi\|_2^{2(1-\alpha)} \leq \alpha \epsilon^{\frac{1}{\alpha}} \|\nabla \phi\|_2^2 + (1-\alpha) \epsilon^{-\frac{1}{(1-\alpha)}} \|\phi\|_2^2$$

for any choice of ϵ, $0 < \epsilon < \infty$. Thus

$$\|\phi\|_p^2 \leq \{(1/\sqrt{n})[(n-1)/(n-2)]\}^{2\alpha} \left[\alpha \epsilon^{\frac{1}{\alpha}} \|\nabla \phi\|_2^2 + (1-\alpha) \epsilon^{-\frac{1}{(1-\alpha)}} \|\phi\|_2^2 \right] .$$

We can recast this expression in a more transparent form as

$$\left(\int |\phi|^p \, d^n x\right)^{2/p} \leq \delta \int (\nabla\phi)^2 \, d^n x + M^2(\delta)\int \phi^2 \, d^n x$$

where

$$\delta \equiv \{(1/\sqrt{n})[(n-1)/(n-2)]\}^{2\alpha} \, \alpha \, \epsilon^{\frac{1}{\alpha}} \, ,$$
$$M^2(\delta) \equiv \{(1/\sqrt{n})[(n-1)/(n-2)]\}^{2\alpha} \, (1-\alpha) \, \epsilon^{-\frac{1}{(1-\alpha)}} \, .$$

The notation is meant to suggest that when $p < 2n/(n-2)$ the interaction term may be bounded by a nonzero but *arbitrarily small* coefficient (δ) of the gradient term at the expense of choosing a large mass (M) term.

If $p = 2n/(n-2)$, on the other hand, then $\alpha = 1$ and consequently

$$\|\phi\|_p^2 \leq \{(1/\sqrt{n})[(n-1)/(n-2)]\}^2 \, \|\nabla\phi\|_2^2$$

to which we add a small term involving a mass (μ). Thus we see that

$$\left(\int |\phi|^p \, d^n x\right)^{2/p} \leq C_n \int (\nabla\phi)^2 \, d^n x + \mu^2 \int \phi^2 \, d^n x \, ,$$

where $\mu^2 > 0$ is as small as desired and

$$C_n \equiv \left[\frac{1}{\sqrt{n}}\frac{(n-1)}{(n-2)}\right]^2 \, ,$$

all of which holds when

$$p = \frac{2n}{(n-2)} \, .$$

Furthermore we observe that the C_n are monotonically decreasing as n becomes larger, i.e., $C_n \leq C_3$, and from the value of C_3 we see that

$$C_n \leq \frac{4}{3} \, , \qquad n \geq 3 \, .$$

Thus by choosing $\mu^2 = 4/3$, and noting for $\epsilon = 1$ that $\delta = C_n^\alpha \alpha \leq 4/3$ and $M^2(\delta) = C_n^\alpha(1-\alpha) \leq 4/3$, it follows that we can assert that

$$\left(\int |\phi|^p \, d^n x\right)^{2/p} \leq (4/3)\int [(\nabla\phi)^2 + \phi^2] \, d^n x$$

holds for all $n \geq 3$ whenever

$$p \leq \frac{2n}{n-2} \, .$$

The bounds described above all have the effect of stating that the interaction action is controlled by the free action whenever $p \leq 2n/(n-2)$; finiteness of the free action implies finiteness of the interaction action. The condition on p for which this condition holds is especially interesting since, according to renormalized perturbation theory for a massive covariant scalar quantum field – as discussed in Chapter 5 – the interval $p < 2n/(n-2)$ coincides *exactly* with the superrenormalizable models and the situation $p = 2n/(n-2)$ coincides *exactly* with the strictly renormalizable models. Furthermore, we reemphasize the additional fact that for the strictly renormalizable models there is a *minimum* coefficient required for the gradient term, while for the superrenormalizable models this coefficient can be made as small as desired. This behavior suggests a possible connection with the need for field-strength renormalization in the strictly renormalizable models, i.e., a need for renormalization of the gradient term itself – a need which is absent in the case of all superrenormalizable models.

Exercises

8.1 Make a sketch of several paths for the imaginary-time propagator as a path integral for the pseudofree harmonic oscillator that shows how various paths cancel leading to the formula

$$P_0'(x'', T; x', 0) \equiv \theta(x''x')[P_0(x'', T; x', 0) - P_0(x'', T; -x', 0)] \ .$$

8.2 Given the definition of local time

$$t^*(y) \equiv \int_0^T \delta(x(u) - y) \, du$$

expressed in terms of a Brownian motion path $x(u)$, $0 \leq u \leq T$, determine, for some c, $0 < c < \infty$, that

$$\int [t^*(y) - t^*(z)]^2 \, d\mu(x) \leq c|y - z| \ , \qquad |y - z| < 1$$

with the help of a representation of the δ-function by its Fourier transform, where $\int (\cdot) \, d\mu(x)$ denotes an average in Wiener measure.

8.3 For $m > 0$ and $k \in \{1, 2, 3, \ldots\}$, let

$$W_{0,k} \equiv \tfrac{1}{2} \int \{\Sigma_j [\partial_j^k \phi(x)]^2 + m^2 \phi(x)^2\} \, d^n x \ .$$

Consider the functional integral

$$C_k\{h\} \equiv N_k \int e^{i \int h(x)\phi(x) \, d^n x} e^{-W_{0,k}(\phi)} \, \mathcal{D}\phi$$

$$= \int e^{i \int h(x)\phi(x) \, d^n x} \, d\mu_k(\phi) \ .$$

Discuss the support properties of the measures $\mu_k(\phi)$ as a function of the positive integer k (see, e.g., [Hi 70]). In particular, show that as k increases, the distributions in the support of the measure become increasingly smoother in some sense. For what value of k does the underlying stochastic process change from a generalized process that is not point-wise defined to a process that is point-wise defined? Compare the support properties of the measure μ_k with the criterion of finiteness of the Euclidean action $W_{0,k}$.

8.4 For $n \geq 3$, consider the Sobolev inequality

$$[\int |\phi(x)|^p \, d^n x]^{2/p} \leq (4/3) \int \{[\nabla \phi(x)]^2 + \phi(x)^2\} \, d^n x \,,$$

proved to hold when $p \leq 2n/(n-2)$ for functions $\phi \in C_0^\infty$. Discuss the extension of this inequality to more general functions based on the use of suitably convergent Cauchy sequences.

9

Independent-Value Models

WHAT TO LOOK FOR

Some model problems are more singular than others, and in this chapter we examine models which, in some sense, are the most singular of all. Conventional treatments are unsatisfactory, and without prejudice toward the formulation, we successfully appeal to symmetry and general principles and obtain a fully satisfactory answer that has all the features of a discontinuous perturbation alluded to in the previous chapter. As has been the norm for several chapters we choose units in which $\hbar = 1$.

9.1 Introduction

From a mathematical point of view, it is no exaggeration to claim that the principal equation to analyze for a (Euclidean) self-interacting scalar quantum field theory is the formal functional integral given for a ϕ_n^p (p even) interaction by the now familiar formula

$$\mathcal{N} \int \exp[\int (ih(x)\phi(x) - \tfrac{1}{2}\{[\nabla\phi(x)]^2 + m^2\phi(x)^2\} - g\phi(x)^p)\, d^n x]\, \mathcal{D}\phi \,,$$

where we have introduced an imaginary source term for later convenience. As customary, we assume this expression is normalized so that for $h = 0$ it is unity. Despite our good intentions, this expression is not well-defined as it stands, and to proceed further it is necessary to give it meaning. This is the role of regularization and renormalization; however, these twin aspects have many potential realizations, and for a number of theories we are still looking for a suitable realization. In this chapter we study a different and noncovariant mathematical model with the aim of sharpening our tools and instincts for the more challenging covariant problems.

The independent-value models may be characterized most simply as covariant models in which the gradient term has been dropped. In other words, the model we study in this chapter is characterized by the formal functional integral

$$\mathcal{N} \int \exp\{\int [ih(x)\phi(x) - \tfrac{1}{2}m^2\phi(x)^2 - g\phi(x)^p]\, d^n x\}\, \mathcal{D}\phi \,,$$

which differs from the previous equation by the absence of the term $-\tfrac{1}{2}[\nabla\phi(x)]^2$. No physical justification is offered for discarding the gradient terms; it is simply a mathematical model useful to discuss the problems encountered in functional integrals of this type. If the presence of gradients tends to soften the distributional character of the fields involved in the functional integral, as we have stressed in the preceding chapter, then their absence in the independent-value models implies that for the models considered in this chapter the nature of the fields involved is even more unruly than in the covariant model with the gradients present. This remark is only to say that by giving up the gradients we haven't made the problem easier for ourselves from the point of view of the distributional nature of the fields; if anything, we have made it harder for ourselves. If we are able to solve this more singular model, then it stands to reason that a solution for the less singular covariant model should also exist. Nothing is proved by these remarks, of course, but the implication seems reasonable enough.

Although the fields are formally more singular without the gradients, the advantage that we have won involves a great increase in the symmetry of the model, a symmetry we will soon exploit to our advantage. But before we outline our solution to these models, let us pursue the consequences of a conventional approach. Since we have no gradients, and in particular no gradient term that could serve as a time derivative in going from Euclidean to Minkowski space, we must confine our "conventional approach" to these models to what would be done simply on the basis of a functional integral analysis of the problem. However, we start even more simply, namely with the noninteracting, i.e., free independent-value model.*

Free independent-value model

The free independent-value model is expressed in terms of the formal functional integral

$$C_0(h) \equiv \mathcal{N} \int \exp\{\int [ih(x)\phi(x) - \tfrac{1}{2}m_0^2\phi(x)^2]\, d^n x\}\, \mathcal{D}\phi \,,$$

where $\mathcal{N}$ is chosen so that $C_0(0) = 1$. We adopt the natural interpretation of this expression as a formal, mean zero, Gaussian functional integral, the evaluation of which may be readily obtained. Based on our earlier discussions, the answer

---

* Our discussion in this chapter is largely based on [Kl 75a].

is given by

$$C_0(h) = \exp[-\tfrac{1}{2}m_0^{-2}\int h(x)^2\,d^n x]\ .$$

Such an expression is well defined for all $h \in L^2(\mathbb{R}^n)$ and for no other h; in particular, we cannot introduce the sharp-time independent-value free field because to do so would require that $h(t, \mathbf{x}) = \delta(t)\,f(\mathbf{x})$ was an acceptable test function. We can also obtain the same answer for $C_0(h)$ by means of a natural lattice formulation in which we interpret the formal functional integral as

$$C_0(h) = \lim_{a \to 0} N_0 \int \exp\{\Sigma[ih_k\phi_k a^n - \tfrac{1}{2}m_0^2\phi_k^2 a^n]\}\,\Pi\,d\phi_k\ .$$

Here a is the lattice spacing, and $k = (k^0, \ldots, k^{n-1})$, $k^j \in \{0, \pm 1, \ldots\}$, labels a lattice site. Since the integrals are independent of each other the result is given quite simply by

$$C_0(h) = \lim_{a \to 0} \exp[-\tfrac{1}{2}m_0^{-2}\Sigma\,h_k^2\,a^n] = \exp[-\tfrac{1}{2}m_0^{-2}\int h(x)^2\,d^n x]\ .$$

Thanks to completing the square, it is worth emphasizing that the only integral encountered in this computation was the simple Gaussian integral $I = \int \exp(-x^2/2)\,dx$, which, as well known, is evaluated by the "trick" of considering

$$I^2 = \int e^{-(x^2+y^2)/2}\,dx\,dy = \int e^{-r^2/2}\,r\,dr\,d\theta = -2\pi\int de^{-u} = 2\pi\ ,$$

in which we emphasize that no "real" integral is involved. In fact, from an analytical point of view, no one ever directly "computes" an integral; integrals are evaluated by indirect means!

9.2 Conventional viewpoint

We may study the functional integral for interacting independent-value models by two different approaches. As attempts to give meaning to the formal functional integral, we can, on the one hand, use perturbation theory, while, on the other hand, we can use a straightforward lattice regularization. Let us look at both approaches.

Perturbation theory

For the sake of the present discussion we assume that the interaction power $p = 4$ so that we are dealing with a quartic independent-value model; other even powers may be discussed in a similar fashion. The perturbation theory for such a model can be carried out in much the same way as was outlined in Chapter 5 for the covariant quartic interaction, and the contributions may be catalogued by means of a graphical identification that is – from a graphical point of view at

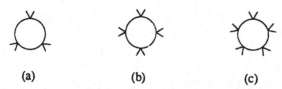

Fig. 9.1. Several one-loop graphs contributing, respectively, to the (a) six-point, (b) eight-point, and (c) ten-point amplitudes.

least – identical to that for the covariant case. What changes is the contribution of the individual graphs, and, in particular, the form of the propagator. Just as in the covariant case we assume that the interaction is normal ordered which reduces the number of graphs that must be considered.

In Fig. 9.1 we display several one-loop graphs that arise for quartic models, be they covariant or independent-value (or even something else as well). For the independent-value models the propagator in real space is given by

$$\Delta_0(x - y) = \frac{1}{m_0^2}\,\delta(x - y) = \frac{1}{(2\pi)^n m_0^2}\int e^{ip\cdot(x-y)}\,d^n p\;,$$

while for the relativistic (Euclidean) field the propagator reads

$$\Delta_E(x - y) = \frac{1}{(2\pi)^n}\int \frac{e^{ip\cdot(x-y)}}{p^2 + m^2}\,d^n p\;;$$

the difference in these two expressions is just the absence or presence of the contribution from the gradient term, respectively. For either model the contribution of the six-, eight-, and ten-point graphs of Fig. 9.1 effectively involves, respectively, the expressions

$$g^3\int_C \Delta(x - y)\Delta(y - z)\Delta(z - x)\,d^n x\,d^n y\,d^n z\;,$$

$$g^4\int_C \Delta(w - x)\Delta(x - y)\Delta(y - z)\Delta(z - w)\,d^n w\,d^n x\,d^n y\,d^n z\;,$$

$$g^5\int_C \Delta(v - w)\Delta(w - x)\Delta(x - y)\Delta(y - z)\Delta(z - v)\,d^n v\,d^n w\,d^n x\,d^n y\,d^n z\;,$$

integrated *not* over all of $\mathbb{R}^n$ but only over a compact region C, for instance, due to the implicit presence of suitable test functions. For the independent-value models the result for the three examples is given, respectively, by

$$g^3 m_0^{-6}\int_C \delta(x - x)\,d^n x \simeq g^3 m_0^{-6}\Lambda^n\;,$$

$$g^4 m_0^{-8}\int_C \delta(x - x)\,d^n x \simeq g^4 m_0^{-8}\Lambda^n\;,$$

$$g^5 m_0^{-10}\int_C \delta(x - x)\,d^n x \simeq g^5 m_0^{-10}\Lambda^n\;,$$

where on the right side we have indicated the result of a high momentum cutoff (Λ) for which in each case the result is proportional to Λ^n. On the other hand, in the covariant case the result of a similar calculation leads, qualitatively speaking, to factors of the order of

$$g^3 \Lambda^{n-6} , \qquad g^4 \Lambda^{n-8} , \qquad g^5 \Lambda^{n-10} ,$$

a set of relations which should be interpreted just as they read if the exponent of Λ is positive, interpreted as a logarithm if the exponent is zero, and interpreted as a finite number if the exponent is negative. In short, in the covariant case, as the number of interactions is increased in a one-loop diagram, the divergence grows ever weaker, eventually becoming nondivergent for sufficiently many interactions for any space-time dimension. For the independent-value models, on the other hand, the contributions of a one-loop graph are all equally divergent, no matter how many interaction terms are present. To renormalize such divergences requires a counterterm for each one that corresponds to a local interaction of a power equal to the number of external legs. For the covariant case, if $n = 4$, for example, the only graph of this kind that is divergent has four legs, and we already have a local fourth-power interaction; that is the strictly renormalizable case. For the independent-value case, if $n = 4$ again, then each of these graphs is divergent, and a local counterterm is needed for each one (e.g., ϕ_4^6, ϕ_4^8, ϕ_4^{10}, etc.), which means that the independent-value models are perturbatively nonrenormalizable. Clearly, the same conclusion for the independent-value models holds for *any* space-time dimension $n \geq 1$; each and every one of them is perturbatively nonrenormalizable thanks to our dropping of all the gradient terms.

Even more can be said if we look at selected infinite-order graph resummations. Consider the graphs illustrated in Fig. 5.2. For models with a quartic interaction the contribution to the amplitude of the set of graphs illustrated in Fig. 5.2 is given by $F(w, x; y, z)$ which is a solution to the integral equation

$$F(w, x; y, z) = g \int \Delta(w - u)\Delta(x - u)\Delta(y - u)\Delta(z - u)\, d^n u$$

$$- g \int \Delta(w - u)\Delta(x - u)\, F(u, u; y, z)\, d^n u .$$

For the independent-value models this equation reduces to

$$F(w, x; y, z) = g m_0^{-8} \delta(w - x)\delta(x - y)\delta(y - z)$$
$$- g m_0^{-4} \delta(w - x) F(w, x; y, z) ,$$

which with the same high momentum cutoff as before, i.e., $\delta(0) \to \Lambda^n$, leads to the solution

$$F_\Lambda(w, x; y, z) = \frac{g m_0^{-8}}{1 + g m_0^{-4} \Lambda^n} \delta(w - x)\delta(x - y)\delta(y - z) .$$

In summary, the contribution of this set of graphs has the form of the first contribution itself with a renormalized coefficient. And, moreover, as the cutoff is removed, i.e., as $\Lambda \to \infty$, it follows that the contribution of this particular set of graphs *vanishes!* It is not difficult to imagine other sets of graphs which, combined with analogously iterated sets, also lead to a vanishing contribution for the renormalized four-point coupling constant. Thus this argument, often referred to in the literature as due to a Landau pole (or Landau ghost) [Zi 96], suggests that far from being nonrenormalizable the theory is actually trivial. Support for this conclusion also follows from a lattice approach to the functional integral.

Lattice regularization for the functional integral

Let us focus on the quartic interaction and consider the conventional lattice approximation to the functional integral for the independent-value models. In particular, let us consider

$$C(h) \equiv N_0 \int \exp[i\Sigma \, h_k \phi_k \Delta - \tfrac{1}{2} m_0^2 \Sigma \, \phi_k^2 \Delta - g_0 \Sigma \, \phi_k^4 \Delta] \, \Pi d\phi_k \; ,$$

where $\Delta \equiv a^n$, the lattice cell volume, and N_0 is chosen so that $C(0) = 1$. Implicitly, we let both m_0 and $g_0 \geq 0$ depend on the cutoff, specifically on Δ, and we seek to choose these parameters in such a way as the cutoff is removed, i.e., as $\Delta \to 0$, that $C(h)$ converges to something of interest. Fortunately, this problem is sufficiently simple that we can analyze the continuum limit in its entirety.

First, since there are no gradients there are no correlations between the field ϕ_k at one lattice site and the field $\phi_{k'}$ at any other lattice site, $k \neq k'$. Thus we can rewrite

$$C(h) = \Pi_k \, N \int \exp(ih_k \phi \Delta - \tfrac{1}{2} m_0^2 \phi^2 \Delta - g_0 \phi^4 \Delta) \, d\phi$$

$$= \Pi_k \, N' \int \exp(ih_k u - \tfrac{1}{2} m_0^2 u^2 \Delta^{-1} - g_0 u^4 \Delta^{-3}) \, du$$

$$\equiv \Pi_k \, \langle \exp(ih_k u) \rangle \; ,$$

where we have introduced the change of variables $\phi \equiv u/\Delta$ and expressed the result in terms of a self-evident average in a Δ-dependent probability distribution. Assuming the necessary moments to exist – which only amounts to having $g_0 > 0$, or when $g_0 = 0$ having $m_0^2 > 0$ – it follows that

$$\Pi_k \langle \exp(ih_k u) \rangle = \Pi_k [1 - \tfrac{1}{2} h_k^2 \langle u^2 \rangle + \tfrac{1}{24} h_k^4 \langle u^4 \rangle - \cdots] \; .$$

In order for this expression to be well defined in the continuum limit it is necessary that $\langle u^2 \rangle \propto \Delta$. Intuitively, this means that the distribution for u is narrow and highly peaked; this picture holds whether $m_0^2 \geq 0$ and so the distribution is singly peaked, or $m_0^2 < 0$ and so the distribution is doubly peaked. By the very nature of a narrow distribution it follows that $\langle u^4 \rangle \propto \Delta^2$, and indeed that $\langle u^{2r} \rangle \propto \Delta^r$ for all $r = 1, 2, 3, \ldots$. As a consequence, besides the factor 1, only the next term in the series, proportional to $\langle u^2 \rangle$, contributes and leads to

$$C\{h\} \equiv \lim_{\Delta \to 0} C(h)$$
$$= \exp[-\tfrac{1}{2} \lim_{\Delta \to 0} \Sigma h_k^2 \langle u^2 \rangle]$$
$$= \exp[-\tfrac{1}{2} A \int h(x)^2 \, d^n x] \,,$$

where $A \equiv \lim_{\Delta \to 0} \langle u^2 \rangle / \Delta$. The limit for A necessarily satisfies the condition $0 \le A \le \infty$; we reject those parameterizations that lead to the values $A = 0$ or $A = \infty$ and retain only those choices for which $0 < A < \infty$. Explicitly, and with $v = u\sqrt{\Delta}$,

$$A = \lim_{\Delta \to 0} \frac{\int v^2 \exp(-\tfrac{1}{2} m_0^2 v^2 - g_0 v^4 \Delta^{-1}) \, dv}{\int \exp(-\tfrac{1}{2} m_0^2 v^2 - g_0 v^4 \Delta^{-1}) \, dv} \,,$$

an expression that makes it clear that g_0 may contribute to the final result for A (e.g., if $g_0 \propto \Delta$). Although the coefficient g_0 of the quartic interaction may contribute to the covariance parameter A, it can not resist the overwhelmingly attractive power of the central limit theorem [Hi 70] in determining that the ultimate result is Gaussian and thus equivalent to some particular free theory.

In summary, we have shown that a perturbation theory leads to a nonrenormalizable theory while a continuum limit of a natural and straightforward lattice formulation leads to a free ($\equiv$ Gaussian) theory, or as one often terms such a result, a "trivial" theory. In our opinion *both* of these results are unsatisfactory, and the purpose of the next section is to show how the functional integral may be alternatively defined and evaluated so as to yield a *nontrivial* result.

9.3 Nontrivial solution and its properties

The characterization of the independent-value models is complete if we are able to evaluate the functional integral

$$E\{h\} = \mathcal{N} \int \exp[i \int h\phi \, d^n x - \tfrac{1}{2} m_0^2 \int \phi^2 \, d^n x - g_0 \int \phi^p \, d^n x] \, \mathcal{D}\phi$$

where p is even, $\mathcal{N}$ is chosen to ensure the normalization $E\{0\} = 1$, and as before m_0^2 and g_0 are bare parameters chosen so that the generating functional E is well defined. However, the rest of the expression is formal indeed, e.g., what does ϕ^2 mean, what does ϕ^p mean, and especially what does $\mathcal{D}\phi$ mean? Although such questions seem to be out of place here, they are in an important sense the Main Questions. It is *absolutely imperative* that we keep an open mind regarding these issues and be guided by the solution as it develops. If too many preconceptions are introduced without justification, the flexibility needed to realize a nontrivial solution will be lacking.

We shall first present the solution itself before undertaking the task of its derivation in the next section. The result of the functional integral for $E\{h\}$ is given by

$$E\{h\} = \exp(-\tfrac{1}{2}b\int d^n x \int \{1 - \cos[uh(x)]\}e^{-\frac{1}{2}bm^2u^2 - gb^{p-1}u^p}\, du/|u|) \,,$$

where b is an arbitrary positive constant with the dimensions (length)$^{-n}$ which will be needed in constructing the solution. Units may always be chosen in which the parameter $b = 1$, in which case

$$E\{h\} = \exp(-\tfrac{1}{2}\int d^n x \int \{1 - \cos[uh(x)]\}e^{-\frac{1}{2}m^2u^2 - gu^p}\, du/|u|) \,,$$

and this has commonly been the case in the prior literature regarding these models [Kl 75a, KlN 76a, KlN 76b]. However, it is useful to retain the explicit dependence on b. The finite parameters m^2 and g are related to their bare counterparts by infinite, multiplicative renormalizations formally defined by

$$m_0^2 = bm^2/\delta(0) \,,$$
$$g_0 = b^{p-1}g/\delta(0)^{p-1} \,,$$

and thus thanks to b, m^2 and g carry the same dimensions as m_0^2 and g_0, respectively.

Pseudofree solution

It is evident that as $g \to 0$ *all* of the interacting theory solutions (i.e., all p) pass to the *pseudofree* solution characterized by the generating functional

$$E_0'\{h\} \equiv \exp(-\tfrac{1}{2}b\int d^n x \int \{1 - \cos[uh(x)]\}e^{-\frac{1}{2}bm^2u^2}\, du/|u|) \,,$$

which may be given a convergent power series representation, say for $h \in C_0^\infty$, in the form

$$E_0'\{h\} = \exp\{-m^{-2}\Sigma_{l=1}^\infty (-2/bm^2)^{l-1}[(l-1)!/(2l)!] \int h(x)^{2l}\, d^n x\} \,.$$

We refer to this solution as a pseudofree solution since it is different from the free solution

$$E_0\{h\} = \exp[-\tfrac{1}{2}m_0^{-2}\int h(x)^2\, d^n x]$$

whatever the chosen value for the bare mass m_0. Observe that the pseudofree solution does *approximately* reduce to the free solution with $m_0 = m$ when b is regarded as a suitably large parameter, e.g., for C_0^∞ functions such that

$$\sup_x h(x)^2 \ll bm^2 \,.$$

The property that *all* models with local nonlinear interactions are continuously connected to the pseudofree theory can be given a pictorial representation as shown in Fig. 9.2.

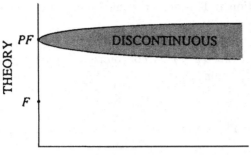

COUPLING CONSTANT

Fig. 9.2. Graphical representation of Theory vs. Coupling Constant for the independent-value models. For zero coupling there are two theories shown, F for free and PF for pseudofree. Every local nonlinear interacting theory represents a discontinuous perturbation of the free theory; hence, once the coupling is turned on, the theory jumps to a curve in the shaded region and stays there as the coupling constant increases. As that coupling is decreased to zero, the theory passes, by continuity, to the pseudofree theory. (Of course, models exist that are continuously connected to the free theory, but none of them involve local interactions.)

9.4 Derivation of the solution for independent-value models

In the absence of any gradients it is evident that the field $\phi(x)$ is statistically independent at every point $x \in \mathbb{R}^n$. This property implies that these models exhibit independent values at each point x, and suggests – as we have already done – that we call these models "independent-value models".

The symmetry implicit in this statistically independent behavior is captured by the assertion that for some function L the result of such a functional integral must have the form

$$E\{h\} \equiv e^{-\int L[h(x)]\,d^n x} \, ,$$

normalized so that $L[0] = 0$. As the characteristic functional of a probability distribution, it is necessary that $E\{th\}$ must be continuous in the real scale factor t. To achieve this continuity it is necessary and sufficient that L be a continuous function. We insist that acceptable renormalizations do not destroy the basic statistical independence of the field at distinct points. Moreover, we expect that renormalizations still lead to the fact that $E\{h\}$ is the Fourier transform of a probability measure, and therefore $E\{h\}$ is a positive-definite functional which satisfies the property that

$$\sum_{j,k=1}^{N} \alpha_j \alpha_k^* E\{h_j - h_k\} \geq 0$$

for any choice of the test functions $\{h_j\}$, complex numbers $\{\alpha_j\}$, and $N < \infty$. Indeed, since $E\{h\}$ is normalized, $E\{0\} = 1$, the functional $E\{h\}$ is a characteristic functional. This property enables us to develop a useful canonical form for the function L.

Let us specialize to functions h of the form $h(x) \equiv s\chi_\Upsilon(x)$, where $\chi_\Upsilon(x) \equiv 1$ if $x \in \Upsilon \subset \mathbb{R}^n$ and $\chi_\Upsilon(x) \equiv 0$ if $x \notin \Upsilon$. For compact Υ we also set $\Delta \equiv \int \chi_\Upsilon(x)\,d^n x$. It follows for such functions that

$$E\{h\} \to E_\Delta(s) \equiv e^{-\Delta L[s]}$$

which is still a normalized, positive-definite function on $s \in \mathbb{R}$ such that

$$\sum_{j,k=1}^{\infty} \alpha_j \alpha_k^* E_\Delta(s_j - s_k) \geq 0 .$$

According to Bochner's Theorem [Lu 70], such a function is the Fourier transform of a positive, normalized (probability) measure,

$$E_\Delta(s) = \int e^{isu}\,d\mu_\Delta(u)$$

for each Δ, $0 < \Delta < \infty$. Consequently, we can represent L as

$$\begin{aligned}
L[s] &= \lim_{\Delta \to 0} \Delta^{-1}(1 - e^{-\Delta L[s]}) \\
&= \lim_{\Delta \to 0} \int (1 - e^{isu})\Delta^{-1}\,d\mu_\Delta(u) .
\end{aligned}$$

The general form for this limit gives the desired canonical form for L.

To find such a limit let us introduce $d\sigma_\Delta(u) = \Delta^{-1}d\mu_\Delta(u)$ and rewrite the necessary expression as

$$\begin{aligned}
\int [1 - e^{isu} &+ isu/(1 + u^2)]\,d\sigma_\Delta(u) - is\int [u/(1 + u^2)]\,d\sigma_\Delta(u) \\
&= \int_{|u|>\Delta}[1 - e^{isu} + isu/(1 + u^2)]\,d\sigma_\Delta(u) \\
&\quad + \int_{|u|\leq\Delta}[1 - e^{isu} + isu/(1 + u^2)]\,d\sigma_\Delta(u) \\
&\quad - is\int [u/(1 + u^2)]\,d\sigma_\Delta(u) .
\end{aligned}$$

In the limit that $\Delta \to 0$ – a limit which of course is assumed to exist – the first term retains its form, the second term passes to a nonnegative multiple of s^2, while the third term passes to a multiple of is. Thus the general form for L is given by

$$L[s] = -ias + cs^2 + \int_{|u|>0}[1 - e^{isu} + isu/(1 + u^2)]\,d\sigma(u) ,$$

where $-\infty < a < \infty$, $c \geq 0$, and $d\sigma \geq 0$ is a nonnegative measure with the property that

$$\int_{|u|>0}[u^2/(1 + u^2)]\,d\sigma(u) < \infty .$$

This representation formula for $L[s]$ is known as the Levy canonical representation [Lu 70], and it is a fundamental result in the theory of infinitely-divisible characteristic functions, i.e., those characteristic functions $C(s)$ for which $C(s)^\alpha$ for all positive α, $0 < \alpha < \infty$, are also characteristic functions; see Section 6.7.

We are particularly concerned about representations for $L[s]$ in the case of even potentials for which $L[-s] = L[s]$ (nonsymmetric potentials are considered in Exercise 9.1), and thus

$$L[s] = cs^2 + \int_{|u|>0}[1 - \cos(su)]\, d\sigma(u) .$$

In that case

$$E\{h\} = \exp\{-c\int h^2\, d^n x - \int d^n x \int_{|u|>0}[1 - \cos(uh)]\, d\sigma(u)\}$$

where the conditions given earlier for c and $d\sigma$ still hold.

As the product of two factors the field ϕ may be written as

$$\phi(x) = \phi_G(x) + \phi_P(x)$$

implying a split of the field into two statistically independent components, one referring to the Gaussian component (G), and the other referring to the Poisson component (P). The Gaussian portion is appropriate to the free field solution as we have previously noted; let us then turn our attention exclusively to the Poisson component assuming hereafter that $c \equiv 0$. Moreover, we shall further assume that σ is absolutely continuous and therefore there exists a real function $C(u)$ – hereafter called the "model function" – such that

$$d\sigma(u) = C(u)^2\, du$$

and for which

$$\int [u^2/(1 + u^2)]\, C(u)^2\, du < \infty .$$

In addition we assume that $1/C(u)^2 < \infty$ everywhere on the real line. These assumptions impose no real restrictions since any other measure σ could be approached as a weak limit from the given class of absolutely continuous measures of the form $C(u)^2\, du$.

Thus we focus our attention on the putative relationship

$$E\{h\} = \exp(-\int d^n x \int \{1 - \cos[uh(x)]\}\, C(u)^2\, du)$$
$$= \mathcal{N} \int \exp[i\int h(x)\phi(x)\, d^n x - \tfrac{1}{2}m_0^2\int \phi(x)^2\, d^n x - g_0\int \phi(x)^p\, d^n x]\, \mathcal{D}\phi .$$

Our goal is to link the model function $C(u)$ to the specific form of the potential in the formal path integral. For this purpose we shall appeal to a Hilbert space realization of the field ϕ.

Operator realization

We have discussed the Fock representation of field operators in Chapters 5 and 6, and once again we have an occasion to appeal to that important representation.

Let $A(x, u)$ and $A(x, u)^\dagger$ denote a family of annihilation and creation operators, respectively, which satisfy the canonical commutation relation

$$[A(x, u), A(y, v)^\dagger] = \delta(x - y)\delta(u - v) ,$$

where $x, y \in \mathbb{R}^n$ and $u, v \in \mathbb{R}$. We introduce the no-particle state $|0\rangle$ in the Hilbert space and insist that

$$A(x, u) |0\rangle = 0$$

hold for all $(x, u) \in \mathbb{R}^{n+1}$. With the further assumption that, up to a factor, $|0\rangle$ is the only state annihilated by all the operators $A(x, u)$, we have identified the representation of A and $A^\dagger$ as a Fock representation. The Hilbert space is built by repeated application of the creation operator $A(x, u)^\dagger$ for different points (x, u) on the no-particle state. In fact, it is worthwhile to recall more precisely what we have previously shown about annihilation and creation operators for fields. As we have stressed in Chapter 6, although $A(x, u)$ is an operator for each point $(x, u) \in \mathbb{R}^{n+1}$, its adjoint, $A(x, u)^\dagger$, has only the zero vector in its domain. Consequently, it is necessary to smear the creation operator with a function ψ from $L^2(\mathbb{R}^{n+1})$ in order to have a proper operator. Let such a smeared operator be called

$$A^\dagger(\psi) \equiv \int \psi(x, u) \, A(x, u)^\dagger \, d^n x \, du .$$

Then the Hilbert space is spanned by the set of vectors

$$|\psi\rangle = e^{A^\dagger(\psi)} |0\rangle$$

which are just the unnormalized coherent states for the problem at hand. We further recall that the coherent states are eigenvectors for the A operators, and in particular, for continuous ψ,

$$A(x, u)|\psi\rangle = \psi(x, u) |\psi\rangle .$$

The Fock operators will be the basic tool once again in the study of the independent-value models!

Let $C(u)$ denote a real, even $[C(-u) = C(u)]$ nowhere vanishing function for which $\int [u^2/(1 + u^2)]C(u)^2 \, du < \infty$, and which will become the model function referred to above for the functional integral we intend to evaluate. Then it follows that the field operator of interest to us is given by

$$\begin{aligned}
\phi(x) = &\int A(x, u)^\dagger \, u \, A(x, u) \, du \\
&+ \int A(x, u)^\dagger \, u \, C(u) \, du \\
&+ \int C(u) \, u \, A(x, u) \, du .
\end{aligned}$$

Our next task is to verify this claim. To that end it is useful to introduce a pair of translated Fock operators given by

$$\begin{aligned}
B(x, u) &\equiv A(x, u) + C(u) , \\
B(x, u)^\dagger &\equiv A(x, u)^\dagger + C(u) ,
\end{aligned}$$

and to set

$$\int u\, C(u)^2\, du = 0$$

by symmetry, either as an absolutely convergent integral or otherwise as a principal value integral. Then we may also write

$$\phi(x) = \int B(x, u)^\dagger\, u\, B(x, u)\, du \ .$$

Observe that the operators B and $B^\dagger$ satisfy the same canonical commutation relations as do the operators A and $A^\dagger$, namely

$$[B(x, u), B(y, v)^\dagger] = \delta(x - y)\delta(u - v) \ .$$

However, the operators B and $B^\dagger$ are *not* Fock operators since there is no state annihilated by all the B that could serve as a no-particle state. Stated otherwise, the B and $B^\dagger$ operators are unitarily inequivalent to the operators A and $A^\dagger$, and in fact they are even *locally* inequivalent (i.e., within any region of finite, nonzero volume) to a Fock representation whenever the model function $C(u)$ is not square integrable, which in fact will be our ultimate conclusion.

By construction $\phi(x)$ is a local self-adjoint operator, and therefore

$$\exp[i\int h(x)\phi(x)\, d^n x]$$

is a unitary operator for a wide class of real test functions, such as, for example, all real $h \in C_0^\infty$. Now we note the important identity

$$\exp[i\int d^n x \int B(x, u)^\dagger\, u\, h(x)\, B(x, u)\, du]$$
$$= :\exp\{\int d^n x \int B(x, u)^\dagger[e^{iu\, h(x)} - 1]B(x, u)\, du\} : \ ,$$

where the pair of colons : : denotes normal ordering of the expression contained within, namely, all $A^\dagger$ to the left of all A without regard for the fact that these operators do not commute. We may verify this relation as follows. Let us introduce the abbreviation

$$(A, w\, A) \equiv \int A(x, u)^\dagger\, w\, A(x, u)\, d^n x\, du \ ,$$

where w denotes some self-adjoint operator, e.g., acting by multiplication and/or differentiation on functions over the space $\mathbb{R}^{n+1}$. Consider the diagonal, normalized (by a factor $N = \exp[-(\psi, \psi)/2]$) coherent-state matrix elements of the unitary operator generated by such an operator, namely

$$\langle\psi| \exp[i(A, w\, A)]|\psi\rangle = N \sum_{n=0}^{\infty} \langle\psi| \exp[i(A, w\, A)]\, [A^\dagger(\psi)]^n/(n!)\, |0\rangle$$
$$= N \sum_{n=0}^{\infty} \langle\psi| [A^\dagger(e^{iw}\psi)]^n/(n!)\, |0\rangle$$
$$= \langle\psi|e^{iw}\psi\rangle$$
$$= \exp[(\psi, [e^{iw} - 1]\, \psi)] \ .$$

Now from the definition of normally ordered products we also know that

$$\langle\psi| : \exp[(A, [e^{iw} - 1] A)] : |\psi\rangle = \exp[(\psi, [e^{iw} - 1]\psi)]$$

holds as well, and in view of the fact that the diagonal coherent state matrix elements determine such an operator uniquely, it follows that

$$\exp[i(A, w\, A)] \equiv\, : \exp[(A, [e^{iw} - 1] A)] : .$$

Since this relation holds for any Fock representation it holds for any operators that are unitarily equivalent to Fock operators, namely to the operators $A(x, u) \to \overline{A}(x, u) \equiv A(x, u) + \lambda(x, u)$ and $A(x, u)^\dagger \to \overline{A}(x, u)^\dagger \equiv A(x, u)^\dagger + \lambda(x, u)^*$ for any $\lambda \in L^2(\mathbb{R}^{n+1})$. In other words, we have the identity

$$\exp[i(\overline{A}, w\, \overline{A})] \equiv\, : \exp[(\overline{A}, [e^{iw} - 1] \overline{A})] : .$$

Although this relation holds whenever $\lambda \in L^2$, it can also be extended to additional $\lambda \notin L^2$. Approaching such a λ through a sequence, an acceptable extension holds provided that $\lambda(x, -u) = \lambda(x, u)$ and $uh(x)\lambda(x, u)/\sqrt{1 + u^2}$ remains square integrable. We choose a sequence such that $\lambda(x, u) \to C(u)$, in which case we need only require that $\int[u^2/(1 + u^2)]\, C(u)^2\, du < \infty$. Since this is our standard requirement on the model function, it follows that we have established the needed identity.

Now armed with this identity consider the expectation functional

$$\begin{aligned}
E\{h\} &= \langle 0| \exp[i\int h(x)\phi(x)\, d^n x] |0\rangle \\
&= \langle 0| \exp[i\int d^n x \int B(x, u)^\dagger\, uh(x)\, B(x, u)\, du] |0\rangle \\
&= \langle 0| : \exp\{\int d^n x \int B(x, u)^\dagger [e^{iuh(x)} - 1] B(x, u)\, du\} : |0\rangle \\
&= \exp\{-\int d^n x \int C(u)[1 - e^{iuh(x)}]\, C(u)\, du\} \\
&= \exp(-\int d^n x \int \{1 - \cos[uh(x)]\}\, C(u)^2\, du) .
\end{aligned}$$

In the last expression we have made use of the fact that C is an even function. With this relation we have established that

$$\phi(x) = \int B(x, u)^\dagger\, u\, B(x, u)\, du ,$$

and according to the GNS Theorem [Na 64, Em 72] any other realization of this local operator is unitarily equivalent to the one we have just developed. As was our intention, the resultant functional is exactly the characteristic functional for the independent-value models. Therefore we have achieved a valid realization of the independent-value model fields, and have done so in terms of a bilinear combination of translated Fock operators with which we shall find it convenient to work.

Local products

Every field operator representation determines its own rules for the definition of *local products*. One of the principal tools for the study of local products is the operator product expansion [ItZ 80], that is, the development of the product of two local operators into a series of local operators multiplied by c-number coefficients having distinct rates of divergence as the separation between the spatial arguments of the two local operators is reduced to zero. Let us examine such an expansion with the aid of the representation of the field operator given previously. To that end we note that

$$
\begin{aligned}
\phi(x)\,\phi(y) &= \int B(x,u)^\dagger\, u\, B(x,u)\, du \int B(y,v)^\dagger\, v\, B(y,v)\, dv \\
&= \int B(x,u)^\dagger\, u\, [B(x,u),\, B(y,v)^\dagger\,]\, v\, B(y,v)\, du\, dv \\
&\quad + \int B(x,u)^\dagger\, B(y,v)^\dagger\, u\, v\, B(x,u)\, B(y,v)\, du\, dv \\
&= \delta(x-y) \int B(x,u)^\dagger\, u^2\, B(x,u)\, du + :\, \phi(x)\,\phi(y)\, :\ .
\end{aligned}
$$

In the last expression we have achieved a decomposition in which the first term is a local operator multiplied by a singular c-number coefficient, i.e., $\delta(x-y)$, and the second term is likewise a local operator which is without singularity in the vicinity of $x = y$. In other words, the first term is the one with the most singular coefficient (indeed the only one) and is the term from which we extract the local product. However, before doing so we need to take precautions not to lose proper field dimensionality. Let us introduce a positive constant b having the dimensions $(\text{length})^{-n}$ so that

$$
\phi(x)\,\phi(y) = [b^{-1}\delta(x-y)]b\int B(x,u)^\dagger\, u^2\, B(x,u)\, du + :\, \phi(x)\,\phi(y)\, :
$$

In this form the singular prefactor in brackets has been rendered *dimensionless* and therefore its factor,

$$
\phi_R^2(x) \equiv b\int B(x,u)^\dagger\, u^2\, B(x,u)\, du\ ,
$$

has the square of the dimensions of ϕ, whatever they may be. It is not really critical what numerical value we choose for b so long as $0 < b < \infty$; what is most important to recognize is that *some* choice must be made if dimensions are to be preserved. Of course, one could always choose units such that the numerical value of b is unity, but such a choice is in no way required.

The formal way in which to extract the finite part of the local product is simply to divide by the singular c-number coefficient. Thus we formally have

$$
\phi_R^2(x) = [b^{-1}\delta(0)]^{-1}\,\phi(x)\,\phi(x) \equiv Z\,\phi(x)\,\phi(x) \equiv Z\,\phi(x)^2
$$

where Z denotes a formal "zero" factor, $Z = b/\delta(0)$, which extracts out the finite part of the local product. Note formally that the division by infinity has eliminated the remaining less singular terms, here simply $:\, \phi(x)\phi(y)\, :$. The proper way to extract the finite part is by means of a test function sequence.

Consider the operator

$$\phi_x = \Delta^{-1} \int_{z \in \Upsilon} \phi(x+z) \, d^n z$$

where Υ is a (hyper)cube defined by the requirement that each coordinate $|z^j| < a/2$, and $\Delta \equiv a^n$. The smeared field ϕ_x is an operator for all Δ, $0 < \Delta < \infty$, and it represents the average of the field at x over a (hyper)cubic cell of edge length a. From the previously given formula for $\phi(x)\phi(y)$ it follows that

$$\begin{aligned}
(b\Delta)\phi_x^2 &= (b\Delta)\Delta^{-2}[\int_{z \in \Upsilon} \phi(x+z) \, d^n z]^2 \\
&= \Delta^{-1} \int_{z \in \Upsilon} \phi_R^2(x+z) \, d^n z \\
&\quad + b\Delta^{-1} \int_{y,z \in \Upsilon} : \phi(x+y) \, \phi(x+z) : \, d^n y \, d^n z \; .
\end{aligned}$$

In the limit $\Delta \to 0$ only the first term on the right-hand side contributes since the second term is $O(\Delta)$. Thus we learn the result that

$$\begin{aligned}
\phi_R^2(x) &\equiv \lim_{\Delta \to 0} (b\Delta)[\Delta^{-1} \int_{z \in \Upsilon} \phi(x+z) \, d^n z]^2 \\
&= b \int B(x,u)^\dagger \, u^2 \, B(x,u) \, du \; .
\end{aligned}$$

Note that the subscript "R" denotes renormalized product.

The operation of forming the local product of two operators *each* of which is bilinear in creation and annihilation operators (i.e., $B^\dagger$ and B) has resulted in a local operator that is *also* bilinear in creation and annihilation operators. Consequently we can repeat the process as in the case

$$\phi(x) \, \phi_R^2(y) = [b^{-1}\delta(x-y)]b^2 \int B(x,u)^\dagger \, u^3 \, B(x,u) \, du + \; : \phi(x) \, \phi_R^2(y) : \; .$$

This formula leads to the result that

$$\phi_R^3(x) \equiv b^2 \int B(x,u)^\dagger \, u^3 \, B(x,u) \, du \equiv Z \, \phi(x) \, \phi_R^2(x) \; ,$$

where again $Z = b/\delta(0)$ formally. We may also define the renormalized cube by the relation

$$\begin{aligned}
\phi_R^3(x) &\equiv \lim_{\Delta \to 0} (b\Delta)^2 [\Delta^{-1} \int_{z \in \Upsilon} \phi(x+z) \, d^n z]^3 \\
&= b^2 \int B(x,u)^\dagger \, u^3 \, B(x,u) \, du \; .
\end{aligned}$$

Note that we have used the *same* factor b in maintaining dimensions in the cubic power as was used for the square. In principle, this equality of b factors is not forced on us here; but it does make the algebra of the local powers self consistent if we choose them to be the same. That is, by choosing the same b factor it follows that

$$\phi_R^3(x) = Z \, \phi(x) \, \phi_R^2(x) = Z^2 \, \phi(x) \, \phi(x) \, \phi(x) \equiv Z^2 \, \phi(x)^3$$

holds. Note that the local cube $\phi_R^3(x)$ is *again* bilinear in $B^\dagger$ and B. Higher-order local powers may be defined analogously by the formula

$$\phi_R^p(x) \equiv \lim_{\Delta \to 0} (b\Delta)^{p-1} [\Delta^{-1} \int_{z \in \Upsilon} \phi(x+z)\, d^n z]^p$$
$$= \lim_{\Delta \to 0} b^{p-1} \Delta^{-1} [\int_{z \in \Upsilon} \phi(x+z)\, d^n z]^p$$
$$= b^{p-1} \int B(x,u)^\dagger\, u^p\, B(x,u)\, du$$

which, with $\phi_R(x) \equiv \phi(x)$, holds for all $p \geq 1$. The use of one and the same b factor throughout leads to desirable algebraic relations of the kind

$$\phi_R^{p+m+l}(x) = Z\, \phi_R^{p+m}(x)\, \phi_R^l(x)$$
$$= Z\, \phi_R^p(x)\, \phi_R^{m+l}(x)$$
$$= Z^2\, \phi_R^p(x)\, \phi_R^m(x)\, \phi_R^l(x)\,,$$

where p, m, and l are each ≥ 1, and further relations of a similar nature hold as well. In each case Z is formally given by $b/\delta(0)$. *All* local powers are bilinear in $B^\dagger$ and B!

Adding an interaction

Based on the operator realization of the independent-value model fields and the definition of local products just completed, let us evaluate the expression

$$\langle 0| \exp[i \int h(x)\phi(x)\, d^n x - \int G(x)\phi_R^r(x)\, d^n x] |0\rangle$$

where $r \geq 2$ is an even integer and G is a nonnegative, bounded function of compact support. It follows that this expression is given by

$$\langle 0| \exp\{\int d^n x \int B^\dagger(x,u)[ih(x)u - G(x)b^{r-1}u^r] B(x,u)\, du\} |0\rangle$$
$$= \langle 0| : \exp\{\int d^n x \int B^\dagger(x,u)[e^{ih(x)u - G(x)b^{r-1}u^r} - 1] B(x,u)\, du\} : |0\rangle$$
$$= \exp\{-\int d^n x \int [1 - e^{ih(x)u - G(x)b^{r-1}u^r}]\, C(u)^2\, du\}\,.$$

We next seek to take a limit in which $G(x) \to K$ where K is a *positive constant*. As it stands the result will vanish in such a limit, but that is not too surprising, and what we need to do first is to *renormalize* the expression of interest so that it is a characteristic functional for all G. Then the desired limit may well exist, and if so it will correspond to a weak limit of the underlying probability measures.

Therefore, to study convergence of $G \to K$, we first pass to the renormalized expression

$$F_G = N_G \langle 0| e^{i \int h\phi\, d^n x - \int G\phi_R^r\, d^n x} |0\rangle$$
$$= \frac{\langle 0| e^{i \int h\phi\, d^n x - \int G\phi_R^r\, d^n x} |0\rangle}{\langle 0| e^{-\int G\phi_R^r\, d^n x} |0\rangle}$$
$$= \frac{\exp\{-\int d^n x \int [1 - e^{ihu - Gb^{r-1}u^r}]\, C(u)^2\, du\}}{\exp\{-\int d^n x \int [1 - e^{-Gb^{r-1}u^r}]\, C(u)^2\, du\}}$$
$$= \exp(-\int d^n x \int \{1 - \cos[uh(x)]\} e^{-G(x)b^{r-1}u^r}\, C(u)^2\, du)\,.$$

In this last expression we can readily take the limit $G \to K > 0$, and the result becomes

$$F_K = \lim_{G \to K} F_G$$

$$= \exp(-\int d^n x \int \{1 - \cos[uh(x)]\} \, e^{-Kb^{r-1}u^r} C(u)^2 \, du) \, .$$

The result we have obtained can also be restated formally as $(K_0 = Z^{r-1} K)$

$$F_K = \mathcal{N}' \int \exp[i\int h\phi \, d^n x - \tfrac{1}{2} m_0^2 \int \phi^2 \, d^n x - g_0 \int \phi^p \, d^n x - K_0 \int \phi^r \, d^n x] \, \mathcal{D}\phi \, ,$$

namely as the evaluation of the formal path integral in the presence of an *additional* interaction term. Therefore, a change of the Euclidean action such as

$$W \to W' = W + K_0 \int \phi^r \, d^n x$$

is accounted for by a change of the model function in the manner

$$C(u)^2 \to C'(u)^2 = e^{-Kb^{r-1}u^r} C(u)^2 \, .$$

Consequently, we can account for the *original* term $g_0 \int \phi^p \, d^n x$ in the action if we set

$$C(u)^2 \equiv e^{-gb^{p-1}u^p} C_0(u)^2 \, ,$$

where C_0 is the model function formally associated with the functional integral

$$E_0'\{h\} = \mathcal{N}_0' \int e^{i\int h\phi \, d^n x - \tfrac{1}{2} m_0^2 \int \phi^2 \, d^n x} \, \mathcal{D}\phi$$

$$= \exp(-\int d^n x \int \{1 - \cos[uh(x)]\} \, C_0(u)^2 \, du) \, .$$

We caution the reader to still keep an open mind and, despite appearances, *not* to jump to conclusions that the first line of this relation signifies a Gaussian integral.

Determination of the pseudofree model

In order to determine C_0 we may consider a *mass insertion*, that is, we consider again the introduction of an additional interaction, but this time we let $r = 2$. Consequently, we find that

$$F_K = \mathcal{N} \int e^{i\int h\phi \, d^n x - \tfrac{1}{2} m_0^2 \int \phi^2 \, d^n x - K_0 \int \phi^2 \, d^n x} \, \mathcal{D}\phi$$

$$= \exp\{-\int d^n x \int [1 - \cos(uh)] e^{-Kbu^2} C_0(u)^2 \, du\} \, .$$

However, we can calculate this expression another way when we observe that the two quadratic terms are of the *same type*. Thus, let us rescale ϕ in the functional integral according to the rule

$$\phi(x) \to \phi(x)/\sqrt{1 + 2K/m^2}$$

where we have used the relation $K_0/m_0^2 = K/m^2$. In that case

$$F_K = \mathcal{N} \int e^{i \int h\phi / \sqrt{1+2K/m^2} \, d^n x - \frac{1}{2} m_0^2 \int \phi^2 \, d^n x} \, \mathcal{D}\phi$$

where $\mathcal{N}$ has been adjusted to absorb any formal constant factor arising in the change of integration variables. But now we recognize that F_K is given alternatively by

$$F_K = \exp\{-\int d^n x \int [1 - \cos(uh/\sqrt{1+2K/m^2})] \, C_0(u)^2 \, du\} \,,$$

namely, the functional $E_0'\{h\}$ evaluated for a *rescaled argument*. Equality of the two expressions involved requires that

$$\int [1 - \cos(uh)] \, e^{-Kbu^2} \, C_0(u)^2 \, du$$
$$= \int [1 - \cos(uh/\sqrt{1+2K/m^2})] \, C_0(u)^2 \, du$$
$$= \sqrt{1+2K/m^2} \int [1 - \cos(uh)] \, C_0(\sqrt{1+2K/m^2} \, u)^2 \, du$$

holds for all values of the smooth function h. This requirement in turn entails equality of the integrands in the form

$$e^{-Kbu^2} \, C_0(u)^2 = \sqrt{1+2K/m^2} \, C_0(\sqrt{1+2K/m^2} \, u)^2$$

save perhaps at the point $u = 0$. Set $u = 1$ and it follows that

$$C_0(\sqrt{1+2K/m^2})^2 = \frac{e^{-Kb}}{\sqrt{1+2K/m^2}} \, C_0(1)^2 \,.$$

We now use the fact that K may be freely chosen, and we reintroduce the variable u as

$$u \equiv \sqrt{1+2K/m^2}, \qquad u > 0$$

or alternatively stated as

$$K = \tfrac{1}{2} m^2 (u^2 - 1) \,.$$

Then it follows for $u > 0$ that

$$C_0(u)^2 = \frac{1}{u} e^{-\frac{1}{2} m^2 b u^2} e^{\frac{1}{2} m^2 b} \, C_0(1)^2 \,.$$

Since C_0, like C, is an even function we finally obtain

$$C_0(u)^2 = \frac{\kappa}{|u|} e^{-\frac{1}{2} m^2 b u^2} \,,$$

where κ is a positive constant – an overall scale factor – that cannot be determined by this argument. Inspection of the form of $E\{h\}$ shows that the dimensions of κ are $(\text{length})^{-n}$, which are exactly the same dimensions as the arbitrary positive constant b. We arbitrarily choose $\kappa = b/2$; note that this choice is equivalent to a particular normalization of the two-point function. Specifically, the choice $\kappa = b/2$ leads to

$$\tfrac{1}{2}b \int u^2 \, e^{-\frac{1}{2}m^2 bu^2} \, du/|u| = m^{-2}$$

as the coefficient of a δ-function for the two-point function.

Summary of the solution

When we combine the facts we have obtained so far it follows that

$$E\{h\} = \mathcal{N} \int \exp[i\int h\phi \, d^n x - \tfrac{1}{2}m_0^2 \int \phi^2 \, d^n x - g_0 \int \phi^p \, d^n x] \, \mathcal{D}\phi$$

$$= \exp(-\tfrac{1}{2}b \int d^n x \int \{1 - \cos[uh(x)]\} \, e^{-\frac{1}{2}m^2 bu^2 - gb^{p-1}u^p} \, du/|u|)$$

where

$$m_0^2 = bm^2/\delta(0) \,, \qquad g_0 = b^{p-1}g/\delta(0)^{p-1} \,,$$

as noted at the outset of this derivation.

It is important to emphasize that *no actual integration has been used to obtain this result;* instead, we have used only *symmetry, general principles, and self consistency.* In particular, symmetry led to the functional form of the answer; the Levy canonical formula followed from general principles, which in turn determined the nature and meaning of local operator products; and self-consistency involving a mass insertion determined the rest, at least up to an overall factor that we have chosen arbitrarily. In the final section of this chapter we shall exploit the arbitrariness in this parameter as well.

9.5 Path integral formulation of independent-value models

In Section 9.2 we argued that a conventional path integral, or more correctly a conventional functional integral treatment of local nonlinear independent-value models leads to triviality, i.e., a Gaussian theory. Nevertheless, general arguments, not using a conventional path integral, have led to a nontrivial result. Our goal now is to take the nontrivial answer we have found and see how the same result could have been obtained by starting with an *un*conventional path integral formulation.

To achieve this goal we seek to identify a measure μ such that

$$\int \exp[i\phi(h)] \, d\mu(\phi)$$

$$= \exp(-\tfrac{1}{2}b \int d^n x \int \{1 - \cos[uh(x)]\} e^{-\frac{1}{2}bm^2 u^2 - gb^{p-1}u^p} \, du/|u|) \,.$$

Based on the fact that the right-hand side is a continuous, positive-definite functional, we know that such a measure μ exists. We can find μ by taking an inverse functional Fourier transformation. Or better yet, let us first *regularize* this expression and seek an inverse Fourier transformation of the *regularized* expression; this will enable us to more clearly see the needed modifications of a conventional formulation. To that end, consider the relation

$$\int \exp\left[i\Sigma_k \phi_k h_k \Delta\right] \rho(\phi)\, \Pi_k d\phi_k$$

$$= \exp(-\tfrac{1}{2}b\Sigma_k \Delta \int \{1 - \cos[uh_k]\} e^{-\frac{1}{2}bm^2 u^2 - gb^{p-1}u^p}\, du/|u|)$$

$$= \prod_k \exp(-\tfrac{1}{2}b\Delta \int \{1 - \cos[uh_k]\} e^{-\frac{1}{2}bm^2 u^2 - gb^{p-1}u^p}\, du/|u|)$$

$$\equiv \prod_k \int \exp[ih_k \phi \Delta - S(\phi, \Delta)]\, d\phi \,,$$

where Δ is the cell volume just as was the case before. The final two expressions have made use of the statistical independence of the fields at distinct points, which results in a product of identical Fourier transformations. Unfortunately, as it stands, the explicit inverse Fourier transform of the given expression is unknown and therefore we cannot find $S(\phi, \Delta)$ as defined. Fortunately, we do not need the explicit transform since all we really care about is reproducing the desired result in the continuum limit $\Delta \to 0$.

It is in fact easiest to present the answer and to prove that it achieves the desired goal. We initially focus on a quartic interaction, and in particular, we study the expression

$$\prod_k N_\Delta \int \exp[ih_k \phi \Delta - \tfrac{1}{2}(b\Delta)m^2 \phi^2 \Delta - g(b\Delta)^3 \phi^4 \Delta]\, |\phi|^{-(1-b\Delta)}\, d\phi \,.$$

Here N_Δ denotes a normalization factor chosen so that when $h_k = 0$, for all k, the result of the integral is unity. Besides the rather conventional terms in the exponent, note as well the appearance of an unexpected weighting introduced by the factor $|\phi|^{-(1-b\Delta)}$. This "minor" modification, as we shall soon see, makes all the difference!

By a suitable change of variables ($\phi = u/\Delta$) the previous expression is brought to the form

$$\prod_k M_\Delta \int \exp[ih_k u - \tfrac{1}{2}bm^2 u^2 - gb^3 u^4]\, |u|^{-(1-b\Delta)}\, du \,,$$

where M_Δ represents the new normalization factor. Besides the normalization factor, observe that now the only place the cutoff Δ occurs is in the auxiliary factor $|u|^{-(1-b\Delta)}$. To proceed further we need to estimate the leading dependence of M_Δ on the cutoff. To that end first observe that as $\Delta \to 0$ the integral diverges at the origin and so to compensate for this divergence, it is necessary that $M_\Delta \to 0$ as $\Delta \to 0$. Hence, we can estimate the leading behavior of M_Δ

from a study of

$$M \int_{-B}^{B} |u|^{-(1-b\Delta)}\, du = \frac{2M}{b\Delta} B^{(b\Delta)} \,,$$

which as Δ becomes small tends to $2M/(b\Delta)$ independently of the value of B, which has been taken as a simple representative of the fact that the integrand rapidly vanishes for large $|u|$. To achieve normalization, therefore, it is sufficient to let $M_\Delta = b\Delta/2$.

We now return to our principal calculation which assumes the form

$$\lim_{\Delta \to 0} \prod_k (\tfrac{1}{2}b\Delta) \int \exp[ih_k u - \tfrac{1}{2}bm^2 u^2 - gb^3 u^4] \frac{du}{|u|^{(1-b\Delta)}}$$

$$= \lim_{\Delta \to 0} \prod_k \left(1 - (\tfrac{1}{2}b\Delta) \int [1 - e^{ih_k u}] e^{-\tfrac{1}{2}bm^2 u^2 - gb^3 u^4} \frac{du}{|u|^{(1-b\Delta)}} \right)$$

$$= \lim_{\Delta \to 0} \prod_k \left(1 - (\tfrac{1}{2}b\Delta) \int [1 - e^{ih_k u}] e^{-\tfrac{1}{2}bm^2 u^2 - gb^3 u^4} \frac{du}{|u|} \right)$$

$$= \exp(-\tfrac{1}{2}b \int d^n x \int \{1 - \cos[uh(x)]\} e^{-\tfrac{1}{2}bm^2 u^2 - gb^3 u^4}\, du/|u|) \,.$$

The procedure just carried out for a quartic interaction may also be repeated for a general p^{th}-power interaction and leads to the general expression given by

$$\exp(-\tfrac{1}{2}cb \int d^n x \int \{1 - \cos[uh(x)]\} e^{-\tfrac{1}{2}bm^2 u^2 - gb^{p-1} u^p}\, du/|u|)$$

$$= \lim_{\Delta \to 0} \prod_k N_\Delta \int \exp[ih_k \phi\Delta - \tfrac{1}{2}(b\Delta)m^2 \phi^2 \Delta - g(b\Delta)^{p-1}\phi^p \Delta] \frac{d\phi}{|\phi|^{(1-cb\Delta)}} \,.$$

In this expression we have inserted a dimensionless constant c representing the only parameter that cannot be determined on general grounds; we now see how it would arise within the regularized functional integral formulation itself.

With this calculation we have obtained an analytic expression for the lattice action within a regularized Euclidean functional integral, the result of which, in the continuum limit, passes to the appropriate nontrivial functional expression obtained on the basis of symmetry and general arguments. We can recognize the essentials of the original lattice action for the independent-value model – modulo two important modifications. The first modification has involved a very special multiplicative renormalization for each of the model parameters; this sort of modification should not be unexpected. However, the major modification has been a change of the *measure* away from its naive form. Such a modification of the measure could not have been anticipated on the basis of classical arguments (such as they are for this model) because if the factor that represents the modification of the measure is put into the exponent, it is seen that it amounts to an $O(\hbar)$ modification of the classical action. Note well that it has been this relatively "minor" change in the measure, including its dependence on the cutoff parameter Δ, that has allowed us to beat the central limit theorem and obtain a non-Gaussian answer.

9.6* Many-component independent-value models

The extension of the single-component independent-value model to an $O(N)$-symmetric, N-component independent-value model, $N > 1$, is a relatively straightforward exercise. Our goal is to evaluate the functional integral

$$\mathcal{N} \int \exp[i\int h(x) \cdot \phi(x)\, d^n x - \tfrac{1}{2} m_0^2 \int \phi(x)^2\, d^n x - g_0 \int \phi(x)^p\, d^n x]\, \mathcal{D}\phi \,,$$

where both h and ϕ are now N-component vectors. Here, $\phi^2 = \phi \cdot \phi$, and ϕ^p is understood to mean $(\phi^2)^{p/2}$, p even. In addition to the stochastic independence of the fields at each distinct point x, we assume that the models also exhibit an invariance under rotation of the N-component field ϕ. In particular, this invariance implies that the functional integral necessarily has the analytical form given by

$$\exp\{-\int L[h(x)]\, d^n x\} \,,$$

where the function L depends only on the length of the vector h. Arguments completely analogous to those presented for the single-component case then lead to the fact that

$$\mathcal{N} \int \exp[i\int h(x) \cdot \phi(x)\, d^n x - \tfrac{1}{2} m_0^2 \int \phi(x)^2\, d^n x - g_0 \int \phi(x)^p\, d^n x]\, \mathcal{D}\phi$$

$$= \exp(-\tfrac{1}{2} cb \int d^n x \int \{1 - \cos[h(x) \cdot u]\} e^{-\frac{1}{2} bm^2 u^2 - g b^{p-1} u^p}\, d^N u / |u|^N) \,.$$

Here, $u \in \mathbb{R}^N$, and the measure is clearly invariant with respect to rotations of the function h. Indeed, since there are no gradients in the formal definition of these models, the functional integral is actually invariant under arbitrary, point-dependent $O(N)$ rotations of the vector field h.

For completeness, we may also offer a lattice-limit formulation of the N-component independent-value models. In particular, it readily follows that

$$\lim_{\Delta \to 0} \prod_k N_\Delta \int \exp[ih_k \cdot \phi\Delta$$

$$- \tfrac{1}{2} m^2 (b\Delta)\phi^2 \Delta - g(b\Delta)^{p-1}\phi^p \Delta] \frac{d^N\phi}{|\phi|^{N(1-c'b\Delta)}}$$

$$= \exp(-\tfrac{1}{2} cb \int d^n x \int \{1 - \cos[h(x) \cdot u]\} e^{-\frac{1}{2} bm^2 u^2 - g b^{p-1} u^p}\, d^N u / |u|^N) \,.$$

In this expression, $c = N\Gamma(N/2)c'/\pi^{N/2}$, which incorporates a factor arising from the volume of a unit $(N-1)$-dimensional sphere. As written, this expression actually applies for all finite $N \geq 1$ [KlN 76a].

---

* This section lies somewhat outside the main line of discussion and may be omitted on a first reading.

9.7* Infinite-component independent-value models

The N-component independent-value models discussed in the preceding section can also be studied in the limit in which $N \to \infty$. First of all we rewrite the expression for the N-component case in a way that makes clear which parameters are at our disposal, i.e. those we may need to choose as functions of the parameter N. In particular, we note that

$$\mathcal{N} \int \exp[i \int h(x) \cdot \phi(x) \, d^n x - \tfrac{1}{2} m_0^2 \int \phi(x)^2 \, d^n x - g_0 \int \phi(x)^p \, d^n x] \, \mathcal{D}\phi$$
$$= \exp(-\tfrac{1}{2} c_N b \int d^n x \int \{1 - \cos[h(x) \cdot u]\} e^{-\frac{1}{2} b m_N^2 u^2 - g_N b^{p-1} u^p} \, d^N u / |u|^N),$$

where c_N, m_N, and g_N have been introduced as new parameters. To determine the N-dependence of these parameters we proceed in a manner rather similar to that used in discussing the rotationally symmetric model in Section 7.2. In particular, for fixed x, we observe that

$$c_N \int \{1 - \cos[h(x) \cdot u]\} e^{-\frac{1}{2} b m_N^2 u^2 - g_N b^{p-1} u^p} \, d^N u / |u|^N$$
$$= c_N \int \{1 - \cos[|h(x)||u| \cos(\theta)]\}$$
$$\times e^{-\frac{1}{2} b m_N^2 u^2 - g_N b^{p-1} u^p} |u|^{-1} d|u| \, \sin(\theta)^{N-2} \, d\theta \, d\Omega_{N-2},$$

where the integral over $d\Omega_{N-2}$ gives the surface of the unit sphere in $N-1$ dimensions. Using the standard expression for this surface area, as well as an approximation to the integration over θ valid for very large N, we are led to a result for the previous expression given by

$$c_N \frac{\sqrt{2\pi}}{\sqrt{N-2}} \frac{2\pi^{(N-1)/2}}{\Gamma((N-1)/2)} \int \{1 - \exp[-\tfrac{1}{2}(N-2)^{-1} h(x)^2 u^2]\}$$
$$\times e^{-\frac{1}{2} b m_N^2 u^2 - g_N b^{p-1} u^p} |u|^{-1} d|u|.$$

A change of integration variables from $|u|$ to $a = |u|^2/(N-2)$ leads next to the expression

$$c_N \frac{\sqrt{2\pi}}{\sqrt{N-2}} \frac{\pi^{(N-1)/2}}{\Gamma((N-1)/2)} \int \{1 - \exp[-\tfrac{1}{2} h(x)^2 a]\}$$
$$\times e^{-\frac{1}{2} b m_N^2 (N-2) a - g_N b^{p-1} (N-2)^{p/2} a^{p/2}} a^{-1} da.$$

It now becomes clear how to proceed. We first choose (say)

$$c_N \frac{\sqrt{2\pi}}{\sqrt{N-2}} \frac{\pi^{(N-1)/2}}{\Gamma((N-1)/2)} = \frac{1}{2},$$

---

† This section lies somewhat outside the main line of discussion and may be omitted on a first reading.

independent of N, and we also rescale the mass and coupling constant so that $M^2 \simeq (N-2)m_N^2$ and $G \simeq (N-2)^{p/2}g_N$ are both independent of N, at least for large N. In particular, we may explicitly choose

$$M^2 \equiv Nm_N^2 , \qquad\qquad G \equiv N^{p/2}g_N ,$$

for all $N \geq 1$. If we adopt these renormalizations, then it follows that, according to the symmetry and general principles which we have discussed, the infinite-component, independent-value model formally presented in terms of the ill-defined functional integral

$$\mathcal{N} \int \exp[i\int h(x)\cdot\phi(x)\,d^nx - \tfrac{1}{2}m_0^2\int\phi(x)^2\,d^nx - g_0\int\phi(x)^p\,d^nx]\,\mathcal{D}\phi ,$$

is evaluated as

$$\exp(-\tfrac{1}{2}b\int d^nx\int\{1-\exp[-\tfrac{1}{2}h(x)^2a]\}\,e^{-\frac{1}{2}bM^2a - Gb^{p-1}a^{p/2}}\,da/a) .$$

Evidently these solutions are not free ($\equiv$ Gaussian) nor do they reduce to any free theory as the nonlinear coupling is turned off, $G \to 0$. In particular, all such local, nonlinear, infinite-component, independent-value models pass to one and the same pseudofree theory as the coupling constant vanishes. That pseudofree theory is determined by the expression

$$\exp(-\tfrac{1}{2}b\int d^nx\int\{1-\exp[-\tfrac{1}{2}h(x)^2a]\}\,e^{-\frac{1}{2}bM^2a}\,da/a) .$$

It is of some interest that this expression for the pseudofree generating functional can be evaluated as

$$\exp\{-\tfrac{1}{2}b\int d^nx\ln[1+h^2(x)/(bM^2)]\} ,$$

which, of course, is significantly different from the usual infinite-component, independent-value free model for which the generating functional reads

$$\exp[-\tfrac{1}{2}\mathsf{M}^{-2}\int h(x)^2\,d^nx] ,$$

for some mass parameter M, $0 < \mathsf{M} < \infty$ [KlN 76b].

Exercises

9.1 Repeat the steps necessary to determine the functional integral for a single-component ($N = 1$), *non*symmetric independent-value model, say, in particular,

$$\mathcal{N} \int \exp\{i\int h(x)\phi(x)\,d^nx - \int[\tfrac{1}{2}m_0^2\phi(x)^2 + \tau_0\phi(x)^3 + g_0\phi(x)^4]\,d^nx\}\,\mathcal{D}\phi ,$$

with $\tau_0 \neq 0$ and $g_0 > 0$.

9.2 How would you proceed to set up a perturbation theory to introduce, within a Euclidean framework, the missing gradients that would turn an independent-value model into a relativistic model; see [Kö 76]? Determine whether the perturbation in question should be regarded as a continuous perturbation or a discontinuous perturbation in the sense of Chapter 8.

9.3 Given the form of the expectation functional for a quartic independent-value model, show that the expectation functional admits a meaningful perturbation series about the pseudofree theory as an expansion in the quartic coupling constant.

9.4 Find the leading $1/N$ correction term to the large N behavior of a quartic, $O(N)$-symmetric, independent-value model; see [KlN 76b].

9.5 For a single-component field $(N = 1)$, the functional given by

$$C\{h\} \equiv \exp(-a \int d^n x \int \{1 - \cos[h(x)u]\} \, e^{-\frac{1}{2}bm^2u^2 - gb^3u^4} \, du/u^2) ,$$

for a a positive constant with appropriate dimensions, defines a positive-definite functional and therefore admits a representation as the functional Fourier transform of a measure, namely,

$$C\{h\} = \int e^{i \int h(x)\Lambda(x) \, d^n x} \, d\mu(\Lambda) .$$

Observe that the integral in the exponent in the present case involves u^{-2} rather than $|u|^{-1}$ which is appropriate to independent-value models. In turn, this functional integral may be written as the limit of a lattice regularized expression such as

$$C\{h\} = \lim_{\Delta \to 0} \prod_k \int \exp[ih_k \phi \Delta - S_\Delta(\phi)] \, d\phi .$$

Find a suitable lattice action S_Δ so that in the continuum limit the first quoted expression for $C\{h\}$ emerges.

10
Ultralocal Models

WHAT TO LOOK FOR

Simply put, the models of the present chapter are local relativistic field theories with all space derivatives discarded but the temporal derivative retained. Such models can be discussed both from an operator and a functional integral point of view. Viewed conventionally these models are trivial (Gaussian). However, alternative, nonperturbative methods demonstrate that non-Gaussian solutions are possible which are consistent with the nontrivial classical theory. Moreover, as the nonlinear coupling is turned off, the interacting solutions pass continuously to a pseudofree solution rather than to any free solution, suggesting the interpretation that the interaction acts as a discontinuous perturbation of the free theory. Although kinematically simplified, these models possibly may offer a hint of how to treat more interesting models (see Chapter 11). Throughout this chapter we generally choose units so that $\hbar = 1$.

10.1 Classical formulation

Consider the quartic self-interacting scalar field model, defined for an s-dimensional space, $s \geq 1$, and characterized by the classical action

$$I = \int \{ \tfrac{1}{2}[\dot{g}(t,\mathbf{x})^2 - m^2 g(t,\mathbf{x})^2] - \kappa\, g(t,\mathbf{x})^4 \}\, d^s x\, dt \ .$$

There are (at least) three ways to look at this model, each of which offers a special viewpoint. From the first point of view, this model differs from a covariant quartic self-interacting scalar field by the term $-\tfrac{1}{2}[\boldsymbol{\nabla} g(t,\mathbf{x})]^2$ in the integrand involving the spacial gradients. It is due to this viewpoint that we refer to such models as *ultralocal models* inasmuch as the only causal influence of the field behavior at a point $(t,\mathbf{x})$ is effective just along the time line $(t + s, \mathbf{x})$,

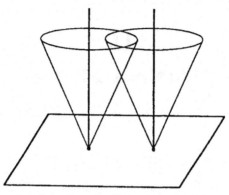

Fig. 10.1. Schematic space-time diagram of the causal influence zone for a covariant (conical lines) and an ultralocal (vertical lines) theory. Observe for covariant theories that for any two distinct spatial points, the future-directed light cones from each of them – within which their causal zone of influence is contained – will always intersect sufficiently far in the future. For an ultralocal theory, on the other hand, the causal zone of influence is confined to a future directed one-dimensional *line*, and as a consequence, the field behavior associated with two distinct spatial points is always statistically independent for any given temporal values.

$0 < s < \infty$, rather than in a future directed light cone as is the case for a relativistically covariant theory; see Fig. 10.1.

The second viewpoint begins with the classical equation of motion arising from such an action, namely

$$\ddot{g}(t, \mathbf{x}) = -m^2 g(t, \mathbf{x}) - 4\kappa\, g(t, \mathbf{x})^3$$

subject to suitable initial conditions. The solution $g(t, \mathbf{x}) \equiv g_{\mathbf{x}}(t)$, where $\mathbf{x}$ plays only a parametric role, and where

$$\ddot{g}_{\mathbf{x}}(t) = -m^2 g_{\mathbf{x}}(t) - 4\kappa\, g_{\mathbf{x}}(t)^3 \,,$$

illustrates that $\mathbf{x}$ merely serves as an auxiliary parameter in the formulation and solution of the classical model. The full classical solution is then obtained by solving the anharmonic oscillator and extending the solution for different parametric values. Such theories have been termed "diastrophic" [Kl 73a], a term chosen to indicate a *continuous extension* of a base theory (here just the anharmonic oscillator, no $\mathbf{x}$) into new directions with various "levels" labelled continuously by $\mathbf{x}$, and where the behavior at any level is statistically independent of the behavior at any other level.

A third way to look at the action functional for the quartic ultralocal model is as an independent-value model, as studied in the last chapter, to which we have added an extra term, namely, the square of the time derivative of the field. It is useful to view the present models, therefore, as the way a gradient-squared term should be added to an independent-value model. It is plausible that if we learn enough about how such terms can be added to preexisting local theories, it may be possible to add further gradient-squared terms corresponding to spatial derivatives, leading, eventually, to a relativistically covariant theory. Unfortunately, at the present time, this program remains just a dream; see, however, the remarks in Section 11.3.

Whether one arrives at ultralocal models by adding a temporal derivative to an independent-value model, by discarding spatial derivatives in a relativistic model, or by adding continuous parameter labels of a Euclidean space $\mathbb{R}^s$ to a base theory, the classical result is the same, a theory of the form indicated. Since the anharmonic oscillator is nontrivial, it follows that the classical solutions of the equations of motion for ultralocal models are generally nontrivial as well, provided of course that the coupling constant $\kappa \neq 0$, or more particularly $\kappa > 0$ to ensure positive energy as represented by the classical Hamiltonian

$$H(f,g) = \int \{\tfrac{1}{2}[f(\mathbf{x})^2 + m^2 g(\mathbf{x})^2] + \kappa\, g(\mathbf{x})^4\}\, d^s x \;.$$

The quartic self-interacting ultralocal model is but one of many that may be considered. For example, we may also consider the model determined by the classical Hamiltonian

$$H(f,g) = \int \{\tfrac{1}{2}[f(\mathbf{x})^2 + m^2 g(\mathbf{x})^2] + V[g(\mathbf{x})]\}\, d^s x \;,$$

where V is a real potential chosen to satisfy $m^2 g^2 + 2V[g] \geq 0$, with $g = 0$ the only solution of $m^2 g^2 + 2V[g] = 0$. It suffices to restrict attention to polynomial potentials, and interest centers on potentials that involve field powers higher than two (and, in addition, have no linear term). If $V[-g] = V[g]$, for all g, then we say the potential is symmetric or that it is even; if, instead, $V[-g] \neq V[g]$, then we say the potential has no symmetry. As will become apparent – and apart from the issue of symmetry of the potential – the analysis of any interacting theory is qualitatively the same as the quartic interacting theory. Except for an occasional remark, we shall generally confine our discussion to the special case of a quartic interaction.[*]

Although the dynamics does not connect behavior at distinct spatial points, we nevertheless insist that the topology of the underlying space ($\mathbb{R}^s$) is maintained, and in particular, that there exist classical spatial translation generators $\mathbf{P}(f,g)$ given as usual by

$$\mathbf{P}(f,g) = -\int f(\mathbf{x}) \boldsymbol{\nabla} g(\mathbf{x})\, d^s x$$

for any spatial dimension $s \geq 1$. Moreover, if $s \geq 2$, then we assume that classical rotation group generators given by

$$J_{lm}(f,g) = \int f(\mathbf{x})(x_m \partial_l - x_l \partial_m)\, g(\mathbf{x})\, d^s x$$

also exist. By design, the Poisson brackets of the Hamiltonian with the trans-

---

[*] The discussion in this chapter takes [Kl 70a] as its starting point.

lation and rotation generators vanish. Thus the models all enjoy full Euclidean covariance, rather than the Minkowski covariance of a relativistic theory.

We do not claim any special physical relevance for ultralocal models; their principal interest lies in the fact of their complete quantum solubility, as explained in detail later in this chapter. However, the nontrivial solution we shall obtain does not follow from conventional quantization procedures which instead – as we next show – invariably lead to a trivial (i.e., Gaussian) result; see, e.g., [ZhK 94, ZhK 95].

10.2 Conventional quantization

The conventional quantization of the *free* ultralocal theories ($\kappa \equiv 0$) for any mass m is given simply by a collection of independent harmonic oscillators all having the same angular frequency (m). The quantization in this case is straightforward, and we prefer to defer further discussion of the free ultralocal theories until later. Instead, let us turn our immediate attention to a brief discussion of the conventional quantization of the quartic self-interacting ultralocal model with $\kappa > 0$.

It is convenient to analyze the conventional quantization of a quartic ultralocal model with the help of a formal Euclidean-space functional integral with an imaginary source. In particular, we focus on the expression

$$\mathcal{N} \int \exp(\int \{ih(x)\phi(x) - \tfrac{1}{2}[\dot{\phi}(x)^2 + m^2\phi(x)^2] - \kappa\phi(x)^4\} \, d^n x \,) \, \mathcal{D}\phi \,,$$

normalized to unity when $h \equiv 0$, and which we interpret, following conventional procedures, as the continuum limit of the formal *spatial* lattice expression

$$\prod_k \mathcal{N}' \int \exp(\int \{ih_k(t)\phi_k(t)a^s - \tfrac{1}{2}[\dot{\phi}_k(t)^2 + m_a^2\phi_k(t)^2]a^s$$
$$- \kappa_a \phi_k(t)^4 a^s\} \, dt \,) \, \mathcal{D}\phi_k \,,$$

within which the temporal integration extends from $-\infty$ to ∞. Here, $k = (k^1, k^2, \ldots, k^s)$, $k^j \in \{0, \pm 1, \pm 2, \ldots\}$, labels a lattice site, a^s is the size of a single spatial lattice cell, and m_a and κ_a denote a possibly cutoff-dependent mass and coupling constant, respectively. The functional form of this expression is evidently given by

$$S\{h\} = \lim_{a \to 0} \prod_k F_a\{h_k a^{s/2}\}$$

for some functional F_a determined (for fixed a) by the path integral of a single, conventional anharmonic oscillator, and normalized so that $F_a\{0\} = 1$.

We next invoke the central limit theorem. Expansion in powers of h_k – and retaining only the even powers on the basis of symmetry – leads to

$$F_a\{h_k a^{s/2}\} = 1 - \frac{a^s}{2!} \int h_k(t_2) h_k(t_1)\, C_a^{(2)}(t_2 - t_1)\, dt_2 dt_1$$

$$+ \frac{a^{2s}}{4!} \int h_k(t_4) h_k(t_3) h_k(t_2) h_k(t_1)$$

$$\times C_a^{(4)}(t_4 - t_1, t_3 - t_1, t_2 - t_1)\, dt_4 dt_3 dt_2 dt_1 + \cdots ,$$

where $C_a^{(n)}$ denotes the nth-order correlation function of the anharmonic oscillator. It is straightforward to conclude, even with general cutoff-dependent coefficients, that if $C_a^{(2)}(t_2 - t_1) \to C^{(2)}(t_2 - t_1)$, as $a \to 0$ – a limit that is presumed to exist – then it follows that $C_a^{(4)}(t_4 - t_1, t_3 - t_1, t_2 - t_1) \to C^{(4)}(t_4 - t_1, t_3 - t_1, t_2 - t_1)$, or at least is bounded. This result for the 4th-order correlation holds because it is uniformly bounded [GlJ 87]. A similar bound applies to all higher-order even correlation functions. As a consequence, the continuum limit for the generating functional leads to

$$S\{h\} = \exp[-\tfrac{1}{2} \textstyle\int h(t, \mathbf{x})\, C^{(2)}(t - s)\, h(s, \mathbf{x})\, d^s x\, dt\, ds] ,$$

which is strictly a Gaussian result. Any effect of the quartic interaction that existed on the lattice has been subsumed into the two-point correlation function $C^{(2)}$; see Exercise 10.1. Such a result does not manifest any non-Gaussian behavior, and it therefore may be regarded as trivial.

An analysis in real time fares no better, and an analog of the central limit theorem again directly leads to a Gaussian distribution.

Lastly, we observe that a conventional operator formulation of the problem, based on the introduction of canonical operators that satisfy canonical commutation relations, also leads to a trivial theory. We do not prove this point here; rather, we shall implicitly establish this fact in the course of our detailed analysis and construction in the next several sections.

Free ultralocal theory

The free ultralocal theory ($\kappa \equiv 0$) is similar to a collection of independent, identical harmonic oscillators. In particular, the local field $\varphi_F(\mathbf{x})$ and momentum $\pi_F(\mathbf{x})$ operators (F for "free") at time $t = 0$ and pertaining to the mass value m, may be conveniently described in terms of a set of Fock annihilation and creation operators $a(\mathbf{k})$ and $a(\mathbf{k})^\dagger$. For all arguments, these operators fulfill the canonical commutation relation

$$[a(\mathbf{k}), a(\mathbf{q})^\dagger] = \delta(\mathbf{k} - \mathbf{q})$$

as well as the condition $a(\mathbf{k})\,|0\rangle = 0$, where $|0\rangle$, the no-particle state, is the only state up to a factor that fulfills this relation. The Hamiltonian operator and the

Euclidean generators are given, respectively, by

$$\mathcal{H} = m \int a(\mathbf{k})^\dagger a(\mathbf{k}) \, d^s k \,,$$
$$\mathcal{P} = \int a(\mathbf{k})^\dagger \, \mathbf{k} \, a(\mathbf{k}) \, d^s k \,,$$
$$\mathcal{J}_{lm} = i \int a(\mathbf{k})^\dagger (k_m \partial_l - k_l \partial_m) \, a(\mathbf{k}) \, d^s k \,.$$

In turn, the time-dependent field operator reads

$$\varphi(t, \mathbf{x}) = \frac{1}{(2\pi)^{s/2}} \int [a(\mathbf{k}) e^{i(\mathbf{k}\cdot\mathbf{x} - mt)} + a(\mathbf{k})^\dagger e^{-i(\mathbf{k}\cdot\mathbf{x} - mt)}] \, \frac{d^s k}{\sqrt{2m}} \,.$$

We identify $\varphi(\mathbf{x}) \equiv \varphi(0, \mathbf{x})$, and along with $\pi(\mathbf{x}) \equiv \dot\varphi(0, \mathbf{x})$, these local self-adjoint operators satisfy the canonical commutation relation

$$[\varphi(\mathbf{x}), \pi(\mathbf{y})] = i\delta(\mathbf{x} - \mathbf{y}) \,.$$

Finally, the normalized coherent states $|f, g\rangle = \exp\{i[\varphi(f) - \pi(g)]\}|0\rangle$, defined for a dense set of functions f and g, provide the connection between the classical and quantum generators through the weak correspondence principle. In particular, $H(f, g) = \langle f, g | \mathcal{H} | f, g \rangle$, $\mathbf{P}(f, g) = \langle f, g | \mathcal{P} | f, g \rangle$, and $J_{lm}(f, g) = \langle f, g | \mathcal{J}_{lm} | f, g \rangle$, as needed. All multipoint functions may be derived from the fact that

$$\langle 0 | e^{i \int h(x)\varphi(x) \, d^n x} | 0 \rangle = \exp[-(1/4m) \int h(t, \mathbf{x}) \, e^{-im(t-s)} \, h(s, \mathbf{x}) \, d^s x \, dt \, ds] \,.$$

For different values of the mass m, the local field operators φ and π at a sharp time characterize inequivalent representations of the canonical commutation relations, as is clear from the use of tags (see Section 6.5), and the fact that the sharp-time expectation functionals are distinct for different m values. For any m, the free theory is said to be trivial since it has a Gaussian distribution.

10.3 Results of the nontrivial quantization

Before presenting our detailed derivation of the nontrivial quantization, let us first illustrate and discuss the result that we will find in the following sections. For illustrative purposes we again mainly use the quartic model classically given by the action functional

$$\int \{\tfrac{1}{2}[\dot{g}(x)^2 - m_o^2 \, g(x)^2] - \kappa_o \, g(x)^4\} \, d^n x \,,$$

where, for convenience, we have added a subscript (o) to the parameters involved to distinguish them from similar parameters that enter in the quantum solution.

First we offer a few general remarks. In our present discussion we assume that the sharp-time field is a local self-adjoint operator; relaxation of that assumption will be discussed subsequently. On the other hand, it is noteworthy that for all interacting solutions, the sharp-time momentum is a Hermitian form, but it is *never* a local operator. As a consequence, and unlike the free ultralocal theories, it follows that *interacting ultralocal quantum theories do not satisfy canonical commutation relations!*

In the following sections we shall establish for the quartic self-coupled theory that the real-time T-product generating functional is given by

$$Z\{h\} = \langle 0|Te^{i\int h(x)\varphi(x)\,d^n x}|0\rangle$$
$$= \exp\{-\int d^s x \int c(\lambda)[1 - Te^{i\int h(t,\mathbf{x})\lambda(t)\,dt}]\,c(\lambda)\,d\lambda\}\,.$$

Here, in this expression, several terms need definition.* In particular,

$$\lambda(t) = e^{i\hbar t}\,\lambda\,e^{-i\hbar t}\,,$$

where

$$\hbar = -\frac{1}{2b}\frac{\partial^2}{\partial\lambda^2} + \frac{1}{2b}\frac{\gamma(\gamma+1)}{\lambda^2} + \frac{b}{2}m^2\lambda^2 + \kappa b^3\lambda^4 - \text{const.}\,,$$

$$c(\lambda) = \frac{Ke^{-y(\lambda)}}{|\lambda|^\gamma}\,,\qquad 1/2 \le \gamma < 3/2\,,\qquad y(0) = 0\,,$$

$$\hbar c(\lambda) = 0\,,$$

where the nonvanishing function $c(\lambda)$ – called the "model function" – is real, b is a positive constant with dimensions $(\text{length})^{-s}$, and K is a factor that may be set by choosing the normalization of the two-point function. The magnitude of the parameter γ – called the "singularity parameter" – is initially undetermined; all that is established is an acceptable range of values. An argument will be given later that suggests both a physical interpretation and possible proper values for γ. The term const. is chosen so that the last equation holds. The relation between m_o and m, and also that between κ_o and κ, is derived below.

Analogous results hold for alternative ultralocal models, that is for other than quartic coupling; for example, if the term κg^4 of the classical theory is replaced by the term κg^p, p even, then the only change is that in $\hbar$ the term $\kappa b^3\lambda^4$ is replaced by $\kappa b^{p-1}\lambda^p$. Of course, the function c also changes in view of the fact that $\hbar c = 0$ must always hold. As a consequence, the functional expression for the generating functional $Z\{h\}$ will generally be different for each distinct classical Hamiltonian. This difference carries over to the expectation functional for the time-zero field, when it exists. When it does exist, as we assume at present, the desired result may be obtained simply by setting $h(t,\mathbf{x}) = \delta(t)\,f(\mathbf{x})$. In particular, the time-zero expectation functional is given by

$$E(f) = \langle 0|e^{i\int f(\mathbf{x})\varphi(\mathbf{x})\,d^s x}|0\rangle = \exp\{-\int d^s x \int [1 - e^{if(\mathbf{x})\lambda}]\,c(\lambda)^2\,d\lambda\}\,.$$

---

* For ultralocal models we use λ as a real variable and not as a coupling constant. The parameter b is analogous to but different from the parameter b introduced in Chapter 9.

For symmetric potentials, it follows that $c(-\lambda) = c(\lambda)$, and so the expectation functional is unchanged if we replace the factor $\exp[if(\mathbf{x})\lambda]$ by $\cos[f(\mathbf{x})\lambda]$. For a potential without symmetry, e.g., the classical interaction $\sigma g^3 + \kappa g^4$, $\sigma \neq 0$ and $\kappa > 0$, then $c(-\lambda) \neq c(\lambda)$ and the more general form indicated is required. As already remarked, we will discuss later the assumption that the smeared time-zero field operator $\varphi(f) = \int f(\mathbf{x})\,\varphi(\mathbf{x})\,d^s x$ exists, and we shall show how to generalize the given solution in the case that it does not exist. On the other hand, for suitable test functions h the space-time smeared field operator $\varphi(h) = \int h(x)\,\varphi(x)\,d^n x$ is always a well-defined operator even when the sharp-time field is not a local operator.

It is evident that for sufficiently many well-chosen test functions above (i.e., h or f, respectively), the indicated expressions for the form of the solution of a general ultralocal model are reduced to well-defined quadratures. It is in this sense that we say that we are able to solve the quantum field theory for ultralocal models.

Pseudofree ultralocal theory

Consider the class of models covered by various potentials $V[g]$ introduced in place of the quartic interaction in our illustrative example. Let each model be given the coupling constant $\kappa > 0$. Then we make the important observation, for any of these models, that as the coupling constant $\kappa \to 0$, all such theories pass continuously to a *single pseudofree theory*, the field for which we denote by $\bar{\varphi}$, and which is characterized by the expression

$$\langle 0|\mathsf{T}e^{i\int h(x)\bar{\varphi}(x)\,d^n x}|0\rangle = \exp\{-\int d^s x \int \bar{c}(\lambda)\,[1 - \mathsf{T}e^{i\int h(t,\mathbf{x})\lambda(t)\,dt}]\,\bar{c}(\lambda)\,d\lambda\}\,.$$

Here

$$\lambda(t) = e^{i\bar{h}t}\,\lambda\,e^{-i\bar{h}t}\,,$$

and

$$\bar{h} = -\frac{1}{2b}\frac{\partial^2}{\partial\lambda^2} + \frac{1}{2b}\frac{\gamma(\gamma+1)}{\lambda^2} + \frac{b}{2}m^2\lambda^2 + (\gamma - \tfrac{1}{2})m\,,$$

$$\bar{c}(\lambda) = \frac{Ke^{-bm\lambda^2/2}}{|\lambda|^\gamma}\,,\qquad 1/2 \leq \gamma < 3/2\,,$$

$$\bar{h}\bar{c}(\lambda) = 0\,.$$

In turn, the expectation functional for the smeared, time-zero, pseudofree ultralocal field is given by

$$\bar{E}(f) = \exp(-K^2 \int d^s x \int \{1 - \cos[f(\mathbf{x})\lambda]\}\,e^{-bm\lambda^2}\,d\lambda/|\lambda|^{2\gamma})\,.$$

Of course, the pseudofree theory depends on γ, and for general γ it is not possible to explicitly evaluate the integral over λ in terms of elementary functions;

nevertheless, the existence of the integral, e.g., for any bounded function f with compact support, is evident.

Let us recall the discussion in Chapter 8 regarding continuous and discontinuous perturbations. On the heuristic grounds discussed previously, we are led to consider whether or not a finite constant C exists such that

$$[\int g(x)^4 \, d^n x]^{1/2} \leq C \int [\dot{g}(x)^2 + m^2 g(x)^2] \, d^n x$$

for any $s \geq 1$. The answer is clearly no, as illustrated by any function of the form $g(x) = z(t)f(\mathbf{x})$ where $z \in C_0^\infty$, say, and $f \in L^2(\mathbb{R}^s)$ but $f \notin L^4(\mathbb{R}^s)$, e.g., $f(\mathbf{x}) = |\mathbf{x}|^{-r}(1 + |\mathbf{x}|^2)^{-s}$, where $s/4 \leq r < s/2$. Thus the fact that the zero-coupling limit of any nonlinear, interacting, ultralocal theory is a pseudofree theory may be expected simply because there are many fields for which the free (Euclidean) action is finite while the nonlinear interaction term diverges. This expectation is in fact confirmed by the exact results summarized above inasmuch as the unique ultralocal pseudofree theory for any mass parameter (and any γ) is inequivalent to any ultralocal free theory. As a consequence we may conclude that the interacting ultralocal models represent *discontinuous perturbations* of the associated ultralocal free theory. Indeed, the relation of any of the local, interacting (nonlinear) theories to the free and to the pseudofree theories is exactly as depicted in Fig. 9.1 for the independent-value models.*

Recall for the independent-value models there was no circumstance for which the sharp-time field was a local operator, not even for the free independent-value models. This situation is to be contrasted with that for the ultralocal models for which under suitable conditions the sharp-time field is a local operator, this certainly being the case for the free models. Arguing from a formal functional integral point of view, it follows that the introduction of the temporal gradient term has – relative to the independent-value models – "softened" the nature of the distributions involved in the integral, indeed, softened them so much that the set of acceptable test functions may be sufficiently enlarged to allow sharp-time test functions, and hence sharp-time local fields. This result is in accord with the heuristic expectation that as the set of functions for which the classical (Euclidean) action is finite is decreased – and therefore the required distributions in the support of the functional integration measure are correspondingly softened – then the set of allowed test functions for the quantum field operator is correspondingly enlarged.

---

* Just as was the case for the independent-value models, there are nonlinear models that are continuously connected to an ultralocal free theory, but they do not involve strictly *local* interactions.

10.4 Construction of the nontrivial solution

Our goal is to establish the validity of the nontrivial solution introduced and discussed above. We start with the assumption that the sharp-time field is a local, self-adjoint operator, and therefore that $\varphi(f)$ is self adjoint for a wide class of real test functions. Let $\mathcal{H} \geq 0$ denote the nonnegative, self-adjoint Hamiltonian operator, and let $|0\rangle$ denote the ground state of the system for which $\mathcal{H}|0\rangle = 0$, and which we assume to be nondegenerate. Let us analyze the expectation functional $E(f) \equiv \langle 0|\exp[i\varphi(f)]|0\rangle$. Whatever else the various renormalizations end up doing, we insist that the basic symmetry of the ultralocal models should be preserved. That symmetry asserts that there is complete statistical independence for all times between the field values at one spatial point and those at any other spatial point. As a consequence it is necessary that

$$E(f) = e^{-\int L[f(\mathbf{x})]\,d^s x}$$

for some function L which depends only on the field value $f(\mathbf{x})$ at the point $\mathbf{x}$ (and not, for example, on any derivatives). This functional form is simply a mathematical restatement of the statistical independence of the field values for every spatial point. Moreover, the operator fields at a fixed time all mutually commute with each other, and so it follows that a probability measure μ on field values exists such that

$$E(f) = e^{-\int L[f(\mathbf{x})]\,d^s x} = \int e^{i\Lambda(f)}\,d\mu(\Lambda)\,.$$

The realization of the sharp-time local field operator as Λ and the associated introduction of the measure μ simply incorporates the Schrödinger representation for the field.

We have encountered this kind of expression for probability measures before, first, implicitly, in Section 6.7 in connection with a general discussion of infinitely-divisible probability distributions, and second, explicitly, in Section 9.3 in connection with the expectation functional for the independent-value models. We may immediately adopt the result of those studies to learn, for real p, that it is necessary that L has a general canonical form given by

$$L[p] = -i\alpha p + \beta p^2 + \int [1 - e^{ip\lambda} + ip\lambda/(1+\lambda^2)]\,d\sigma(\lambda)\,,$$

where α is real, $\beta \geq 0$, and the nonnegative measure σ satisfies

$$\int [\lambda^2/(1+\lambda^2)]\,d\sigma(\lambda) < \infty\,.$$

The general expression for L leads to a Gaussian distribution for the field (determined by β, and part of α), and an independent Poisson distribution for the field (determined by σ, and the rest of α); more specifically, for the latter part, we deal with a compound Poisson distribution whenever $0 < \int d\sigma(\lambda) < \infty$, and a generalized Poisson distribution whenever $\int d\sigma(\lambda) = \infty$. Just as in Chapter 9, our interest will again focus on the case of generalized distributions. If, for convenience, we restrict the measure σ so that $\int |\lambda|/(1+|\lambda|)\, d\sigma(\lambda) < \infty$, then it follows that the canonical form for L simplifies and is given by

$$L[p] = \beta p^2 + \int [1 - e^{ip\lambda}]\, d\sigma(\lambda) \ .$$

This expression is often enough to establish most desired results. Note that the limit of a convergent sequence of the simpler expression for $L[p]$ has the form of the most general canonical form.

If the potential is symmetric, i.e., $V[-g] = V[g]$, then we expect that $E(-f) = E(f)$, or in other words, for symmetric potentials that we can choose

$$L[p] = \beta p^2 + \int [1 - \cos(p\lambda)]\, d\sigma(\lambda) \ ;$$

in this case we shall assume as well that the measure σ possesses the correct symmetry itself. Finally, we recognize the free model solutions in this form, i.e., those for $\alpha = 0$, $\beta = 1/4m$, and $\sigma(\lambda) = 0$. Having exhausted the potential of the free theories, we will concentrate on a Poisson solution for the interacting cases, and *hereafter we choose* $\beta = 0$. Finally, we shall suppose that the measure σ is absolutely continuous, and more specifically that there exists a nonvanishing, nonnegative weight function $c(\lambda)^2$ such that $d\sigma(\lambda) = c(\lambda)^2\, d\lambda$.

As a consequence of the preceding discussion, we are led to an expression for the expectation functional for general lower-bounded potentials without symmetry given by

$$\langle 0|e^{i\varphi(f)}|0\rangle = \exp(-\int d^s x \int \{1 - e^{i[f(\mathbf{x})\lambda]}\}\, c(\lambda)^2\, d\lambda) \ ,$$

which in this simpler form may be interpreted, if necessary, as a principal value integral. The corresponding expression in the case of symmetric (i.e., even) lower-bounded potentials is given by

$$\langle 0|e^{i\varphi(f)}|0\rangle = \exp(-\int d^s x \int \{1 - \cos[f(\mathbf{x})\lambda]\}\, c(\lambda)^2\, d\lambda) \ .$$

In both cases, the expressions involve the real function $c(\lambda)$ such that

$$\int [\lambda^2/(1+\lambda^2)]\, c(\lambda)^2\, d\lambda < \infty \ ,$$

although, in either case, we shall invariably be led to the fact that

$$\int c(\lambda)^2\, d\lambda = \infty$$

for all our examples in the study of ultralocal models.

Unless stated otherwise, we shall focus hereafter on *even* potentials, i.e., $V[-g] = V[g]$, for which we may assume that $c(-\lambda) = c(\lambda)$ without loss of generality.

Operator realization of the sharp-time field

Having determined that the expectation functional is functionally similar to that for the independent-value models, we can follow the construction developed in

Section 9.4 to find an operator realization for the sharp-time ultralocal field. In particular, let us introduce a Fock representation of annihilation and creation operators given, in the present case, by the operators $A(\mathbf{x}, \lambda)$ and $A(\mathbf{x}, \lambda)^\dagger$ which, for all arguments, satisfy the commutation relation

$$[A(\mathbf{x}, \lambda), A(\mathbf{y}, \lambda')^\dagger] = \delta(\mathbf{x} - \mathbf{y})\delta(\lambda - \lambda') .$$

In addition, we introduce the no-particle state $|0\rangle$ for which

$$A(\mathbf{x}, \lambda)|0\rangle = 0$$

holds for all arguments. The assertion that up to multiples the state $|0\rangle$ is the unique no-particle state establishes that we are dealing – once again – with a Fock representation of basic operators, operators which in the present case depend on a spatial point $\mathbf{x} \in \mathbb{R}^s$ and an additional parameter $\lambda \in \mathbb{R}$. As usual, a dense set of states in the Hilbert space is given by repeated action of the creation operators on the no-particle state. Since the Fock representation is irreducible, it follows that all other Hilbert space operators are obtained as functions of the set of annihilation and creation operators. In particular, for a specific ultralocal model with an even potential and characterized by the model function c, we claim that the sharp-time field operator is given by

$$\varphi(\mathbf{x}) = \int A(\mathbf{x}, \lambda)^\dagger \lambda A(\mathbf{x}, \lambda)\, d\lambda + \int c(\lambda)\, \lambda A(\mathbf{x}, \lambda)\, d\lambda + \int A(\mathbf{x}, \lambda)^\dagger \lambda c(\lambda)\, d\lambda .$$

In view of the symmetry $c(-\lambda) = c(\lambda)$, let us agree that $\int \lambda c(\lambda)^2\, d\lambda = 0$, either as an absolutely convergent integral or as a principal value integral. In that case a more convenient form for the field operator can be given. Let us introduce the translated operators

$$B(\mathbf{x}, \lambda) \equiv A(\mathbf{x}, \lambda) + c(\lambda) ,$$
$$B(\mathbf{x}, \lambda)^\dagger \equiv A(\mathbf{x}, \lambda)^\dagger + c(\lambda) ,$$

then we may also write

$$\varphi(\mathbf{x}) = \int B(\mathbf{x}, \lambda)^\dagger \lambda B(\mathbf{x}, \lambda)\, d\lambda ,$$

which, for future reference, we emphasize is *bilinear* in the operators $B^\dagger$ and B. We add that the operators $B^\dagger$ and B satisfy the same commutation relations as $A^\dagger$ and A, namely

$$[B(\mathbf{x}, \lambda), B(\mathbf{y}, \lambda')^\dagger] = \delta(\mathbf{x} - \mathbf{y})\delta(\lambda - \lambda') ,$$

but they do not constitute a Fock representation of the canonical commutation relations since there is no state in the Hilbert space that is annihilated by all B operators.

The proof that the field operator may be so represented follows the pattern used for the independent-value models. In particular, we first make the general observation that

$$\exp[i \int d^s x \int B(\mathbf{x}, \lambda)^\dagger f(\mathbf{x}) \lambda B(\mathbf{x}, \lambda) \, d\lambda]$$
$$= \,: \exp\{\int d^s x \int B(\mathbf{x}, \lambda)^\dagger [e^{if(\mathbf{x})\lambda} - 1] B(\mathbf{x}, \lambda) \, d\lambda\} :$$

holds for a dense set of functions f, where as usual $: :$ denotes normal ordering, i.e., all $A^\dagger$s to the left of all As. The proof of this relation is identical to the proof given in Section 9.4 and is not repeated here. As a consequence, it follows that

$$\langle 0 | e^{i\varphi(f)} | 0 \rangle = \langle 0 | : \exp\{\int d^s x \int B(\mathbf{x}, \lambda)^\dagger [e^{if(\mathbf{x})\lambda} - 1] B(\mathbf{x}, \lambda) \, d\lambda\} : | 0 \rangle$$
$$= \exp\{\int d^s x \int c(\lambda) [e^{if(\mathbf{x})\lambda} - 1] c(\lambda) \, d\lambda\} \,,$$

which directly yields the desired result for $E(f)$. The GNS Theorem [Na 64, Em 72] guarantees us that any other representation of the field operator will be unitarily equivalent to the one given.

Local products

We raise the question as to how one should define formal expressions like $\varphi(\mathbf{x})^2$, $\varphi(\mathbf{x})^4$, etc. This question has no universal answer inasmuch as every operator representation determines its own rules to define local products. A useful tool to study this question is the operator product expansion [ItZ 80, HeK 74]. The analysis here again follows the pattern used for the independent-value models. Consider the expression

$$\varphi(\mathbf{x})\varphi(\mathbf{y}) = \int B(\mathbf{x}, \lambda)^\dagger \lambda B(\mathbf{x}, \lambda) \, d\lambda \int B(\mathbf{y}, \lambda')^\dagger \lambda' B(\mathbf{y}, \lambda') \, d\lambda'$$
$$= \delta(\mathbf{x} - \mathbf{y}) \int B(\mathbf{x}, \lambda)^\dagger \lambda^2 B(\mathbf{x}, \lambda) \, d\lambda + \,: \varphi(\mathbf{x})\varphi(\mathbf{y}) : \,.$$

The first term is the most singular as $\mathbf{y} \to \mathbf{x}$; in fact, the second, normal-ordered term, is nonsingular in this limit. We will define the local field product by the coefficient of the singular term, but before doing so we introduce an arbitrary parameter $b > 0$ with the dimensions of (length)$^{-s}$ so that

$$\varphi(\mathbf{x})\varphi(\mathbf{y}) = [b^{-1}\delta(\mathbf{x} - \mathbf{y})] \, b \int B(\mathbf{x}, \lambda)^\dagger \lambda^2 B(\mathbf{x}, \lambda) \, d\lambda + \,: \varphi(\mathbf{x})\varphi(\mathbf{y}) : \,,$$

and we introduce the local *renormalized* (subscript R) product

$$\varphi_R^2(\mathbf{x}) = b \int B(\mathbf{x}, \lambda)^\dagger \lambda^2 B(\mathbf{x}, \lambda) \, d\lambda \,.$$

This result may also be obtained through the limit

$$\varphi_R^2(\mathbf{x}) = \lim_{\Delta \to 0} b \Delta^{-1} [\int_{\mathbf{z} \in \Upsilon} \varphi(\mathbf{x} + \mathbf{z}) \, d^s z]^2 \,,$$

where $\Delta \equiv \int_{\mathbf{z} \in \Upsilon} d^s z$. In a formal sense we may state that

$$\varphi_R^2(\mathbf{x}) = [b/\delta(\mathbf{0})] \, \varphi(\mathbf{x})\varphi(\mathbf{x}) \,,$$

the formal division by $\delta(\mathbf{0})$ effectively annulling the normal-ordered term. Since the formal factor $b/\delta(\mathbf{0})$ is dimensionless, it follows that φ_R^2 has the square of the ordinary dimensions of the field φ itself. Observe that the renormalized square φ_R^2 is bilinear in $B^\dagger$ and B, just like the field φ itself.

Higher-order local powers may be introduced in a similar fashion. For example,

$$\varphi_R^2(\mathbf{x})\varphi(\mathbf{y}) = b\,\delta(\mathbf{x} - \mathbf{y}) \int B(\mathbf{x}, \lambda)^\dagger\, \lambda^3\, B(\mathbf{x}, \lambda)\, d\lambda + :\varphi_R^2(\mathbf{x})\varphi(\mathbf{y}): \ .$$

Thus we may define

$$\begin{aligned}
\varphi_R^3(\mathbf{x}) &= b^2 \int B(\mathbf{x}, \lambda)^\dagger\, \lambda^3\, B(\mathbf{x}, \lambda)\, d\lambda \\
&= [b/\delta(\mathbf{0})]\, \varphi_R^2(\mathbf{x})\varphi(\mathbf{x})\,,
\end{aligned}$$

where the last line is heuristic. The renormalized cube then has the cube of the dimension of the field itself. Of course, there is no requirement that the factor "b" used in the latter expression has to be the same as the original "b", only that the dimensions are identical. However, there is some algebraic utility as well as economy in using the same b throughout. In particular, with the same b it follows that we have such natural relations as $\varphi_R^3 = [\varphi_R^2\,\varphi]_R = [\varphi\,\varphi\,\varphi]_R$.

The pattern for additional local renormalized powers is now evident, and we may introduce directly the pth renormalized field power as

$$\varphi_R^p(\mathbf{x}) \equiv b^{p-1} \int B(\mathbf{x}, \lambda)^\dagger\, \lambda^p\, B(\mathbf{x}, \lambda)\, d\lambda\,.$$

If we identify $\varphi_R \equiv \varphi$, then this definition holds for all integer $p \geq 1$. Furthermore, polynomial functions, and through them, convergent limits to more general expression can also be given. In particular, for a general function W, such that $W[0] = 0$, we define

$$W[\varphi(\mathbf{x})]_R \equiv b^{-1} \int B(\mathbf{x}, \lambda)^\dagger\, W[b\lambda]\, B(\mathbf{x}, \lambda)\, d\lambda\,.$$

As an example we may introduce a family of local field operators defined by

$$|\varphi(\mathbf{x})|_R^p \equiv b^{p-1} \int B(\mathbf{x}, \lambda)^\dagger\, |\lambda|^p\, B(\mathbf{x}, \lambda)\, d\lambda\,,$$

for any real $p > 0$ such that $\int |\lambda|^p c(\lambda)^2\, d\lambda < \infty$, which is a well-defined expression for a local operator for a wide class of model functions. *Observe well that in every case the renormalized field product has remained bilinear in $B^\dagger$ and B!*

Introduction of dynamics

We turn our attention from kinematics to dynamics. Like the field operator, the Hamiltonian $\mathcal{H}$ must be constructed out of the irreducible set of basic annihilation and creation operators A and $A^\dagger$ (or B and $B^\dagger$). In view of the ultralocal symmetry, we expect that the nonnegative, self-adjoint operator

$$\mathcal{H} = \int \mathcal{H}(\mathbf{x})\, d^s x\,,$$

where $\mathcal{H}(\mathbf{x})$ denotes a local, nonnegative self-adjoint operator. It is also reasonable to suppose that $\mathcal{H}(\mathbf{x})$ is constructed solely out of the field operators $A(\mathbf{x}, \cdot)^\dagger$ and $A(\mathbf{x}, \cdot)$, all taken at the same point $\mathbf{x}$. Thus we can expect that

$$\mathcal{H}(\mathbf{x}) = \Sigma_{p \geq 0, q \geq 0} \int A(\mathbf{x}, \lambda_1)^\dagger \cdots A(\mathbf{x}, \lambda_p)^\dagger \, h(\lambda_1, \ldots, \lambda_p; \lambda_1', \ldots, \lambda_q')$$
$$\times A(\mathbf{x}, \lambda_1') \cdots A(\mathbf{x}, \lambda_q') \, d\lambda_1 \cdots d\lambda_p \, d\lambda_1' \cdots d\lambda_q' \, .$$

Insisting that $\mathcal{H}$ be a local Hermitian operator implies not only that

$$h(\lambda_1', \ldots, \lambda_q'; \lambda_1, \ldots, \lambda_p) = h(\lambda_1, \ldots, \lambda_p; \lambda_1', \ldots, \lambda_q')^* \, ,$$

but that the sum includes only those terms for which $0 \leq p \leq 1$ and $0 \leq q \leq 1$. This result follows because, as discussed in Section 6.3, the expression $A(\mathbf{x}, \lambda_1)^\dagger A(\mathbf{x}, \lambda_2)^\dagger$ is *not* a local operator; in particular, even after smearing with a test function, the domain of the resultant expression consists of only the zero vector. It follows, therefore, that

$$\mathcal{H}(\mathbf{x}) = \int A(\mathbf{x}, \lambda)^\dagger \, h(\lambda; \lambda') A(\mathbf{x}, \lambda') \, d\lambda \, d\lambda'$$
$$+ \int A(\mathbf{x}, \lambda)^\dagger Y(\lambda) \, d\lambda + \int Y(\lambda')^* A(\mathbf{x}, \lambda') \, d\lambda' + C(\mathbf{x}) \, .$$

Insisting on $\mathcal{H}|0\rangle = 0$ leads to the fact that the c-number functions $Y = C = 0$, and so

$$\mathcal{H}(\mathbf{x}) = \int A(\mathbf{x}, \lambda)^\dagger \, h(\lambda; \lambda') A(\mathbf{x}, \lambda') \, d\lambda \, d\lambda'$$

necessarily holds. A general integral kernel $h(\lambda; \lambda')$ may be replaced by an operator expression $\hbar(-i\partial/\partial\lambda, \lambda)$, and thus

$$\mathcal{H}(\mathbf{x}) = \int A(\mathbf{x}, \lambda)^\dagger \, \hbar(-i\partial/\partial\lambda, \lambda) A(\mathbf{x}, \lambda) \, d\lambda \, .$$

Furthermore, we note that $\mathcal{H} \geq 0$ implies that $\hbar(-i\partial/\partial\lambda, \lambda) \geq 0$, and if we insist that $|0\rangle$ be the unique state of zero energy, then it is in fact necessary that $\hbar(-i\partial/\partial\lambda, \lambda) > 0$.

Our goal now is to determine a reasonable choice for $\hbar(-i\partial/\partial\lambda, \lambda)$. We first note, formally speaking, that the Hamiltonian density has the form

$$\text{``} \mathcal{H}(\mathbf{x}) = \tfrac{1}{2}[\pi(\mathbf{x})^2 + m_o^2 \varphi(\mathbf{x})^2] + \kappa_o \varphi(\mathbf{x})^4 \text{''} \, .$$

From this formal expression we take the idea that the potential part of this expression may be constructed from the local field powers introduced earlier, all of which are bilinear in $B^\dagger$ and B. To that end, we therefore shall require that

$$\mathcal{H}(\mathbf{x}) = \int A(\mathbf{x}, \lambda)^\dagger \, \hbar(-i\partial/\partial\lambda, \lambda) A(\mathbf{x}, \lambda) \, d\lambda$$
$$= \int B(\mathbf{x}, \lambda)^\dagger \, \hbar(-i\partial/\partial\lambda, \lambda) B(\mathbf{x}, \lambda) \, d\lambda \, ,$$

a connection which demands that

$$\hbar(-i\partial/\partial\lambda, \lambda) \, c(\lambda) = 0 \, .$$

This differential equation connects the *kinematics* of the ultralocal models (embodied in c) with the *dynamics* (embodied in $\hbar$). Moreover, since $\hbar > 0$, it follows that $c(\lambda)$ *cannot* be square integrable. Otherwise, if $c(\lambda)$ were square integrable, there would exist infinitely many unequal states with zero energy, contrary to our initial hypothesis; for example, any unnormalized coherent state of the form

$$|u, c\rangle \equiv \exp[\int A(\mathbf{x}, \lambda)^\dagger u(\mathbf{x}) c(\lambda) d^s x \, d\lambda]|0\rangle ,$$

for any $u \in L^2(\mathbb{R}^s)$, would satisfy $\mathcal{H}|u, c\rangle = 0$. To exclude this possibility, it is necessary, therefore, that

$$\int c(\lambda)^2 \, d\lambda = \infty ,$$

as we have anticipated earlier. Since $\int[\lambda^2/(1 + \lambda^2)] c(\lambda)^2 \, d\lambda < \infty$ it follows that c has the (sufficiently) general form

$$c(\lambda) = \frac{K \, e^{-y(\lambda)}}{|\lambda|^\gamma} , \qquad 1/2 \le \gamma < 3/2 ,$$

for some function $y(\lambda)$, and some parameters K and γ. We choose the normalization $y(0) = 0$ so as to remove any ambiguity in the choice of the scale factor K. Nevertheless, observe that the connection between $\hbar$ and c is *entirely independent of the scale factor K*.

Our final step in determining $\hbar(-i\partial/\partial\lambda, \lambda)$ comes from an analogy with the usual quantization procedures. In particular, let us choose

$$\hbar = -\frac{1}{2b}\frac{\partial^2}{\partial\lambda^2} + W(\lambda) .$$

Since we have assumed that $\hbar c = 0$ and since $c(\lambda)$ is nowhere vanishing, it follows in fact that

$$\hbar = -\frac{1}{2b}\left[\frac{\partial^2}{\partial\lambda^2} - \frac{c''(\lambda)}{c(\lambda)}\right] .$$

Another useful representation of the same expression is given by

$$\hbar = -\frac{1}{2b}\frac{1}{c(\lambda)}\frac{\partial}{\partial\lambda}c(\lambda)^2\frac{\partial}{\partial\lambda}\frac{1}{c(\lambda)} .$$

Since we have assumed a functional form for $c(\lambda)$, it follows from either of these latter expressions that

$$\hbar = -\frac{1}{2b}\frac{\partial^2}{\partial\lambda^2} + \frac{1}{2b}\left[\frac{\gamma(\gamma+1)}{\lambda^2} + \frac{2\gamma}{\lambda}y'(\lambda) + y'(\lambda)^2 - y''(\lambda)\right] .$$

As an example, when $y(\lambda) = \bar{y}(\lambda) \equiv bm\lambda^2/2$, it follows that

$$\bar{\hbar} = -\frac{1}{2b}\frac{\partial^2}{\partial\lambda^2} + \frac{\gamma(\gamma+1)}{2b\lambda^2} + \frac{b}{2}m^2\lambda^2 + (\gamma - \tfrac{1}{2})m .$$

This latter expression characterizes the Hamiltonian for the pseudofree theory, and it is useful to discuss this special example further at this point.

Pseudofree Hamiltonian

Let us study the eigenvalue problem

$$\left[-\frac{1}{2b}\frac{\partial^2}{\partial\lambda^2} + \frac{\gamma(\gamma+1)}{2b\lambda^2} + \frac{b}{2}m^2\lambda^2 + (\gamma-\tfrac{1}{2})m\right]\bar{u}_r(\lambda) = \bar{\mu}_r\bar{u}_r(\lambda)$$

and seek solutions $\bar{u}_r(\lambda)$ which are square integrable in λ. From the point of view of ordinary quantum mechanics it is evident that a complete set of normalizable eigenfunctions exist [BlEH 94]. Due to the fact that $\gamma \geq 1/2$ each eigenvalue for the ultralocal pseudofree model is doubly degenerate, and in fact the two eigenfunctions associated with each pair of degenerate eigenvalues are functionally identical save for the fact that one of them is an *even* function of λ while the other one is an *odd* function of λ; by taking suitable linear combinations these eigenfunctions can also be arranged to vanish for $\lambda < 0$ or for $\lambda > 0$, respectively. Indeed, every eigenfunction vanishes for $\lambda = 0$. (These general remarks actually hold for the interacting models as well as for the pseudofree model.) The eigenvalues for the pseudofree Hamiltonian are given by $\bar{\mu}_r = 2([r/2]+\gamma+1/2)m$ for $r \in \{0,1,2,\ldots\}$. In this expression $[r/2]$ denotes the integral part of $r/2$, and therefore the eigenvalues are given by $\bar{\mu}_0 = \bar{\mu}_1 = (1+2\gamma)m$, $\bar{\mu}_2 = \bar{\mu}_3 = (3+2\gamma)m$, $\bar{\mu}_4 = \bar{\mu}_5 = (5+2\gamma)m$, etc. In terms of these quantities the ultralocal pseudofree Hamiltonian can be expressed as

$$\begin{aligned}\mathcal{H} &= \int\mathcal{H}(\mathbf{x})\,d^s x\\ &= \Sigma_r\bar{\mu}_r\int d^s x\int A(\mathbf{x},\lambda)^\dagger\bar{u}_r(\lambda)\,d\lambda\int\bar{u}_r(\lambda')^* A(\mathbf{x},\lambda')\,d\lambda'\\ &\equiv \Sigma_r\bar{\mu}_r\mathsf{N}_r\,,\end{aligned}$$

where

$$\mathsf{N}_r \equiv \int d^s x\int A(\mathbf{x},\lambda)^\dagger\bar{u}_r(\lambda)\,d\lambda\int\bar{u}_r(\lambda')^* A(\mathbf{x},\lambda')\,d\lambda'$$

denotes the number operator for excitation "r" defined in terms of the normalized eigenfunction $\bar{u}_r(\lambda)$. Since the eigenvalues for any such number operator are $0,1,2,\ldots$, it follows that the spectrum for the pseudofree ultralocal Hamiltonian is of the form

$$E = \Sigma_r n_r\bar{\mu}_r\,, \qquad n_r \in \{0,1,2,\ldots\}\,.$$

Since each $\bar{\mu}_r \geq (1+2\gamma)m \geq 2m > 0$, it follows that the ground state is unique and is determined by $n_r = 0$ for all $r \geq 0$. Due to the uniform spacing of the energy levels for the pseudofree Hamiltonian, the spectrum of the pseudofree ultralocal Hamiltonian may also be written in the compact form

$$E = (N_1 + 2N_2\gamma)m\,, \qquad N_1\,, N_2 \in \{0,1,2,3,\ldots\}\,,$$

where $N_1 \equiv \Sigma_r(1+2[r/2])n_r$ and $N_2 \equiv \Sigma n_r$.

The formulas for the energy spectrum for the pseudofree ultralocal Hamiltonian evidently depend on the singularity parameter γ, and it is clear that the analysis of the Hamiltonian and its spectrum apply for *any* $\gamma \geq 1/2$. Below we shall show the special relevance of cases for which $J \equiv 2\gamma + 1$ is an *even integer*. In that case the spectrum of the pseudofree theory is of the extremely simple form

$$E = Nm , \qquad\qquad N \in \{0, J, J+2, J+4, J+6, \ldots\} .$$

Apart from a certain zero point level, it is noteworthy that the spectrum of the pseudofree Hamiltonian has the structure of a uniformly spaced ladder of energy levels. This property lends credence to our use of the pseudofree theory as some sort of "free" ultralocal theory.

General interacting Hamiltonian

The analysis of the spectrum in the general case of an interaction is qualitatively rather similar to that of the pseudofree Hamiltonian except that in general the uniform level spacing is lost. Let us focus on the quartic ultralocal theory and the corresponding eigenvalue equation given by

$$\left[-\frac{1}{2b}\frac{\partial^2}{\partial\lambda^2} + \frac{1}{2b}\frac{\gamma(\gamma+1)}{\lambda^2} + \frac{bm^2}{2}\lambda^2 \right.$$
$$\left. + b^3\kappa\lambda^4 - \text{const.} \right] u_r(\lambda) = \mu_r u_r(\lambda) .$$

Because $\gamma \geq 1/2$, the eigenvalues are doubly degenerate once again, that is $0 < \mu_0 = \mu_1 < \mu_2 = \mu_3 < \mu_4 = \mu_5 < \cdots$. In turn, the eigenfunctions can be chosen either to have the parity of r or, alternatively, to vanish for $\lambda < 0$ if r is even, or for $\lambda > 0$ if r is odd.

In the case of a potential *without* symmetry, for which $c(-\lambda) \neq c(\lambda)$, it is noteworthy that the degeneracy of the eigenvalues is in general lifted. In particular, consider the case of the nonsymmetric potential where

$$\left[-\frac{1}{2b}\frac{\partial^2}{\partial\lambda^2} + \frac{1}{2b}\frac{\gamma(\gamma+1)}{\lambda^2} + \frac{bm^2}{2}\lambda^2 \right.$$
$$\left. + b^2\sigma\lambda^3 + b^3\kappa\lambda^4 - \text{const.} \right] u_r(\lambda) = \mu_r u_r(\lambda) ,$$

with $\sigma \neq 0$ and $\kappa > \sigma^2/(2m^2)$. In the generic case $0 < \mu_0 < \mu_1 < \mu_2 < \mu_3 < \cdots$, and the corresponding eigenfunctions all vanish either for $\lambda < 0$ or for $\lambda > 0$ with no superposition allowed; any exception occurs in the unlikely event of an accidental degeneracy.

The spectrum of the full Hamiltonian $\mathcal{H}$ follows directly from the spectrum of the operator $\hbar$. In particular, just like the case for the pseudofree theory,

$$\mathcal{H} = \int \mathcal{H}(\mathbf{x}) \, d^s x$$
$$= \Sigma_r \mu_r \int d^s x \int A(\mathbf{x}, \lambda)^\dagger u_r(\lambda) \, d\lambda \int u_r(\lambda')^* A(\mathbf{x}, \lambda') \, d\lambda'$$
$$\equiv \Sigma_r \mu_r \mathsf{N}_r ,$$

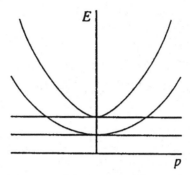

Fig. 10.2. A qualitative depiction of Energy vs. Momentum for a relativistic theory (curved lines) and an ultralocal theory (horizontal lines). The figure corresponds approximately to a free relativistic theory and a free ultralocal theory, although, qualitatively speaking, any polynomially interacting ultralocal theory has the same behavior. Observe that any momentum dependence of the energy levels present in the relativistic case disappears in the ultralocal limit; hence, in the general case, all energy levels for polynomial ultralocal theories are purely discrete.

where

$$N_r \equiv \int d^s x \int A(\mathbf{x}, \lambda)^\dagger u_r(\lambda)\, d\lambda \int u_r(\lambda')^* A(\mathbf{x}, \lambda')\, d\lambda'$$

denotes the number operator for excitation "r" defined in terms of the normalized eigenfunction $u_r(\lambda)$. Observe that in the case of a quartic interaction, the spectrum of the ultralocal models is *discrete*; this result should not be too surprising since we have effectively removed any momentum dependence of a traditional energy-momentum relation. The relation between the momentum dependence of the energy spectrum for a relativistic theory and an ultralocal theory is qualitatively illustrated in Fig. 10.2.

Finally, we note that the eigenvalue equation for the energy eigenvalues $\{\mu_r\}$ is well defined for any value of $\gamma \geq 1/2$. Indeed, we shall implicitly appeal to just these energy levels when we discuss the generalization of the ultralocal models to cases for which $\gamma \geq 3/2$.

Time-dependent field operator

Armed with the sharp-time local field operator

$$\varphi(\mathbf{x}) = \int B(\mathbf{x}, \lambda)^\dagger \, \lambda \, B(\mathbf{x}, \lambda)\, d\lambda$$
$$= \int A(\mathbf{x}, \lambda)^\dagger \, \lambda \, A(\mathbf{x}, \lambda)\, d\lambda$$
$$+ \int A(\mathbf{x}, \lambda)^\dagger \lambda\, c(\lambda)\, d\lambda + \int c(\lambda)\, \lambda\, A(\mathbf{x}, \lambda)\, d\lambda \,,$$

and the Hamiltonian

$$\mathcal{H} = \int d^s x \int A(\mathbf{x}, \lambda)^\dagger \, \hbar \, A(\mathbf{x}, \lambda)\, d\lambda \,,$$

we are ready to define the time-dependent field operator. In particular,

$$\varphi(t, \mathbf{x}) \equiv e^{i\mathcal{H}t}\varphi(\mathbf{x})e^{-i\mathcal{H}t}$$
$$= \int A(\mathbf{x}, \lambda)^\dagger \, e^{i\hbar t}\lambda e^{-i\hbar t} A(\mathbf{x}, \lambda) \, d\lambda$$
$$+ \int A(\mathbf{x}, \lambda)^\dagger \, e^{i\hbar t}\lambda \, c(\lambda) \, d\lambda + \int c(\lambda) \, \lambda \, e^{-i\hbar t} A(\mathbf{x}, \lambda) \, d\lambda$$
$$\equiv \int B(\mathbf{x}, \lambda)^\dagger \, \lambda(t) \, B(\mathbf{x}, \lambda) \, d\lambda \, .$$

In this last expression we have introduced

$$\lambda(t) \equiv e^{i\hbar t} \, \lambda \, e^{-i\hbar t} \, ,$$

and have implicitly used the rule that

$$e^{-i\hbar t} \, c(\lambda) = c(\lambda) \, ,$$

as well as the complex conjugate (with $\hbar$ acting to the left)

$$c(\lambda) \, e^{i\hbar t} = c(\lambda) \, ,$$

which hold as differential identities despite the fact that $c \notin L^2(\mathbb{R})$.

Multipoint functions

The explicit operator construction indicated above enables us to readily determine expressions for multipoint truncated vacuum expectation values. First observe that

$$\langle 0|\varphi(t, \mathbf{x})|0\rangle = \langle 0|\varphi(\mathbf{x})|0\rangle = \int c(\lambda) \, \lambda \, c(\lambda) \, d\lambda = 0 \, .$$

In fact, it is evident that all odd-order vacuum expectation values vanish because of the symmetry $c(-\lambda) = c(\lambda)$; on the other hand, the odd-order vacuum expectation values do *not* necessarily vanish for an asymmetric model function appropriate, for instance, to a potential that includes $\sigma\lambda^3 + \kappa\lambda^4$, with $\sigma \neq 0$ and $\kappa \propto \sigma^2 > 0$. However, for the sake of illustration, we shall confine our present remarks to even potentials.

A direct calculation of the two-point function leads to

$$\langle 0|\varphi(t, \mathbf{x})\varphi(s, \mathbf{y})|0\rangle$$
$$= \langle 0|\int B(\mathbf{x}, \lambda)^\dagger \lambda(t) B(\mathbf{x}, \lambda) \, d\lambda \int B(\mathbf{y}, \lambda')^\dagger \lambda'(s) B(\mathbf{y}, \lambda') \, d\lambda'|0\rangle$$
$$= \langle 0|\int c(\lambda)\lambda(t) B(\mathbf{x}, \lambda) \, d\lambda \int B(\mathbf{y}, \lambda')^\dagger \lambda'(s) \, c(\lambda') \, d\lambda'|0\rangle$$
$$= \delta(\mathbf{x} - \mathbf{y}) \int c(\lambda) \, \lambda(t) \, \lambda(s) \, c(\lambda) \, d\lambda$$
$$= \delta(\mathbf{x} - \mathbf{y}) \int c(\lambda) \, \lambda \, e^{-i\hbar(t-s)} \, \lambda \, c(\lambda) \, d\lambda \, .$$

Rather than proceeding directly to the (truncated) four-point function at this point, let us focus directly on the generating functional for all higher-order functions. Due to the form of the time-dependent local field, it follows that

$$\langle 0|e^{i\int h(x)\varphi(x)\,d^nx}|0\rangle$$
$$= \langle 0|\exp[i\int d^sx \int dt \int B(\mathbf{x},\lambda)^\dagger h(t,\mathbf{x})\lambda(t)\,B(\mathbf{x},\lambda)\,d\lambda]|0\rangle$$
$$= \langle 0| : \exp\{\int d^sx \int B(\mathbf{x},\lambda)^\dagger [e^{i\int h(t,\mathbf{x})\lambda(t)\,dt} - 1]\,B(\mathbf{x},\lambda)\,d\lambda\} : |0\rangle$$
$$= \exp\{-\int d^sx \int c(\lambda)\,[1 - e^{i\int h(t,\mathbf{x})\lambda(t)\,dt}]\,c(\lambda)\,d\lambda\}\,.$$

If we set $\varphi(h) \equiv \int h(x)\varphi(x)\,d^nx$, then the relation between truncated ("T") moments and ordinary moments is implicitly given by

$$\langle 0|[e^{i\varphi(h)} - 1]|0\rangle^T \equiv \ln\left[\langle 0|e^{i\varphi(h)}|0\rangle\right]\,.$$

Therefore, we learn the fundamental fact that

$$\langle 0|[e^{i\int h(x)\varphi(x)\,d^nx} - 1]|0\rangle^T = \int d^sx \int c(\lambda)\,[e^{i\int h(t,\mathbf{x})\lambda(t)\,dt} - 1]\,c(\lambda)\,d\lambda\,.$$

As we will establish below, this expression for the generating functional of truncated vacuum expectation values applies for any value of $\gamma \geq 1/2$.

Let us discuss the four-point function in some detail. In particular, we focus on

$$\langle 0|\varphi(t_1,\mathbf{x_1})\varphi(t_2,\mathbf{x_2})\varphi(t_3,\mathbf{x_3})\varphi(t_4,\mathbf{x_4})|0\rangle^T\,.$$

For completeness, and in an abbreviated notation, we recall the relation between the truncated four-point function and the ordinary four-point function given by

$$\langle 0|\varphi_1\varphi_2\varphi_3\varphi_4|0\rangle^T \equiv \langle 0|\varphi_1\varphi_2\varphi_3\varphi_4|0\rangle - \langle 0|\varphi_1\varphi_2|0\rangle\langle 0|\varphi_3\varphi_4|0\rangle$$
$$- \langle 0|\varphi_1\varphi_3|0\rangle\langle 0|\varphi_2\varphi_4|0\rangle - \langle 0|\varphi_1\varphi_4|0\rangle\langle 0|\varphi_2\varphi_3|0\rangle\,,$$

as follows from the expression above. From the logarithm of the generating functional it is straightforward to show that

$$\langle 0|\varphi(t_1,\mathbf{x_1})\varphi(t_2,\mathbf{x_2})\varphi(t_3,\mathbf{x_3})\varphi(t_4,\mathbf{x_4})|0\rangle^T$$
$$= \delta(\mathbf{x_1} - \mathbf{x_2})\delta(\mathbf{x_2} - \mathbf{x_3})\delta(\mathbf{x_3} - \mathbf{x_4})$$
$$\times \int c(\lambda)\,\lambda(t_1)\,\lambda(t_2)\,\lambda(t_3)\,\lambda(t_4)\,c(\lambda)\,d\lambda\,.$$

The generalization of this kind of relation to any higher-order truncated vacuum expectation values is evident. The point to observe clearly from these calculations is that the *truncated* field operator correlation functions are determined from the *ordinary* correlation functions of the variable λ.

A similar construction applies to the generating functional for time-ordered vacuum expectation values, when they exist. In that case, it follows that

$$Z\{h\} = \langle 0|Te^{i\int h(x)\varphi(x)\,d^nx}|0\rangle$$
$$= \exp\{-\int d^sx \int c(\lambda)\,[1 - Te^{i\int h(t,\mathbf{x})\lambda(t)\,dt}]\,c(\lambda)\,d\lambda\}\,.$$

And therefore we observe that the truncated, time-ordered field operator vacuum expectation values are given simply by the time-ordered ordinary correlation functions of the λ variables.

Absence of canonical commutation relations

We now take up the question of canonical operators and canonical commutation relations. First of all we define what one would expect to be the canonical momentum (at time zero) by

$$
\begin{aligned}
\pi(\mathbf{x}) &\equiv i[\mathcal{H}, \varphi(\mathbf{x})] \\
&= i[\int d^s y \int B(\mathbf{y}, \lambda)^\dagger \not{k} B(\mathbf{y}, \lambda)\, d\lambda,\ \int B(\mathbf{x}, \lambda')^\dagger \lambda' B(\mathbf{x}, \lambda')\, d\lambda'] \\
&= i \int B(\mathbf{x}, \lambda)^\dagger [\not{k}, \lambda]\, B(\mathbf{x}, \lambda)\, d\lambda \\
&= -i b^{-1} \int B(\mathbf{x}, \lambda)^\dagger (\partial/\partial\lambda)\, B(\mathbf{x}, \lambda)\, d\lambda\ .
\end{aligned}
$$

We assert that the resultant expression does *not* generate a local field operator, and it is worth establishing this fact in some detail. Granting, due to symmetry, that the c-number part (proportional to the identity operator) of $\pi(\mathbf{x})$ vanishes, it follows that

$$
\begin{aligned}
b\,\pi(\mathbf{x}) = &-i\int A(\mathbf{x}, \lambda)^\dagger (\partial/\partial\lambda)\, A(\mathbf{x}, \lambda)\, d\lambda \\
&-i\int A(\mathbf{x}, \lambda)^\dagger (\partial/\partial\lambda)\, c(\lambda)\, d\lambda - i\int c(\lambda)\, (\partial/\partial\lambda)\, A(\mathbf{x}, \lambda)\, d\lambda\ .
\end{aligned}
$$

It is noteworthy that the first and last terms in this expression do define local operators, it is only the middle term which fails to do so. In view of the singularity in $c(\lambda)$ at the origin it follows that $(\partial/\partial\lambda)\, c(\lambda) = c'(\lambda)$ is not square integrable near the origin. Stated as an equation, $\int_{|\lambda|<1} |c'(\lambda)|^2\, d\lambda = \infty$. For this discussion, let us assume that the test function $0 \not\equiv g \in L^2(\mathbb{R}^s)$, and set $A(g, \lambda)^\dagger \equiv \int A(\mathbf{x}, \lambda)^\dagger g(\mathbf{x})\, d^s x$. Then it follows that

$$
b^2\, \|\pi(g)|0\rangle\|^2 = \|\int A(g, \lambda)^\dagger c'(\lambda)\, d\lambda|0\rangle\|^2 = (g, g)\int |c'(\lambda)|^2\, d\lambda = \infty\ .
$$

A similar calculation confirms that $\pi(g)$ has only the zero vector in its domain, and in our terminology $\pi(g)$ is not an operator; instead, $\pi(g)$ is a bilinear form. This fact has consequences for the canonical commutation relations.

Let us ignore the fact that $\pi(\mathbf{y})$ is not a local operator. Indeed, let us proceed blindly, and formally calculate

$$
\begin{aligned}
[\varphi(\mathbf{x}),\, \pi(\mathbf{y})] \\
&= -i b^{-1}[\int B(\mathbf{x}, \lambda)^\dagger \lambda B(\mathbf{x}, \lambda) d\lambda,\ \int B(\mathbf{y}, \lambda')^\dagger (\partial/\partial\lambda') B(\mathbf{y}, \lambda') d\lambda'] \\
&= -i\delta(\mathbf{x} - \mathbf{y}) b^{-1} \int B(\mathbf{x}, \lambda)^\dagger [\lambda, (\partial/\partial\lambda)]\, B(\mathbf{x}, \lambda)\, d\lambda \\
&= i\delta(\mathbf{x} - \mathbf{y}) b^{-1} \int B(\mathbf{x}, \lambda)^\dagger B(\mathbf{x}, \lambda)\, d\lambda \\
&= i\delta(\mathbf{x} - \mathbf{y}) b^{-1} \int [c(\lambda)^2 + \cdots]\, d\lambda\ ,
\end{aligned}
$$

which clearly has a diverging c-number component. Consequently, we learn that *no interacting ultralocal model obeys canonical commutation relations!*[*]

---

[*] Even setting dynamical considerations aside, one can show [HeK 72] that there is *no* local self-adjoint operator that could serve as a canonical momentum to the given local self-adjoint field operator $\varphi(\mathbf{x})$.

Cyclicity of the field operators

The Hilbert space in the present case is the Fock space $\mathfrak{H}$ built from repeated action of the creation operators acting on the unique no-particle state $|0\rangle$. As has been the case previously, a total set of vectors is provided by the general set of (normalized) coherent states, namely vectors of the form

$$|\psi\rangle = N \exp[\int d^s x \int d\lambda \, A(\mathbf{x}, \lambda)^\dagger \psi(\mathbf{x}, \lambda)]\,|0\rangle \;,$$

where

$$N = \exp[-\tfrac{1}{2} \int d^s x \int d\lambda \, |\psi(\mathbf{x}, \lambda)|^2] \;,$$

defined for a dense set of elements or indeed for all $\psi \in L^2(\mathbb{R}^{s+1})$. We wish to show that the vector $|0\rangle$ is a cyclic vector for the sharp-time field operator algebra generated by the unitary operators $\exp[i\phi(f)]$ for all real f in some natural space, e.g., all bounded functions of compact support. Recall that a vector $|0\rangle$ is cyclic for an algebra $\mathcal{A}$ provided the set of vectors $\mathcal{A}|0\rangle$ is dense in the Hilbert space [Em 72].

We may establish the cyclicity of $|0\rangle$ by first recalling that local powers of the field operator have already been shown to lead to local operators of the general form

$$W[\phi(\mathbf{x})]_R \equiv b^{-1} \int B(\mathbf{x}, \lambda)^\dagger \, W[b\lambda] \, B(\mathbf{x}, \lambda) \, d\lambda \;.$$

For real, bounded, smooth functions W with $W[0] = 0$, it is straightforward to establish that when the local field operator is smeared with a real test function f, the result is a self-adjoint operator $W[f]_R$ [Kl 71]. Let W_j denote a set of such operators and f_j a set of suitable real test functions, and define

$$\mathcal{J} \equiv \Sigma_j W_j[f_j]_R$$

for an arbitrary finite linear combination such that $\mathcal{J}$ remains self adjoint; note that all such operators commute with each another. Let us consider the unitary operators $\exp[i\mathcal{J}]$ generated by these self-adjoint operators. The action of these unitary operators on the no-particle state leads to coherent states of the form

$$\exp[i\mathcal{J}]|0\rangle = N \exp[\int d^s x \int d\lambda \, A(\mathbf{x}, \lambda)^\dagger J(\mathbf{x}, \lambda)]\,|0\rangle$$

for a set of square-integrable functions each of which is given by

$$J(\mathbf{x}, \lambda) \equiv \{e^{ib^{-1}\Sigma_j f_j(\mathbf{x})W_j[b\lambda]} - 1\} \, c(\lambda) \;.$$

In the present case

$$N = \exp[-\tfrac{1}{2} \int d^s x \int d\lambda \, |J(\mathbf{x}, \lambda)|^2] \;.$$

Evidently such functions J are dense in the set of functions in $L^2(\mathbb{R}^{s+1})$, and so the set of coherent states defined with a general set of functions J are dense in the set of all coherent states and therefore form a total set in $\mathfrak{H}$. Hence, we have determined that the set of vectors $|f\rangle = \exp[i\varphi(f)]|0\rangle$, defined for all bounded functions f with compact support, span the Hilbert space $\mathfrak{H}$.

Failure of the Araki relations for ultralocal models

The fact that the momentum operator is not a local operator, after all, has profound consequences for relating the quantum solution to the motivating classical theory for ultralocal models. In particular, using $|f\rangle = \exp[i\varphi(f)]|0\rangle$, the Araki relations

$$\langle f|\pi(\mathbf{x})|f'\rangle = \tfrac{1}{2}[f(\mathbf{x}) + f'(\mathbf{x})]\langle f|f'\rangle,$$
$$\langle f|\mathcal{H}|f'\rangle = \tfrac{1}{2}(f, f')\langle f|f'\rangle$$

derived in Section 6.9, no longer hold true for ultralocal models since they are based on the canonical commutation relations. In fact, for ultralocal models, it follows for any $f \not\equiv 0$ that $|f\rangle$ is not in the form domain of $\mathcal{H}$, or stated otherwise, $|f\rangle \notin \mathfrak{D}(\mathcal{H}^{1/2})$. This fact may be readily ascertained.

We first observe that

$$\exp[i\varphi(f)]|0\rangle = \exp[i\textstyle\int d^s x \int B(\mathbf{x}, \lambda)^\dagger f(\mathbf{x})\lambda\, B(\mathbf{x}, \lambda)\, d\lambda]|0\rangle$$
$$= :\exp[\textstyle\int d^s x \int B(\mathbf{x}, \lambda)^\dagger [e^{if(\mathbf{x})\lambda} - 1]B(\mathbf{x}, \lambda)\, d\lambda]: |0\rangle$$
$$= \exp[\textstyle\int d^s x \int B(\mathbf{x}, \lambda)^\dagger [e^{if(\mathbf{x})\lambda} - 1]c(\lambda)\, d\lambda]|0\rangle$$
$$= N_f \exp[\textstyle\int d^s x \int A(\mathbf{x}, \lambda)^\dagger [e^{if(\mathbf{x})\lambda} - 1]c(\lambda)\, d\lambda]|0\rangle,$$

where

$$N_f \equiv \exp[\textstyle\int d^s x \int c(\lambda)[e^{if(\mathbf{x})\lambda} - 1]c(\lambda)\, d\lambda]$$

is a normalization factor. From the general structure – qualitatively of the form $\exp[A^\dagger]|0\rangle$ – we recognize $|f\rangle$ as a coherent state. Consequently,

$$\mathcal{H}|f\rangle = \textstyle\int d^s x \int A(\mathbf{x}, \lambda)^\dagger\, \hbar\, A(\mathbf{x}, \lambda)\, d\lambda\, |f\rangle$$
$$= \textstyle\int d^s x \int A(\mathbf{x}, \lambda)^\dagger\, \hbar\, [e^{if(\mathbf{x})\lambda} - 1]\, c(\lambda)\, d\lambda\, |f\rangle.$$

In turn, it readily follows that

$$\|\mathcal{H}^{1/2}|f\rangle\|^2 \equiv \langle f|\mathcal{H}|f\rangle$$
$$= \textstyle\int d^s x \int c(\lambda)[e^{-if(\mathbf{x})\lambda} - 1]\, \hbar\, [e^{if(\mathbf{x})\lambda} - 1]\, c(\lambda)\, d\lambda$$
$$= \textstyle\int d^s x \int c(\lambda)e^{-if(\mathbf{x})\lambda}\, \hbar\, e^{if(\mathbf{x})\lambda}c(\lambda)\, d\lambda$$
$$= \tfrac{1}{2}b^{-1}\textstyle\int d^s x\, f(\mathbf{x})^2 \int c(\lambda)^2\, d\lambda,$$

which diverges for any $f \not\equiv 0$. Thus $|f\rangle$ is not in the form domain of $\mathcal{H}$ for any $f \not\equiv 0$.

10.5 Functional integral formulation

At the beginning of this chapter we showed that a conventional functional integral formulation for ultralocal models resulted in an ultralocal analog of a generalized free field, namely, a Gaussian expectation functional, and thus a trivial answer.

However, using symmetry and general principles, we have arrived at a nontrivial formulation of an ultralocal quantum field theory embodied, for example, in the fact that

$$Z\{h\} = \langle 0|Te^{i\int h(x)\varphi(x)\,d^n x}|0\rangle$$
$$= \exp\{-\int d^s x \int c(\lambda)\,[1 - Te^{i\int h(t,x)\lambda(t)\,dt}]\,c(\lambda)\,d\lambda\}\,,$$

which holds for a wide class of model functions all of which lead to non-Gaussian results. In this section we take up the question of how to determine this nontrivial result from a functional integral, one which evidently must be qualitatively different from the conventional one studied before. We will find that an $O(\hbar^2)$ modification of the interaction potential will achieve our goal.

From a pedagogical point of view it is easier to start with the answer and demonstrate that it leads to the desired result [Kl 94]. Thus, for a quartic self-interacting ultralocal model, we assert that

$$Z\{h\} = \lim_{a\to 0}\lim_{\epsilon\to 0} M \int \exp(i\{\Sigma h_k \phi_k a^s \epsilon$$
$$+ \tfrac{1}{2}(ba^s)\Sigma(\phi_{k^*} - \phi_k)^2 \epsilon^{-2} a^s \epsilon - \tfrac{1}{2}(ba^s)m^2 \Sigma \phi_k^2 a^s \epsilon - (ba^s)^3 g \Sigma \phi_k^4 a^s \epsilon$$
$$- \tfrac{1}{2}b^2(ba^s)^{-3}\Sigma[\gamma(\gamma + 1)\phi_k^2 - \gamma a^{-2s}\delta^2](\phi_k^2 + a^{-2s}\delta^2)^{-2} a^s \epsilon\}) \Pi\,d\phi_k\,.$$

Here a denotes the spatial lattice spacing, a^s is the volume of an s-dimensional spatial lattice cell, ϵ is the temporal lattice spacing, b is the dimensional parameter introduced previously, and $\delta = \delta(a)$ (which, in the present case, is *not* a Dirac δ-function!) will be defined shortly. In addition, here, M is a normalization chosen so that $Z\{0\} = 1$, $k = (k^0, k^1, \ldots, k^s)$, $k^j \in \{0, \pm1, \pm2, \ldots\}$ for all j, and k^* denotes the point next to k advanced one step in time ($k^0 \to k^0 + 1$). The combination ba^s is dimensionless, goes to zero as $a \to 0$, and represents the lattice regularized version of the formal expression $b/\delta(0)$. As was the case in discussing conventional field theories in Chapter 5, the mass implicitly contains a small negative imaginary part to ensure convergence and which is sent to zero after all integrations have been performed; as we have done on previous occasions, we shall treat this small imaginary term and the associated ultimate limit implicitly. Observe that the expected terms in the potential (proportional to m^2 and g, for example) involve ba^s to a positive power, while the additional, auxiliary term (proportional to γ) involves this dimensionless regularization factor to a negative power. Ultimately, we will show that this auxiliary term is proportional to $\hbar^2$, and thus it vanishes in the classical limit where $\hbar \to 0$.

Our next step is to change integration variables from ϕ_k to $\lambda_k \equiv \phi_k a^s$, which leads to

$$Z\{h\} = \lim_{a\to 0}\lim_{\epsilon\to 0} N' \int \exp(i\{\Sigma h_k \lambda_k \epsilon$$
$$+ \tfrac{1}{2}b\Sigma(\lambda_{k^*} - \lambda_k)^2 \epsilon^{-1} - \tfrac{1}{2}bm^2 \Sigma \lambda_k^2 \epsilon - b^3 g \Sigma \lambda_k^4 \epsilon$$
$$- \tfrac{1}{2}b^{-1}\Sigma[\gamma(\gamma + 1)\lambda_k^2 - \gamma\delta^2](\lambda_k^2 + \delta^2)^{-2}\epsilon\}) \Pi\,d\lambda_k\,.$$

Let $p \equiv k^0$ and $q \equiv (k^1, k^2, \ldots, k^s)$. Note that the spatial label q on $\lambda_{p,q}$ may be dropped as it appears merely as a dummy integration variable. Thus we are led to

$$Z\{h\} = \lim_{a \to 0} \lim_{\epsilon \to 0} \prod_q N \int \exp(i\{\Sigma h_{p,q} \lambda_p \epsilon$$
$$+ \tfrac{1}{2} b \Sigma (\lambda_{p+1} - \lambda_p)^2 \epsilon^{-1} - \tfrac{1}{2} bm^2 \Sigma \lambda_p^2 \epsilon - b^3 g \Sigma \lambda_p^4 \epsilon$$
$$- \tfrac{1}{2} b^{-1} \Sigma [\gamma(\gamma+1)\lambda_p^2 - \gamma \delta^2](\lambda_p^2 + \delta^2)^{-2} \epsilon\}) \, \Pi \, d\lambda_p \ .$$

Observe, for each q, that the indicated expression is the lattice form of a one-dimensional quantum mechanical path integral, as discussed, for example, in Section 4.5. The limit $\epsilon \to 0$ may now be taken and the result written in two different but equivalent forms. On the one hand,

$$Z\{h\} = \lim_{a \to 0} \prod_q \mathcal{N} \int \exp(i \int h_q(t) \lambda(t) \, dt$$
$$+ i \int \{ \tfrac{1}{2} b \dot{\lambda}(t)^2 - \tfrac{1}{2} bm^2 \lambda(t)^2 - b^3 g \lambda(t)^4$$
$$- \tfrac{1}{2} b^{-1} [\gamma(\gamma+1)\lambda(t)^2 - \gamma \delta^2](\lambda(t)^2 + \delta^2)^{-2} \} \, dt) \, \mathcal{D}\lambda \ ,$$

while, on the other hand,

$$Z\{h\} = \lim_{a \to 0} \prod_q \int \psi_\delta^*(\lambda) \mathsf{T} e^{i \int [h_q(t)\lambda - \hbar_\delta(-i\partial/\partial\lambda, \lambda)] \, dt} \psi_\delta(\lambda) \, d\lambda \ ,$$

where as usual T denotes time ordering, and ψ_δ denotes the normalized ground state of $\hbar_\delta$ which in turn is defined by

$$\hbar_\delta(-i\partial/\partial\lambda, \lambda) = -\tfrac{1}{2} b^{-1}(\partial^2/\partial\lambda^2) + \tfrac{1}{2} bm^2 \lambda^2 + b^3 g \lambda^4$$
$$+ \tfrac{1}{2} b^{-1}[\gamma(\gamma+1)\lambda^2 - \gamma\delta^2](\lambda^2 + \delta^2)^{-2} - \text{const.} \ ,$$

with the const. chosen so that $\hbar_\delta \psi_\delta = 0$. Observe that $\delta > 0$ provides a cutoff for the singularity of the auxiliary potential at $\lambda = 0$, and therefore $\hbar_\delta$ is a regularized version of $\hbar$. With δ very small it follows that

$$\psi_\delta(\lambda) = B(\lambda^2 + \delta^2)^{-\gamma/2} e^{-y(\lambda, \delta)} \ ,$$

where $y(\lambda, \delta)$ is very little changed by the presence of δ, i.e., $y(\lambda, \delta) \simeq y(\lambda)$, and B denotes a real normalization constant chosen so that

$$1 = \int |\psi_\delta(\lambda)|^2 \, d\lambda = B^2 \int (\lambda^2 + \delta^2)^{-\gamma} e^{-2y(\lambda, \delta)} \, d\lambda \ .$$

First let $\gamma > 1/2$. Then for $\delta \ll 1$ we may estimate B by means of the equation

$$1 = B^2 \int (\lambda^2 + \delta^2)^{-\gamma} \, d\lambda = B^2 G \delta^{1-2\gamma} \ ;$$

here $G \equiv \int (y^2+1)^{-\gamma} dy = \Gamma(\tfrac{1}{2})\Gamma(\gamma - \tfrac{1}{2})/\Gamma(\gamma)$. We now *require* that $B^2 \equiv K^2 a^s$, a demand that determines $\delta = \delta(a)$ to be

$$\delta = (GK^2 a^s)^{1/(2\gamma-1)} \ .$$

Instead, if $\gamma = 1/2$ and $\delta \ll 1$, then to a sufficient accuracy,

$$1 = B^2 \int_0^{\delta_o} (\lambda^2 + \delta^2)^{-1/2} \, d\lambda = B^2 \int_0^{\delta_o/\delta} (y^2 + 1)^{-1/2} \, dy = B^2 |\ln(\delta/\delta_o)| \,,$$

and as a consequence

$$\delta = \delta_o e^{-1/(K^2 a^s)} \,,$$

where $0 < \delta_o < \infty$.

For all $\gamma \geq 1/2$, observe that $\delta(a) \to 0$ as $a \to 0$. Let $\psi_\delta(\lambda) \equiv a^{s/2} c_\delta(\lambda)$, in which case

$$c_\delta(\lambda) = K(\lambda^2 + \delta^2)^{-\gamma/2} \, e^{-y(\lambda,\delta)} \,.$$

Evidently, $c_\delta(\lambda) \to c(\lambda)$ as $\delta \to 0$, and thus also as $a \to 0$. In the new notation

$$Z\{h\} = \lim_{a \to 0} \Pi_q \, a^s \int c_\delta(\lambda) \, \mathsf{T} \, e^{i \int [h_q(t)\lambda - \hbar_\delta] \, dt} \, c_\delta(\lambda) \, d\lambda$$

$$= \lim_{a \to 0} \Pi_q \{1 - a^s \int c_\delta(\lambda) \, \mathsf{T} \, [1 - e^{i \int h_q(t)\lambda \, dt}] \, e^{-i \int \hbar_\delta \, dt} \, c_\delta(\lambda) \, d\lambda\} \,.$$

At this point, the singularity of $c(\lambda)$ at $\lambda = 0$ is rendered harmless thanks to symmetry and the factor $[1 - \exp(i \int h_q(t)\lambda \, dt)]$. Thus, within the last integral, and to the order needed, the limit $\delta \to 0$ can already be taken leading to

$$Z\{h\} = \lim_{a \to 0} \Pi_q \{1 - a^s \int c(\lambda) \, \mathsf{T} \, [1 - e^{i \int h_q(t)\lambda \, dt}] \, e^{-i \int \hbar \, dt} \, c(\lambda) \, d\lambda\} \,,$$

or stated alternatively as

$$Z\{h\} = \lim_{a \to 0} \Pi_q (1 - a^s \int c(\lambda) \, \{1 - \mathsf{T} e^{i \int [h_q(t)\lambda - \hbar] \, dt}\} \, c(\lambda) \, d\lambda) \,.$$

Taking the continuum limit, we find the result

$$Z\{h\} = \exp(- \int d^s x \int c(\lambda) \, \{1 - \mathsf{T} e^{i \int [h(t,\mathbf{x})\lambda - \hbar] \, dt}\} \, c(\lambda) \, d\lambda) \,.$$

The expression we have just derived is presented in the Schrödinger picture. We may also exhibit the same result in the Heisenberg picture by introducing

$$\lambda(t) \equiv e^{i\hbar t} \, \lambda \, e^{-i\hbar t} \,,$$

which leads to the alternative expression

$$Z\{h\} = \exp(- \int d^s x \int c(\lambda) \, \{1 - \mathsf{T} e^{i \int h(t,\mathbf{x})\lambda(t) \, dt}\} \, c(\lambda) \, d\lambda) \,.$$

With the previous derivation, we have illustrated a lattice-regularized functional integral formulation for the quartic self-interacting ultralocal model which in the continuum limit leads to the nontrivial answer we have derived on the grounds of symmetry and general principles. The extension to treat other models is self evident. It is worth emphasizing the important role played in that derivation by the auxiliary potential, which entered as a regularized form of the potential λ^{-2} in $\hbar$. It is also important to emphasize, however, that this auxiliary potential is in reality proportional to $\hbar^2$, and so does not appear in the classical theory in which $\hbar \to 0$.

Let us briefly reinstate the parameter $\hbar$ so as to convince ourselves of the stated dependence. For convenience, let us illustrate this behavior for the pseudofree model. In that case $c(\lambda) = K|\lambda|^{-\gamma} \exp(-bm\lambda^2/2\hbar)$ and therefore

$$\hbar = -\frac{\hbar^2}{2b}\left[\frac{\partial^2}{\partial\lambda^2} - \frac{c''(\lambda)}{c(\lambda)}\right]$$

$$= -\frac{\hbar^2}{2b}\frac{\partial^2}{\partial\lambda^2} + \frac{\hbar^2\gamma(\gamma+1)}{2b\lambda^2} + \frac{1}{2}bm^2\lambda^2 + \hbar(\gamma - \tfrac{1}{2})m .$$

With this expression it is clear that the auxiliary potential is proportional to $\hbar^2$ as stated. This property holds for all ultralocal models. In the classical limit in which $\hbar \to 0$ such a term would not be present. Until further notice, we return to our custom of setting $\hbar = 1$.

10.6 Large singularity parameters

When the singularity parameter $\gamma \geq 3/2$, the sharp-time field is no longer a local field operator. This follows because the basic condition on the model function $c(\lambda)$, i.e., $\int[\lambda^2/(1+\lambda^2)]c(\lambda)^2 d\lambda < \infty$, is no longer valid. One way around this problem is to use a local field power in place of $\varphi(\mathbf{x})$. In particular, let us consider the local power

$$\varphi_R^\theta(\mathbf{x}) \equiv b^{\theta-1}\int B(\mathbf{x}, \lambda)^\dagger \lambda^\theta B(\mathbf{x}, \lambda) d\lambda ,$$

where λ^θ is defined in the present case as an *odd* function for any $\theta > 0$; that is

$$\lambda^\theta \equiv \pm|\lambda|^\theta , \qquad \pm\lambda \geq 0 .$$

It follows that

$$\langle 0|e^{i\varphi_R^\theta(f)}|0\rangle = \exp(-\int d^s x \int\{1 - \cos[f(\mathbf{x})b^{\theta-1}\lambda^\theta]\}c(\lambda)^2 d\lambda) ,$$

which, with $\lambda^{\theta\cdot2} \equiv (\lambda^\theta)^2$, is well defined so long as

$$\int[\lambda^{\theta\cdot2}/(1+\lambda^{\theta\cdot2})]c(\lambda)^2 d\lambda < \infty .$$

Since $c(\lambda) \propto |\lambda|^{-\gamma}$ near $\lambda = 0$, it follows that the generating functional for φ_R^θ is well defined so long as

$$\tfrac{1}{2} \leq \gamma < \tfrac{1}{2} + \theta .$$

Thus it is always possible to choose θ sufficiently large in order that a local power of the field operator exists as an operator that requires only a spatial smearing function.

However, we can also define an operator without resorting to local powers. Specifically, we wish to show that the operator $\varphi(t, \mathbf{x})$ is a well-defined, operator-valued distribution regarding test functions in both $\mathbf{x}$ *and* t for any $\gamma \geq 1/2$. To show this we may argue as follows. For a quartic self-interacting ultralocal model, as Exercise 10.3 shows, the eigenfunctions $u_r(\lambda)$ of $\mathfrak{h}$, satisfying $\mathfrak{h}u_r(\lambda) = \mu_r u_r(\lambda)$, $r \in \{0, 1, 2, \ldots\}$, for any $\gamma \geq 1/2$, fulfill the asymptotic condition given by

$$|k_r| \equiv |\textstyle\int u_r^*(\lambda)\,\lambda\,c(\lambda)\,d\lambda| \propto r^{2\gamma/3-3/2}\,,$$

while the energy levels themselves satisfy $\mu_r \propto r^{4/3}$ for large r. The two-point function, for example, is given by

$$\langle 0|\varphi(t,\mathbf{x})\varphi(s,\mathbf{y})|0\rangle = \delta(\mathbf{x}-\mathbf{y})\textstyle\int c(\lambda)\,\lambda e^{-i\mathfrak{h}(t-s)}\lambda\,c(\lambda)\,d\lambda$$
$$= \delta(\mathbf{x}-\mathbf{y})\Sigma_r|k_r|^2\,e^{-i\mu_r(t-s)}\,.$$

When $\gamma \geq 3/2$ and the sequence $\{|k_r|\}$ is not square summable, we can make this expression finite by smearing in time with a smooth function $v(t)v(s)$ so that the sum of interest becomes proportional to

$$\Sigma_r|k_r|^2\,|\tilde{v}(\mu_r)|^2\,.$$

Evidently, if the Fourier transform $\tilde{v}(\omega)$ of $v(t)$ falls off sufficiently fast, the sum will converge as desired. This argument may be repeated for general coherent states, which leads to the same conclusion. Thus we find that space-time smearing always leads to an operator for any $\gamma \geq 3/2$ and *a fortiori* for any $\gamma \geq 1/2$.

There is another very instructive way to look at these conclusions when $\gamma > 3/2$. Let us again examine

$$\textstyle\int c(\lambda)\,\lambda e^{-i\mathfrak{h}(t-s)}\,\lambda\,c(\lambda)\,d\lambda\,,$$

and note that this distribution can also be expressed by the function defined, for $\epsilon > 0$, by

$$J(t-s) \equiv \textstyle\int c(\lambda)\,\lambda e^{-\epsilon\mathfrak{h}-i\mathfrak{h}(t-s)}\,\lambda\,c(\lambda)\,d\lambda$$

in the limit that $\epsilon \to 0$. We note without proof – see Exercise 10.4 – that for $\epsilon \ll 1$ and for $|t-s| \ll 1$ as well, and for some nonzero constant R, that

$$J(t-s) \simeq \frac{R}{[\epsilon + i(t-s)]^{\gamma-3/2}}\,.$$

This equation represents the short time behavior of the two-point function for any $\gamma > 3/2$, and we draw the important conclusion that γ *controls the short time singularity for ultralocal models.* It is straightforward to show that this expression leads to a proper distribution as $\epsilon \to 0$. To that end, consider

$$\textstyle\int v(t)J(t-s)v(s)\,dt\,ds = \int v(t+s)J(t)v(s)\,dt\,ds = \int w(t)J(t)\,dt\,,$$

for a suitable w, after integrating out the variable s. That this expression is well defined follows from the fact, for $\gamma \neq 5/2$, that

$$\int_{-1}^{1}\frac{1}{[\epsilon+it]^{\gamma-3/2}}\,dt = \frac{i}{(\gamma-5/2)(\epsilon+it)^{\gamma-5/2}}\Bigg|_{-1}^{1}$$

has a well-defined limit as $\epsilon \to 0$. This integral leads to a logarithm when $\gamma = 5/2$, but the conclusion is the same.

In the same line of inquiry, it is interesting to examine the previous question for *time-ordered* operators, i.e., to consider

$$K(t-s) \equiv \int c(\lambda) \, \lambda \, e^{-\epsilon\hbar - i\hbar|t-s|} \, \lambda \, c(\lambda) \, d\lambda \; .$$

By a similar argument, it follows, for some nonzero constant S, that

$$K(t-s) \simeq \frac{S}{[\epsilon + i|t-s|]^{\gamma-3/2}} \; .$$

Next we study

$$\int v(t) K(t-s) v(s) \, dt \, ds = \int w(t) K(t) \, dt \; .$$

In the present case, the existence of this integral is equivalent to the existence of

$$\int_{-1}^{1} \frac{1}{[\epsilon + i|t|]^{\gamma-3/2}} \, dt = \frac{2i}{(\gamma - 5/2)(\epsilon + it)^{\gamma-5/2}} \bigg|_{0}^{1} \; ,$$

which only leads to a uniformly bounded result provided that $\gamma < 5/2$. The same result holds for general coherent state matrix elements as well, and we conclude that the time-ordered operator expressions are well defined space-time distributions only when $\gamma < 5/2$. For larger γ values, they do not represent space-time distributions.

Nevertheless, if we change the test function space to *avoid* coincident time values, then we can speak of bonafide distributions in the time-ordered case as well for any value of $\gamma \geq 1/2$. As an example of what we have in mind, consider the integral

$$\int v(t,s) \, K(t-s) \, dt \, ds \; ,$$

where now the symmetric function $v(t,s)$ denotes a test function that *vanishes* when $t \to s$ faster than any power. It is then clear that this expression is now well defined, and there exists a natural generalization to symmetric, higher-order, time-ordered correlation functions, and suitable test functions that vanish for any pair of coincident points, that leads to well-defined distributions for any time-ordered set of correlation functions and any $\gamma \geq 1/2$. This discussion is patterned after a comparable discussion that holds in quantum field theory [OsS 73, OsS 75].

In summary, we have found that:

(i) For $3/2 > \gamma \geq 1/2$, $\varphi(t,\mathbf{x})$ is an operator with suitable spatial smearing only.

(ii) For $\gamma \geq 1/2$, $\varphi(t,\mathbf{x})$ is an operator with suitable space and time smearing.

(iii) For $5/2 > \gamma \geq 1/2$, $\varphi(t,\mathbf{x})$ is an operator with suitable space and time smearing that may also be used in time-ordered expressions.

(iv) For $\gamma \geq 1/2$, the family of time-ordered correlation functions constitute a family of distributions for special test functions that vanish faster than any power for coincident temporal points.

Renormalized kinetic energy

Now that we have generalized our basic models to ones for which the singularity parameter can assume any value $\gamma \geq 1/2$, it is appropriate to address the question of the kinetic energy. Recall that we have shown for any $\gamma \geq 1/2$, that the momentum operator expression $\pi(\mathbf{x})$ is only a form and not a local operator. This fact has implications for the local kinetic energy. The Hamiltonian operator

$$\mathcal{H} = \int \mathcal{H}(\mathbf{x}) \, d^s x = \int d^s x \int B(\mathbf{x}, \lambda)^\dagger \, \hbar \, B(\mathbf{x}, \lambda) \, d\lambda$$

contains – for an even polynomial interaction – the differential operator

$$\hbar = -\frac{1}{2b} \frac{\partial^2}{\partial \lambda^2} + \frac{\gamma(\gamma + 1)}{2b\lambda^2} + \sum_{l=0}^{L} v_{2l} \lambda^{2l} \, .$$

In this form, the contribution to the potential energy density is evidently given by a weighted sum of terms of the form

$$\phi_R^{2l}(\mathbf{x}) \equiv b^{2l-1} \int B(\mathbf{x}, \lambda)^\dagger \, \lambda^{2l} \, B(\mathbf{x}, \lambda) \, d\lambda \, ,$$

although each such term is strictly a local operator only for $l > \gamma - 1/2$. Let l_o be the least l for which $l > \gamma - 1/2$. Then we collect all such terms together, and introduce the local potential energy density $\mathcal{V}(\mathbf{x})$ as

$$\mathcal{V}(\mathbf{x}) \equiv \Sigma_{l_o}^L v_{2l} \int B(\mathbf{x}, \lambda)^\dagger \, \lambda^{2l} \, B(\mathbf{x}, \lambda) \, d\lambda \, .$$

In turn, we define the remainder of the Hamiltonian density to be the local kinetic energy density

$$\mathcal{K}(\mathbf{x}) \equiv \int B(\mathbf{x}, \lambda)^\dagger \, [-\tfrac{1}{2} b^{-1} \partial_\lambda^2 + \tfrac{1}{2} b^{-1} \gamma(\gamma + 1) \lambda^{-2}$$
$$+ \Sigma_0^{l_o - 1} v_{2l} \lambda^{2l}] \, B(\mathbf{x}, \lambda) \, d\lambda \, .$$

With this division, it follows that

$$\mathcal{H}(\mathbf{x}) = \mathcal{K}(\mathbf{x}) + \mathcal{V}(\mathbf{x}) \, .$$

Since $\mathcal{H}$ may be defined with $A^\dagger$ and A replacing the $B^\dagger$ and B, respectively, $\mathcal{H}(\mathbf{x})$ is manifestly a well-defined local operator; in addition, $\mathcal{V}(\mathbf{x})$ has been chosen to be a well-defined local operator. As a consequence it follows that $\mathcal{K}(\mathbf{x})$ is a well-defined local operator. Observe, however, that the individual terms in $\mathcal{K}(\mathbf{x})$ do *not* generally correspond to local operators. In particular, the expression

$$\pi_R^2(\mathbf{x}) \equiv -b^{-1} \int B(\mathbf{x}, \lambda)^\dagger \, \partial_\lambda^2 \, B(\mathbf{x}, \lambda) \, d\lambda \, ,$$

is not even a local form; the expression

$$\varphi_R^{-2}(\mathbf{x}) \equiv b^{-1} \int B(\mathbf{x}, \lambda)^\dagger \, \lambda^{-2} \, B(\mathbf{x}, \lambda) \, d\lambda \,,$$

has a divergent c-number contribution; and the expression

$$\varphi_R^0(\mathbf{x}) \equiv b^{-1} \int B(\mathbf{x}, \lambda)^\dagger \, B(\mathbf{x}, \lambda) \, d\lambda \,,$$

also has a divergent c-number contribution. First consider, as an example, $\gamma < 3/2$; in that case $l_o = 1$, and therefore

$$\mathcal{K}(\mathbf{x}) = \int B(\mathbf{x}, \lambda)^\dagger \left[-\tfrac{1}{2} b^{-1} \partial_\lambda^2 + \tfrac{1}{2} b^{-1} \gamma(\gamma + 1)\lambda^{-2} + v_0 \right] B(\mathbf{x}, \lambda) \, d\lambda \,.$$

Each of the three terms that make up this expression, taken individually, does *not* lead to a local operator, and indeed, not even a local form. However, the *sum of the three terms* **does** *correspond to a local operator.* Thus we are led to the interpretation that $\varphi_R^{-2}(\mathbf{x})$ and $\varphi_R^0(\mathbf{x})$ *constitute necessary renormalizations of $\pi_R^2(\mathbf{x})$ in order that the kinetic energy density becomes a meaningful local operator.* When $l_o > 1$, then additional terms are needed in order for $\mathcal{K}(\mathbf{x})$ to correspond to a genuine local operator. At the very least, the nonclassical and nonconventional auxiliary potential term $\varphi_R^{-2}(\mathbf{x})$, based on the λ-dependent potential $1/b\lambda^2$, may be understood as a necessary renormalization counterterm to the ill-defined formal expression $\pi_R^2(\mathbf{x})$, based on the λ-dependent operator $-\partial_\lambda^2/b$.

10.7 Scaling and scale transformations

Scaling for the free theory

Recall the ultralocal free theory for a mass m. The sharp-time field expectation functional for the free theory reads

$$\langle 0 | e^{i\varphi(f)} | 0 \rangle = e^{-(1/4m) \int f(\mathbf{x})^2 \, d^s x} \,,$$

which is well defined for all $f \in L^2(\mathbb{R}^s)$. Let us start with a well known invariance. The expectation functional is evidently invariant if the field f is replaced by $f_\mathbf{a}$, where $f_\mathbf{a}(\mathbf{x}) \equiv f(\mathbf{x} - \mathbf{a})$, due to the translation invariance of the integral. Now we introduce

$$\varphi_\mathbf{a}(f) \equiv \varphi(f_\mathbf{a}) \,,$$

which implies that $\varphi_\mathbf{a}(\mathbf{x}) \equiv \varphi(\mathbf{x} + \mathbf{a})$. The equivalence of

$$\langle 0 | e^{i\varphi_\mathbf{a}(f)} | 0 \rangle = \langle 0 | e^{i\varphi(f)} | 0 \rangle \,,$$

guarantees [Na 64, Ar 60a, Ar 60b] the existence of a unitary transformation $U(\mathbf{a})$ with the property that

$$U(\mathbf{a})^\dagger \varphi(\mathbf{x}) U(\mathbf{a}) = \varphi_\mathbf{a}(\mathbf{x}) \,,$$

and which, in addition, leaves the ground state invariant,

$$U(\mathbf{a})|0\rangle = |0\rangle \ .$$

As a family of continuous one-parameter groups, it follows that

$$U(\mathbf{a}) = \exp(i\,\mathbf{a}\cdot\mathcal{P})\ ,$$

where $\mathcal{P}$ is the vector-valued generator of space translations that satisfies $\mathcal{P}|0\rangle = 0$. In fact, the operator $\mathcal{P}$ is readily shown to be

$$\mathcal{P} = \int A(\mathbf{x})^\dagger(-i\boldsymbol{\nabla})\,A(\mathbf{x})\,d^s x\ ,$$

a form which makes the invariance of the state $|0\rangle$ evident. Specifically, recall that the form of $\mathcal{P}$ follows directly from the fact that

$$\langle f|f'_{\mathbf{a}}\rangle = \exp\{-(1/4m)\int[f'_{\mathbf{a}}(\mathbf{x}) - f(\mathbf{x})]^2\,d^s x\}\ ,$$

along with differentiation with respect to $\mathbf{a}$, followed by setting $\mathbf{a} = 0$. Since the set of states $\{|f\rangle\}$ span the Hilbert space the form of $\mathcal{P}$ is thereby determined.

A similar scenario applies to other invariances of the expectation functional. In particular, we observe that the expectation functional for the free ultralocal model is also invariant if we replace f by f_S where $f_S(\mathbf{x}) \equiv S^{-s/2}f(S^{-1}\mathbf{x})$, with $S^{-1}\mathbf{x} = (S^{-1}x^1,\ldots,S^{-1}x^s)$, and S is a positive scale factor. This transformation is called a *scale transformation*. Let us set $\varphi_S(f) \equiv \varphi(f_S)$ from which it follows that $\varphi_S(\mathbf{x}) = S^{s/2}\varphi(S\mathbf{x})$. The power (here $s/2$) of the parameter S in the prefactor for φ_S is said to be the *scale dimension* of the field; for these models, the value $s/2$ is called the *canonical scale dimension*. As we shall see, *not all fields have a scale dimension given by the canonical scale dimension*. The invariance of the expectation functional means that we can introduce a different set of unitary operators $U(S)$ such that

$$U(S)^\dagger\varphi(\mathbf{x})U(S) = \varphi_S(\mathbf{x})\ , \qquad U(S)|0\rangle = |0\rangle\ .$$

As a continuous family of unitary transformations it follows that

$$U(S) = \exp[i\ln(S)D]\ ,$$

where the *dilation operator* D is given by

$$D = -i\tfrac{1}{2}\int A(\mathbf{x})^\dagger(\mathbf{x}\cdot\boldsymbol{\nabla} + \boldsymbol{\nabla}\cdot\mathbf{x})A(\mathbf{x})\,d^s x\ .$$

All of the foregoing discussion applies for the free ultralocal model only. In addition, one is often interested in a field that is not strictly invariant under a scale transformation, but is approximately so. In that case the field operator that has the approximate symmetry has the scale dimension of an associated scale-invariant field, i.e., one that has the same short distance behavior as the one with the approximate transformation properties. Below, we shall encounter explicit examples of approximate scaling behavior.

Scaling for ultralocal models

When it comes to investigate scaling behavior for ultralocal models it soon becomes apparent that all the model field theories discussed so far exhibit only approximate scaling. For example, let us examine a pseudofree model defined by the expectation functional

$$E(f) = \exp(-K^2 \textstyle\int d^s x \int \{1 - \cos[f(\mathbf{x})\lambda]\} \, e^{-bm^2\lambda^2} \, d\lambda/|\lambda|^{2\gamma}) \,.$$

Evidently, if we replace $\varphi(\mathbf{x})$ by $S^d \varphi(S\mathbf{x})$, or equivalently, $f(\mathbf{x})$ by $S^{d-s} f(S^{-1}\mathbf{x})$, where d denotes a scale dimension to be determined, we do not achieve strict invariance of the expectation functional for any d. In fact, using $c \equiv s - d$ for convenience, what we find is given by

$$\begin{aligned}
E(f_S) &= \exp(-K^2 \textstyle\int d^s x \int \{1 - \cos[S^{-c}f(S^{-1}\mathbf{x})\lambda]\} \, e^{-bm^2\lambda^2} \, d\lambda/|\lambda|^{2\gamma}) \\
&= \exp(-S^{s-c(2\gamma-1)} K^2 \textstyle\int d^s x \int \{1 - \cos[f(\mathbf{x})\lambda]\} \, e^{-bS^{2c}m^2\lambda^2} \, d\lambda/|\lambda|^{2\gamma}) \,.
\end{aligned}$$

In arriving at the final expression we have not only rescaled the integration variables $\mathbf{x}$ but the integration variable λ as well. Observe that we could achieve partial invariance of this expression if we set

$$d = s[1 - 1/(2\gamma - 1)] \,,$$

which cancels the S-dependence outside the integral, but this would not eliminate the S dependence in the exponent of the λ integral. However, we *can* eliminate this remaining S dependence if we discuss the physically artificial but mathematically interesting *stable ultralocal models* for which $m = 0$ within a pseudofree model – or, more generally, $y(\lambda) = 0$ – and for which, to achieve convergence of the λ integral, we need to require $1/2 < \gamma < 3/2$. The stable ultralocal models are defined by an expectation functional of the form

$$E_\gamma(f) \equiv \exp(-K^2 \textstyle\int d^s x \int \{1 - \cos[f(\mathbf{x})\lambda]\} \, d\lambda/|\lambda|^{2\gamma}) \,,$$

which is well defined and in fact can be evaluated as

$$E_\gamma(f) = \langle 0|e^{i\varphi_\gamma(f)}|0\rangle = \exp[-c_\gamma \textstyle\int d^s x \, |f(\mathbf{x})|^{2\gamma-1}] \,,$$

for some positive constant c_γ. Observe that the stable models look somewhat like the free model save for the very important difference that the exponent of the field f is not 2 but rather $(2\gamma - 1)$, which, with $3/2 > \gamma > 1/2$, can be any positive number less than 2.*

So far we have been discussing scale transformations that involve spatial variables alone without any consideration of the time variable. In a relativistic theory, for example, space and time enter in a very symmetric fashion, and that

---

* The models in question are called "stable" models because the field distribution involved is that given by a stable infinitely divisible probability distribution; see [Lu 70].

should be reflected in the scaling properties of the theory as well. For example, part of a typical relativistic wave equation reads

$$\frac{\partial^2}{\partial t^2}F(t,\mathbf{x}) = \alpha \sum_{j=1}^{s}\frac{\partial^2}{\partial x^{j\,2}}F(t,\mathbf{x}) + \cdots ,$$

where we have introduced the parameter α for convenience. Clearly, this part of the wave equation is covariant if we transform F to F_S, where

$$F_S(t,\mathbf{x}) = S^{-2}F(St, S\mathbf{x})$$

because of the similar manner in which the independent variables enter. As $\alpha \to 0$, and this part of a typical relativistic wave equation passes to part of a typical equation for an ultralocal theory, the equality of the spatial and temporal scaling remains.

In transforming these ideas to a stable ultralocal quantum field theory let us consider transformations of the various fields φ_R^θ for a general value of θ all at the same time. With $c \equiv s + 1 - d_\theta$ this time, attention is focused on

$$\langle 0|e^{i\varphi_{\gamma,R,S}^\theta(h)}|0\rangle = \langle 0|e^{i\varphi_{\gamma,R}^\theta(h_S)}|0\rangle$$

$$= \exp(-K^2\textstyle\int d^s x$$
$$\times \int |\lambda|^{-\gamma}\{1 - \cos[b^{\theta-1}\textstyle\int e^{i h_\gamma t}\lambda^\theta e^{-i h_\gamma t}S^{-c}h(S^{-1}t, S^{-1}\mathbf{x})dt]\}|\lambda|^{-\gamma}\,d\lambda)$$
$$= \exp(-K^2 S^s\textstyle\int d^s x$$
$$\times \int |\lambda|^{-\gamma}\{1 - \cos[b^{\theta-1}\textstyle\int e^{i h_\gamma St}\lambda^\theta e^{-i h_\gamma St}S^{1-c}h(t,\mathbf{x})dt]\}|\lambda|^{-\gamma}\,d\lambda)$$
$$= \exp(-K^2 S^{s-(2\gamma-1)/2}\textstyle\int d^s x$$
$$\times \int |\lambda|^{-\gamma}\{1 - \cos[b^{\theta-1}\textstyle\int e^{i h_\gamma t}\lambda^\theta e^{-i h_\gamma t}S^{1-c+\theta/2}h(t,\mathbf{x})dt]\}|\lambda|^{-\gamma}\,d\lambda) .$$

In arriving at this last expression we have used the fact that

$$h_\gamma \equiv -\frac{1}{2b}\frac{\partial^2}{\partial\lambda^2} + \frac{\gamma(\gamma+1)}{2b\,\lambda^2} ,$$

which is *homogeneous* of degree -2 in the variable λ, thus permitting a simple change of variables $(\lambda \to \lambda S^{1/2})$ to bring the factor S out of the exponent within the cosine. To achieve invariance of this expression now requires *two* separate conditions, namely, $s = (2\gamma - 1)/2$ and $c - 1 = s - d_\theta = \theta/2$. Alternatively, these relations imply that

$$\gamma = s + 1/2 ,$$
$$d_\theta = s - \theta/2 .$$

These two relations determine both γ and d_θ, and in addition are compatible with the single previously determined relation (including θ)

$$d_\theta = s[1 - \theta/(2\gamma - 1)] .$$

The invariance of the stable models under scale transformations means that there exist unitary operators $U(S)$, $S > 0$, such that

$$U(S)^{\dagger}\varphi^{\theta}_{\gamma,R}(\mathbf{x})U(S) = \varphi^{\theta}_{\gamma,R,S}(\mathbf{x}), \qquad U(S)|0\rangle = |0\rangle.$$

As before $U(S) = \exp[i\ln(S)D]$ where, in the present case, since we had to make transformations in both $\mathbf{x}$ and λ, it is not surprising that dilation transformations in both variables appear; explicitly (see Exercise 10.6) we find that

$$D = -i\tfrac{1}{2}\int d^s x\, A(\mathbf{x},\lambda)^{\dagger}\{\mathbf{x}\cdot\boldsymbol{\nabla} + \boldsymbol{\nabla}\cdot\mathbf{x} + \tfrac{1}{2}[\lambda(\partial/\partial\lambda) + (\partial/\partial\lambda)\lambda]\}A(\mathbf{x},\lambda)\,d\lambda.$$

The determination of γ from the vestiges of relativistic scaling invariance provides the last ingredient for the determination of the proposed solution of ultralocal models. Earlier, when we were discussing the spectrum of ultralocal pseudofree models, we noted that we would eventually find a reason to choose $2\gamma + 1 = 2s + 2 = 2n$ as an even integer, which at the time we called J. Here we now see that $J = 2n$, namely, J is *twice the space-time dimension*. As an example, let us choose $n = 4$; then, based on that choice, the spectrum of four-dimensional relativistically derived pseudofree ultralocal theories is given by $0, 8m, 10m, 12m, \ldots$.

We have now understood a discrete subset of all possible γ values, but not all of them. The interpretation of other parts of the γ domain follows from a slight generalization of the relativistic wave equation discussed earlier. Suppose, instead of the relativistic version of the wave equation, we had started with

$$\frac{\partial^2}{\partial t^2}F(t,\mathbf{x}) = \alpha\sum_{j=1}^{s}\left(\frac{\partial^2}{\partial x^{j\,2}}\right)^{\beta}F(t,\mathbf{x}) + \cdots, \qquad \beta > 0.$$

Although this wave equation is different for different values of β, it is clear that as $\alpha \to 0$, and the wave equation becomes that for an ultralocal theory, the final equation is the same independently of the value of β. Thus, for completeness, we should investigate the scaling properties associated with such a starting point, in particular, relative to a local operator $\phi^{\theta}_{\gamma,R,S}(h) = \phi^{\theta}_{\gamma,R}(h_S)$ with the associated test function transformation given by

$$h_S(t,\mathbf{x}) = S^{d_{\theta}-s-\beta}h(S^{-\beta}t, S^{-1}\mathbf{x}).$$

A calculation analogous to that just concluded shows, in the present case, that $\gamma = s/\beta + 1/2$ and $d_{\theta} = s - \theta\beta/2$. In this fashion we can account for the entire range of $\gamma > 1/2$, as due in part to the fact in space dimension s that the wave equation for any $\beta > 0$ leads, as $\alpha \to 0$, to one or another of the acceptable values for γ. Of course, the choice $\beta = 1$ is physically special, corresponding to a relativistic wave equation, but from a mathematical point of view, all such wave equations are equally valid, and so it is entirely natural that a general mathematical solution of ultralocal models should allow for all of the mathematically consistent possibilities. The dilation generator D for a general β is similar to the result given previously in the case $\beta = 1$ except for the fact that the multiplier in front of the dilation with respect to λ in the integrand is $\beta/2$ rather than being $1/2$; see Exercise 10.6.

Spectral weight function

Let us examine the two-point function is some detail in the case of stable ultralocal models when $\gamma > 3/2$. In particular, let us focus on the function

$$J_\gamma^\epsilon(t - s) \equiv \int c(\lambda)\, \lambda\, e^{-[\epsilon + i(t-s)]\hbar_\gamma}\, \lambda\, c(\lambda)\, d\lambda \, ,$$

and the distribution to which it gives rise in the limit that $\epsilon \to 0$. We recall that $c(\lambda) \propto |\lambda|^{-\gamma}$ and that

$$\hbar_\gamma \equiv -\tfrac{1}{2} b^{-1} \partial_\lambda^2 + \tfrac{1}{2} b^{-1} \gamma(\gamma + 1)\lambda^{-2}$$

is a positive operator, and by a simple change of variables it follows, for some nonzero R, that

$$J_\gamma^\epsilon(t - s) = \frac{R}{[\epsilon + i(t - s)]^{\gamma - 3/2}}$$

holds as an identity for all $\epsilon > 0$ and for all values of $t - s$. In turn, this expression may be given a Fourier representation according to

$$J_\gamma^\epsilon(t - s) \equiv \frac{1}{2} \int_0^\infty \frac{\rho_\gamma(\omega)}{\omega} e^{-[\epsilon + i(t-s)]\omega}\, d\omega \, ,$$

where the form of the integrand is traditional. The function $\rho(\omega)$ is called the *spectral weight function*, and for the case at hand, it follows that

$$\rho(\omega) = \rho_\gamma(\omega) \equiv [2R/\Gamma(\gamma - 3/2)]\, \omega^{\gamma - 3/2} \, .$$

Here, quite explicitly, is the fact that γ determines the short-time behavior of the two-point function, a behavior that is intimately linked with the high-frequency behavior of the spectral weight function $\rho(\omega)$.

A similar analysis can be carried out for the time-ordered two-point function for stable ultralocal models. In that case, we study

$$K_\gamma^\epsilon(t - s) \equiv \int c(\lambda)\, \lambda\, e^{-[\epsilon + i|t-s|]\hbar_\gamma}\, \lambda\, c(\lambda)\, d\lambda \, .$$

By the same token we find that

$$K_\gamma^\epsilon(t - s) = \frac{R}{[\epsilon + i|t - s|]^{\gamma - 3/2}} \, ,$$

which holds for all $\epsilon > 0$ and all $|t - s|$. In turn, this expression is given by

$$K_\gamma^\epsilon(t - s) = \frac{1}{2} \int_0^\infty \frac{\rho_\gamma(\omega)}{\omega} e^{-[\epsilon + i|t-s|]\omega}\, d\omega \, ,$$

but this formula does not represent a Fourier transformation. To obtain such a Fourier transformation we introduce the alternative expression given by

$$K_\gamma^\epsilon(t-s) = \frac{1}{2\pi i} \int_0^\infty \int_{-\infty}^\infty \frac{e^{-\epsilon\omega - i(t-s)\sigma}}{\sigma^2 - \omega^2 + i\delta} \, d\sigma \, \rho_\gamma(\omega) \, d\omega \; ,$$

in the limit that $\delta \to 0$ from positive values. In offering this expression we are making use of the fact, for $t - s > 0$ and $\omega > 0$, that

$$\frac{1}{2\pi i} \int_{-\infty}^\infty \frac{e^{-i(t-s)\sigma}}{\sigma^2 - \omega^2 + i\delta} \, d\sigma$$

$$= \frac{1}{2\pi i} \int_{-\infty}^\infty \frac{e^{-i(t-s)\sigma}}{(\sigma - \omega_-)(\sigma + \omega_-)} \, d\sigma$$

$$= \frac{e^{-i\omega_-(t-s)}}{2\omega_-}$$

on the basis of Cauchy's integral formula; a similar calculation for $s - t > 0$ leads to the same result with with $s - t$ appearing in place of $t - s$. In this expression

$$\omega_- \equiv \sqrt{\omega^2 - i\delta} \equiv \omega - i0^+ \; .$$

10.8* Current algebra formulation

The study of the various possible wave equations that lead to ultralocal models has clarified the role of γ for all $\gamma > 1/2$. However, that analysis does not apply to $\gamma = 1/2$. It is therefore noteworthy that there is an alternative way to investigate the special case $\gamma = 1/2$. For convenience in this alternative investigation, let us choose units so that $b = 1$.

For $\gamma = 1/2$, the local field operator $\varphi(\mathbf{x})$ becomes a self-adjoint operator with only spatial smearing. The expression that has been suggested for the local momentum, namely

$$\pi(\mathbf{x}) = -i \int B(\mathbf{x}, \lambda)^\dagger (\partial/\partial\lambda) B(\mathbf{x}, \lambda) \, d\lambda \; ,$$

is, however, only a local form and is unsuitable as a kinematical operator. Setting the form π aside for the moment, let us introduce κ defined by the expression

$$\kappa(\mathbf{x}) \equiv -i\tfrac{1}{2} \int B(\mathbf{x}, \lambda)^\dagger \left[\lambda(\partial/\partial\lambda) + (\partial/\partial\lambda)\lambda \right] B(\mathbf{x}, \lambda) \, d\lambda \; ,$$

which is seen to involve an ingredient in the dilation operator previously discussed. It is straightforward to show that κ is a local self-adjoint operator. To

---

* This section lies somewhat outside the main line of discussion and may be omitted on a first reading.

that end, let us introduce the abbreviation

$$\sigma \equiv -i\tfrac{1}{2}[\lambda(\partial/\partial\lambda) + (\partial/\partial\lambda)\lambda] .$$

Since, for all real p, the map

$$\chi(\lambda) \to e^{p/2}\chi(e^p\lambda)$$

defines a continuous one-parameter group of norm preserving transformations on $L^2(\mathbb{R})$, it follows that

$$e^{ip\sigma}\chi(\lambda) = e^{p/2}\chi(e^p\lambda) ,$$

which thus defines σ as a self-adjoint operator. Moreover, for $\gamma = 1/2$, and with, e.g., $y(\lambda) \propto \lambda^2$, $\lambda^2 \ll 1$, we have

$$\int c(\lambda) \left[1 - e^{ip\sigma}\right] c(\lambda)\, d\lambda$$
$$= K^2 \int |\lambda|^{-1/2} e^{-y(\lambda)} \left[1 - e^{ip\sigma}\right] |\lambda|^{-1/2} e^{-y(\lambda)}\, d\lambda$$
$$= K^2 \int [\, |\lambda|^{-1} e^{-2y(\lambda)} - |\lambda|^{-1} e^{-y(e^{-p/2}\lambda) - y(e^{p/2}\lambda)}\,]\, d\lambda ,$$

with the only appearance of p being in the exponent y. This expression is continuous in p and vanishes for $p = 0$; these are sufficient conditions for $\kappa(\mathbf{x})$ to exist as a local self-adjoint operator [Kl 71]. This fortuitous circumstance holds only for $\gamma = 1/2$. Finally, it is useful to observe that

$$\kappa(\mathbf{x}) = \tfrac{1}{2}[\varphi(\mathbf{x})\pi(\mathbf{x}) + \pi(\mathbf{x})\varphi(\mathbf{x})]_R ,$$

where, as before, R denotes a "renormalized" product. The implication of this relation – specifically, that the local renormalized product of a local operator and a local form may lead to a local operator – is a perfectly possible result, and in the present case it is a useful way to interpret κ.

The operators $\varphi(\mathbf{x})$ and $\kappa(\mathbf{x})$ constitute the kinematical operators that take the place of the field and the nonexistent momentum. Moreover, these kinematical operators have a closed commutation algebra given by

$$[\varphi(\mathbf{x}), \kappa(\mathbf{y})] = i\delta(\mathbf{x} - \mathbf{y})\, \varphi(\mathbf{x}) .$$

This closed commutation relation is generally referred to as an *affine commutation relation;* the fields that satisfy such commutation relations are often said to form (one example of) a *current algebra* [AdD 68]. If one takes the formal definition of κ in terms of φ and π into account, then the affine commutation relation is seen to be a formal consequence of using the canonical commutation relation. To a considerable extent, as we shall see, the affine commutation relations can effectively take the place of the nonexistent canonical commutation relations when $\gamma = 1/2$.

Besides the local operator $\varphi(\mathbf{x})$ it is important to consider the family of local products of $\varphi(\mathbf{x})$ which are defined in exactly the same manner that we have introduced earlier. In particular, we have

$$\varphi_R^p(\mathbf{x}) \equiv \int B(\mathbf{x}, \lambda)^\dagger \lambda^p B(\mathbf{x}, \lambda)\, d\lambda ,$$

for all $p \in \{1, 2, 3, \ldots\}$. (N.B. The positive parameter b with dimensions of $(\text{length})^{-s}$ introduced earlier to preserve engineering dimensions of the field powers has been here set equal to unity.) Such fields are especially needed when we turn our attention to the discussion of dynamics.

It is sometimes useful to focus on a subalgebra of all the local powers of φ, for example, one which involves only *even powers*, $p = 2l$, $l = 1, 2, 3, \ldots$. It is clear that all even powers are generated as local powers of the basic even field $\varphi_R^2(\mathbf{x})$, which for the sake of brevity we shall hereafter call

$$\tau(\mathbf{x}) \equiv \varphi_R^2(\mathbf{x}) \ .$$

Evidently,

$$[\tau(\mathbf{x}), \kappa(\mathbf{y})] = 2i\delta(\mathbf{x} - \mathbf{y})\, \tau(\mathbf{x}) \ ,$$

which means that τ and κ also form a closed algebra of local fields. This commutation relation is again referred to as an affine commutation relation, and the set of fields that satisfy such a relation are again said to form a current algebra. With regard to applications to ultralocal model quantum field theories, the current algebra viewpoint has been stressed by Newman [Ne 71, Ne 72].

Let us next introduce several postulates that relate to dynamics. From a heuristic point of view, we again assume that the Hamiltonian has the generic form

$$\text{``}\, \mathcal{H} = \tfrac{1}{2}\textstyle\int \pi(\mathbf{x})^2 d^s x + \mathcal{V}(\varphi) \,\text{''},$$

as is customary, despite the fact that for ultralocal models there is no local momentum operator. Nevertheless, using the canonical commutation relations formally, one is led to the relation

$$[\tau(\mathbf{x}), \mathcal{H}] = 2i\,\kappa(\mathbf{x}) \ .$$

Although arrived at on a heuristic basis, we shall adopt this final relation as a fundamental connection of the kinematics (as embodied by τ and κ) with the dynamics (as embodied by $\mathcal{H}$). Let us specialize the dynamics to what we have earlier called even models, namely those for which the potential satisfies $\mathcal{V}(-\varphi) = \mathcal{V}(\varphi)$. In that case the potential is actually a renormalized function of τ, and so we are led to a complete characterization of the model in terms of the local field $\tau(\mathbf{x})$ and the Hamiltonian $\mathcal{H}$. To summarize, we postulate

$$2\,\kappa(\mathbf{x}) \equiv i[\mathcal{H}, \tau(\mathbf{x})] \ ,$$
$$[\tau(\mathbf{x}), \kappa(\mathbf{y})] = 2i\delta(\mathbf{x} - \mathbf{y})\, \tau(\mathbf{x}) \ ,$$

along with the assertion that $\mathcal{H} = \mathcal{K} + \mathcal{V}$ for some (formal) operator $\mathcal{K}$ and where for a polynomial interaction we (formally) have

$$\mathcal{V}(\phi) = \textstyle\int \Sigma_l \, v_l \, \tau(\mathbf{x})_R^l \, d^s x \ ,$$

for suitable constant coefficients v_l, $1 \le l \le L < \infty$. These expressions are formal due to the integral over an unbounded domain, $\mathbb{R}^s$; however, the combination that forms $\mathcal{H}$ leads to a proper self-adjoint operator.

Coherent states

One can introduce normalized *affine coherent states* for the present case, namely,

$$|u, v\rangle \equiv e^{i \int \tau(\mathbf{x}) u(\mathbf{x}) \, d^s x} e^{-i \int \kappa(\mathbf{x}) v(\mathbf{x}) \, d^s x} |0\rangle \ ,$$

defined, for example, for all u and v which are bounded functions of compact support. It is noteworthy that these states are also coherent states in the canonical sense in that

$$|u, v\rangle = N \exp\{ \int d^s x \int A(\mathbf{x}, \lambda)^{\dagger} [e^{i\lambda^2 u(\mathbf{x})} e^{-i\sigma v(\mathbf{x})} - 1] \, c(\lambda) \, d\lambda \} |0\rangle \ ,$$

where N denotes a suitable normalization. The affine coherent states span the *even* Hilbert space $\mathfrak{H}_E$, which may be defined as the span of states of the form $|u\rangle \equiv |u, 0\rangle$ for suitably many functions. Let us next study matrix elements in the states $|u\rangle$.

The set of states $|u\rangle$, and the affine commutation relations, will enable us to offer an analog of the Araki relations for ultralocal models for which $\gamma = 1/2$; see [IsK 90]. In particular, let us first exploit the fact that the no-particle state $|0\rangle$ is time-reversal invariant for the dynamics of an ultralocal theory to help us find an expression for $\langle u | \kappa(\mathbf{x}) | u' \rangle$. It is even easier if we start more simply.

For the example of a single degree of freedom the analogous expectation functional has a Schrödinger representation given by

$$I = -i \tfrac{1}{2} \int \chi(x) e^{-iux^2} (x \partial_x + \partial_x x) e^{iu'x^2} \chi(x) \, dx \ ,$$

where we have used the fact that the wave function $\chi(x)$ is real due to the assumed time-reversal invariance. The complex conjugate of this expression, plus the interchange of the labels u and u', leads to an equivalent expression give by

$$I = i \tfrac{1}{2} \int \chi(x) e^{iu'x^2} (x \partial_x + \partial_x x) e^{-iux^2} \chi(x) \, dx \ ,$$

and, finally, the average of these two expressions leads to

$$I = (u + u') \int \chi(x) e^{-iux^2} x^2 \, e^{iu'x^2} \chi(x) \, dx \ .$$

The moral of this exercise for the field theory is that time-reversal invariance for the state $|0\rangle$ implies that

$$\langle u | \kappa(\mathbf{x}) | u' \rangle = [u(\mathbf{x}) + u'(\mathbf{x})] \langle u | \tau(\mathbf{x}) | u' \rangle \ ;$$

observe that this relation also implies that $\langle 0 | \kappa(\mathbf{x}) | 0 \rangle = 0$. In the preceding general matrix element, the last factor is also given by a functional derivative, i.e.,

$$\langle u|\tau(\mathbf{x})|u'\rangle = i[\delta/\delta u(\mathbf{x})]\langle u|u'\rangle = i[\delta/\delta u(\mathbf{x})]\langle 0|u' - u\rangle \,,$$

which is evidently a function only of the difference variable $u' - u$. Hence, using the connection between $\mathcal{H}$ and κ, and the fact that $\mathcal{H}|0\rangle = 0$, we also determine that

$$\langle u|\mathcal{H}|u'\rangle = 2\int u(\mathbf{x})u'(\mathbf{x})\,\langle u|\tau(\mathbf{x})|u'\rangle\,d^s x \,.$$

Thus a total set of matrix elements of both κ and the Hamiltonian $\mathcal{H}$ are fully determined by the expectation functional $\langle 0|u\rangle$. In turn, this expectation functional may be expressed in terms of the model function $c(\lambda)$ that describes the ultralocal model according to the relation

$$\langle 0|u\rangle \equiv \langle 0|e^{i\tau(u)}|0\rangle = \exp\{-K^2\textstyle\int d^s x \int \left[1 - e^{iu(\mathbf{x})\lambda^2}\right]|\lambda|^{-1}e^{-2y(\lambda)}\,d\lambda\} \,.$$

Expectation values

When dealing with canonical fields and canonical coherent states in Chapters 5 and 7, for example, we argued that diagonal coherent state expectation values of quantum expressions corresponded to their classical counterparts. Let us see to what extent the diagonal matrix elements of quantum operators using *affine* coherent states are related to their classical counterparts. To begin with, and using the affine commutation relations, it follows that

$$\langle u, v|\tau(\mathbf{x})|u, v\rangle = C_2\, e^{2v(\mathbf{x})} \,,$$
$$\langle u, v|\kappa(\mathbf{x})|u, v\rangle = 2C_2\, u(\mathbf{x})\, e^{2v(\mathbf{x})} \,,$$

where

$$C_2 \equiv \textstyle\int \lambda^2\, c(\lambda)^2\, d\lambda = K^2 \int |\lambda|\, e^{-2y(\lambda)}\, d\lambda \,.$$

The value of C_2 can be adjusted by choosing the free parameter K. In the next section, it is argued that properly $C_2 \equiv \hbar/m$; however, for the moment we choose $C_2 \equiv 1 = \int \lambda^2 c(\lambda)^2\, d\lambda$ in order to simplify the following discussion. For example, for the pseudofree model, where $y(\lambda) = m\lambda^2/2$, the latter choice means that

$$\textstyle\int \lambda^2\, c(\lambda)^2\, d\lambda = K^2 \int |\lambda| e^{-m\lambda^2}\, d\lambda = 1 \,,$$

and in this case it follows that $K^2 = m$. We associate the local operator $\tau(\mathbf{x}) = \varphi_R^2(\mathbf{x})$ with the classical combination $g(\mathbf{x})^2$, and the local operator $\kappa(\mathbf{x}) = \frac{1}{2}[\varphi(\mathbf{x})\pi(\mathbf{x}) + \pi(\mathbf{x})\varphi(\mathbf{x})]_R$ with the classical combination $f(\mathbf{x})g(\mathbf{x})$. Therefore, we are led to identify

$$\langle u, v|\tau(\mathbf{x})|u, v\rangle = e^{2v(\mathbf{x})} \equiv g(\mathbf{x})^2 \,,$$
$$\langle u, v|\kappa(\mathbf{x})|u, v\rangle = 2u(\mathbf{x})\, e^{2v(\mathbf{x})} \equiv f(\mathbf{x})g(\mathbf{x}) \,.$$

Stated otherwise, we identify the classical field $g(\mathbf{x}) = e^{v(\mathbf{x})}$ and the classical momentum $f(\mathbf{x}) = 2u(\mathbf{x})\,e^{v(\mathbf{x})}$. [Observe that we could also have chosen $g(\mathbf{x}) = -e^{v(\mathbf{x})}$ and $f(\mathbf{x}) = -2u(\mathbf{x})\,e^{v(\mathbf{x})}$. For the sake of brevity, however, we shall only discuss the case where $g(\mathbf{x}) > 0$.]

We now turn our attention to the Hamiltonian and we focus on the expression $H(u,v) \equiv \langle u, v | \mathcal{H} | u, v \rangle$. To evaluate this expression recall that

$$\mathcal{H} = \int d^s x \int A(\mathbf{x}, \lambda)^\dagger \, \hbar \, A(\mathbf{x}, \lambda)\, d\lambda \,,$$

as well as

$$\hbar = -\frac{1}{2}\left(\frac{\partial^2}{\partial \lambda^2} - \frac{c''(\lambda)}{c(\lambda)} \right) = -\frac{1}{2}\frac{1}{c(\lambda)}\frac{\partial}{\partial \lambda}c(\lambda)^2\frac{\partial}{\partial \lambda}\frac{1}{c(\lambda)} \,.$$

Based on the second relation for $\hbar$ given here, we determine a manifestly Hermitian and nonnegative expression for the Hamiltonian operator in the form

$$\mathcal{H} = \tfrac{1}{2}\int d^s x \int [c(\lambda)(\partial/\partial\lambda)c(\lambda)^{-1}A(\mathbf{x},\lambda)]^\dagger\, [c(\lambda)(\partial/\partial\lambda)c(\lambda)^{-1}A(\mathbf{x},\lambda)]\, d\lambda \,.$$

Since the states $|u,v\rangle$ are canonical coherent states as well, we learn that

$$
\begin{aligned}
H(u,v) &= \tfrac{1}{2}\int d^s x \int | c(\lambda)(\partial/\partial\lambda)c(\lambda)^{-1}[e^{i\lambda^2 u(\mathbf{x})}e^{-i\sigma v(\mathbf{x})} - 1]c(\lambda)|^2\, d\lambda \\
&= \tfrac{1}{2}K^2\int d^s x \int | |\lambda|^{-1/2}e^{-y(\lambda)}(\partial/\partial\lambda)[e^{i\lambda^2 u + y(\lambda) - y(\lambda e^{-v})} - 1]|^2\, d\lambda \\
&= \tfrac{1}{2}K^2\int d^s x \int | 2i\lambda u + y'(\lambda) - e^{-v}y'(\lambda e^{-v})|^2\, e^{-2y(\lambda e^{-v})}\, d\lambda/|\lambda| \\
&= \tfrac{1}{2}K^2\int d^s x \int \{4u^2\lambda^2 + [y'(\lambda) - e^{-v}y'(\lambda e^{-v})]^2\}\, e^{-2y(\lambda e^{-v})}\, d\lambda/|\lambda| \\
&= \tfrac{1}{2}K^2\int d^s x \int \{4u(\mathbf{x})^2\, e^{2v(\mathbf{x})}\, \lambda^2 \\
&\qquad + [y'(\lambda e^{v(\mathbf{x})}) - e^{-v(\mathbf{x})}y'(\lambda)]^2\}\, e^{-2y(\lambda)}\, d\lambda/|\lambda| \,.
\end{aligned}
$$

Expressed in terms of the fields f and g, we obtain

$$H(u,v) = \tfrac{1}{2}\int f(\mathbf{x})^2\, d^s x + \int W(g(\mathbf{x}))\, d^s x \,,$$

where

$$W(g(\mathbf{x})) \equiv \tfrac{1}{2}K^2\int [y'(\lambda g(\mathbf{x})) - g(\mathbf{x})^{-1}y'(\lambda)]^2\, e^{-2y(\lambda)}\, d\lambda/|\lambda| \,.$$

It is clear that the result obtained for $H(u,v)$ satisfies the general criteria that (i) $H(0,0) = 0$, (ii) $H(u,v) \geq 0$, and

$$H(u,0) = 2\int u(\mathbf{x})^2\, d^s x$$

as required by the previous argument. If we assume

$$y(\lambda) = \Sigma_{l=1}^{L}(2l)^{-1}y_{2l}\lambda^{2l} \,,$$

then it readily follows that

$$W(g) = \tfrac{1}{2}\int\{\Sigma_{l=1}^{L}y_{2l}[g^{2l} - 1]g^{-1}\lambda^{2l-1}\}^2\, c(\lambda)^2\, d\lambda \,.$$

As a more specific example, consider, once again, the pseudofree Hamiltonian for which $y(\lambda) = m\lambda^2/2$. In that case,

$$H(u,v) = \tfrac{1}{2}\int\{f(\mathbf{x})^2 + m^2[g(\mathbf{x}) - 1/g(\mathbf{x})]^2\}\,d^s x \; .$$

This expression somewhat resembles the original Hamiltonian, but it cannot coincide with it since it is constrained to vanish when $u = v = 0$, i.e., when $f = 0$ and $g = 1$ (or $g = -1$) rather than $g = 0$.

The fact that $H(0,v) = 0$ when $g = 1$ (or $g = -1$), and not some other value, is a consequence of our present choice of C_2. If we adopt its proper value of $\hbar/m$, then the expectation value for the pseudofree model is given by

$$H(u,v) = \tfrac{1}{2}\int\{f(\mathbf{x})^2 + m^2[g(\mathbf{x}) - \hbar/m\,g(\mathbf{x})]^2\}\,d^s x \; .$$

In this case the classical Hamiltonian has the expected form apart from $O(\hbar)$ corrections. However, there is a qualitative change brought about by the term $\hbar^2/g(\mathbf{x})^2$ in the Hamiltonian density. So long as $\hbar > 0$ this term forces $g(\mathbf{x}) \neq 0$, and thus in solving the classical equations of motion, g will always remain positive (or always remain negative). Indeed, even in the limit where $\hbar \to 0$, any one of these classical solutions does not lead to any of the classical solutions of the model in which $\hbar = 0$ from the very beginning [Ka 81]. This unexpected modification of the classical theory derives from the exact quantum theory which has left a significant imprint on the classical theory. Apart from this important qualification, therefore, we have offered a suggestive argument that the quantum theory described by the current algebra in the case $\gamma = 1/2$ does indeed represent the quantum field theory of an ultralocal model.

When $\gamma > 1/2$, however, a current algebra does not exist, and therefore, we must appeal to alternative methods to demonstrate that the quantum theory we have developed in earlier sections of this chapter does indeed represent the quantum theory of the associated classical ultralocal theory. In the following discussion, we shall establish a fully satisfactory classical limit for all $\gamma \geq 1/2$.

10.9* Classical limit for ultralocal models

The nonexistence of either canonical or affine commutation relations in the general case causes certain complications in showing that a given nontrivial solution of the ultralocal quantum field models leads, in the classical limit, to an acceptable classical solution. Specifically, even when $\varphi(\mathbf{x})$ is a local field operator, the expression $\pi(\mathbf{x}) = \dot{\varphi}(\mathbf{x})$ is only a form and does not define a local operator. Thus we can not make use of the weak correspondence principle to help us connect

---

* This section lies somewhat outside the main line of discussion and may be omitted on a first reading.

quantum and classical expressions as we have previously done, particularly in Chapter 7 for the rotationally symmetric models. As a result, we need to find an alternative method to show that the classical limit of the ultralocal models obeys the expected classical equations of motion. In turn, the alternative approach to which we are led will have more to do with the affine commutation relations than with the canonical commutation relations. In the discussion that follows in this section, we will generally set the parameter $b = 1$, and we will reintroduce $\hbar$ explicitly in appropriate places.

In order to discuss an alternative way to characterize the classical limit, let us introduce a family of unit vectors $|(f, g)\rangle$, each of which is in fact a normalized coherent state defined by

$$|(f, g)\rangle \equiv N \exp[\int d^s x \int d\lambda \, A(\mathbf{x}, \lambda)^\dagger \Upsilon(\mathbf{x}, \lambda)] |0\rangle \,,$$

where Υ implicitly depends on f and g and is given by

$$\Upsilon(\mathbf{x}, \lambda) \equiv M e^{(i/\hbar)\lambda f(\mathbf{x})} g_\pm(\mathbf{x})^{-1/2} \chi_\pm(\lambda/g_\pm(\mathbf{x})) \,, \qquad |\mathbf{x}| \leq R \,,$$
$$\equiv 0 \,, \qquad\qquad\qquad\qquad\qquad\qquad\quad |\mathbf{x}| > R \,,$$

which is defined for all bounded, smooth $f(\mathbf{x})$ and $g(\mathbf{x}) \neq 0$. Before defining the several terms involved, let us first observe that M is chosen so the function Υ satisfies

$$\int |\Upsilon(\mathbf{x}, \lambda)|^2 \, d\lambda = 1 \,, \qquad |\mathbf{x}| \leq R \,,$$
$$= 0 \,, \qquad |\mathbf{x}| > R \,.$$

Consequently,

$$\int d^s x \int |\Upsilon(\mathbf{x}, \lambda)|^2 \, d\lambda = \int_{|\mathbf{x}|<R} d^s x = k_s R^s / s \,,$$

where k_s is the volume of the unit s-dimensional sphere, the explicit value of which is not important; what is important is that for all finite R the L^2 norm of Υ is finite. Accordingly, the normalization factor in the definition of the state $|(f, g)\rangle$ is $N = \exp(-k_s R^s / 2s)$.

Now we introduce several required definitions:

$$g_\pm(\mathbf{x}) \equiv \tfrac{1}{2}[\,|g(\mathbf{x})| \pm g(\mathbf{x})\,] \,,$$
$$\chi(\lambda) \equiv N' \exp[-\lambda^{-2} - (\lambda^2 - 1)^2 / 8\hbar] \,,$$
$$\chi_\pm(\lambda) \equiv \theta(\pm\lambda)\chi(\lambda) \,.$$

Here, $g(\mathbf{x}) \equiv g_+(\mathbf{x}) - g_-(\mathbf{x})$, $g_+(\mathbf{x})g_-(\mathbf{x}) \equiv 0$, both terms of which are non-negative, $\theta(\pm\lambda) \equiv 1$ if $\pm\lambda > 0$ and zero otherwise, and N' is chosen so that $\int \chi_\pm(\lambda)^2 \, d\lambda = 1$. The choice of $\chi(\lambda)$ is dictated by symmetry, infinite differentiability, and especially, as expressed in terms of the notation

$$\langle\langle (\cdot) \rangle\rangle_\pm \equiv \int (\cdot) \chi_\pm(\lambda)^2 d\lambda \,,$$

by the property that

$$\langle \lambda^{2p+1}\rangle_{\pm} = \pm 1 + O(\hbar)\,,$$
$$\langle \lambda^{2p}\rangle_{\pm} = 1 + O(\hbar)\,,$$

for all $p \in \{0, \pm 1, \pm 2, \pm 3, \ldots\}$.

On the basis of these definitions, we note that

$$\int \Upsilon(\mathbf{x}, \lambda)^* \, \lambda^p \, \Upsilon(\mathbf{x}, \lambda)\, d\lambda = g_+(\mathbf{x})^p \langle \lambda^p\rangle_+ + g_-(\mathbf{x})^p \langle \lambda^p\rangle_-$$
$$= g_+(\mathbf{x})^p + (-1)^p\, g_-(\mathbf{x})^p + O(\hbar)$$
$$= g(\mathbf{x})^p + O(\hbar)\,;$$

observe that this evaluation is valid for all integer p, $-\infty < p < \infty$. In addition, we note that

$$-i\hbar \int \Upsilon(\mathbf{x}, \lambda)^* \, \partial_\lambda \, \Upsilon(\mathbf{x}, \lambda)\, d\lambda$$
$$= f(\mathbf{x}) + O(\hbar)\,,$$
$$-\hbar^2 \int \Upsilon(\mathbf{x}, \lambda)^* \, \partial_\lambda^2 \, \Upsilon(\mathbf{x}, \lambda)\, d\lambda$$
$$= \int |\hbar(\partial/\partial\lambda)\, \Upsilon(\mathbf{x}, \lambda)|^2\, d\lambda$$
$$= f(\mathbf{x})^2 + O(\hbar)\,.$$

The states in question are also canonical coherent states in the sense that

$$A(\mathbf{x}, \lambda)|(f, g)\rangle = \Upsilon(\mathbf{x}, \lambda)|(f, g)\rangle\,.$$

The new states $|(f, g)\rangle$ evidently share the property with the coherent states previously introduced that they are both labelled by a pair of real fields, f and g. However, these states should *not* be confused with the coherent states denoted by $|f, g\rangle$ and introduced in Chapter 5. To make that point clear we use the notation $|(f, g)\rangle$ rather than $|f, g\rangle$. In accordance with the expectation values given above, it follows that

$$\langle (f, g)| \int A(\mathbf{x}, \lambda)^\dagger \, w[-i\hbar(\partial/\partial\lambda), \lambda]\, A(\mathbf{x}, \lambda)\, d\lambda\, |(f, g)\rangle$$
$$= \int \Upsilon(\mathbf{x}, \lambda)^* \, w[-i\hbar(\partial/\partial\lambda), \lambda]\, \Upsilon(\mathbf{x}, \lambda)\, d\lambda$$
$$= w[f(\mathbf{x}), g(\mathbf{x})] + O(\hbar)\,.$$

The construction of expectation values in the set of special coherent states $\{|(f, g)\rangle\}$ is patterned after the standard properties normally used for affine coherent states for a single degree of freedom [AsK 69]. (Instead, basing things on canonical coherent states would be unsatisfactory because of the nonclassical, auxiliary term in $\hbar$ proportional to λ^{-2}.) The use of the specifically chosen function $\chi(\lambda)$ is not critical; what is critical is symmetry, sufficient differentiability, and that the moments of positive and negative powers of λ in the distribution $\chi(\lambda)^2$ satisfy the indicated properties. Other acceptable functions that replace the given $\chi(\lambda)$ will influence only the $O(\hbar)$ terms and *not* the $\hbar$-independent term $w[f(\mathbf{x}), g(\mathbf{x})]$. In consequence, therefore, it follows in the classical limit that

$$\lim_{\hbar \to 0} \langle (f,g)| \int A(\mathbf{x}, \lambda)^\dagger \, w[-i\hbar(\partial/\partial\lambda), \lambda] \, A(\mathbf{x}, \lambda) \, d\lambda \, |(f,g)\rangle = w[f(\mathbf{x}), \, g(\mathbf{x})].$$

Observe, moreover, that for the class of operators investigated, nothing is lost in the classical limit; namely, provided there are no $\hbar$-terms other than those explicitly indicated, only the λ-operator $w[-i\hbar(\partial/\partial\lambda), \lambda)] = 0$ has the classical limit $w[f(\mathbf{x}), g(\mathbf{x})] = 0$. The converse is also true; that is, provided there are no $\hbar$-terms other than those explicitly expected, the classical expression $w[f(\mathbf{x}), g(\mathbf{x})]$ uniquely determines the quantum expression $w[-i\hbar(\partial/\partial\lambda), \lambda)]$, modulo a possible ordering ambiguity. Such connections are relevant for our analysis of the classical limit of ultralocal quantum field theories.

Significantly, the local operator of primary interest, namely $\varphi(t, \mathbf{x})$, is not precisely of the form studied in the preceding several paragraphs. In order to discuss the classical limit of the desired operator, it is also necessary to analyze the $\hbar$-dependence of a certain integral involving the model function. Let us deduce that dependence by analogy. First of all, recall for a harmonic oscillator of angular frequency ω and ground state $|0\rangle$ that $\langle 0|Q^2|0\rangle = \hbar/(2\omega)$. Likewise for an ultralocal free model with mass m and ground state $|0\rangle$, we have $\langle 0|\varphi(f)^2|0\rangle = \hbar(f,f)/(2m)$, which has a similar dependence on $\hbar$. For an ultralocal pseudofree model with mass m and $\gamma < 3/2$, we find (restoring b for the rest of this paragraph) that

$$\langle 0|\varphi(f)^2|0\rangle = \int d^s x \int f(\mathbf{x})^2 \lambda^2 K^2 e^{-mb\lambda^2/\hbar} \, d\lambda/|\lambda|^{2\gamma}$$
$$= (f,f) \int \lambda^2 c(\lambda)^2 d\lambda.$$

It is natural to demand that this expression be approximately equal to $\hbar(f,f)/2m$. To achieve this dependence we conclude that

$$\int \lambda^2 c(\lambda)^2 \, d\lambda \approx \hbar/m;$$

indeed, the required dependence is already dictated by the fact that $b^{-1}c(\lambda)^2 \, d\lambda$ is dimensionless for any γ, along with the dimensionless form of $y(\lambda)$ – namely, $y(\lambda) = mb\lambda^2/2\hbar$ – for pseudofree ultralocal models. Dimensional arguments of the kind presented are not specific to the pseudo-free model, and they evidently hold for any well defined model function. Moreover, when $\gamma \geq 3/2$, time smearing as well as space smearing is essential to obtain an operator; however, that too does not change the dimensions of the expressions at all, and as a consequence, one learns that

$$\lambda^2 c(\lambda)^2 \, d\lambda \propto \hbar$$

for any model function $c(\lambda)$, and particularly, for any value of $\gamma \geq 1/2$. This dependence on $\hbar$ is the basic fact to take away from the discussion in this paragraph.

Our next task is to show that $|(f,g)\rangle$ lies in the diagonal form domain of the unsmeared field expression $\varphi(t,\mathbf{x})$, and so we focus on the expression

$$G(t,\mathbf{x}) \equiv \langle (f,g)|\varphi(t,\mathbf{x})|(f,g)\rangle \,,$$

for $t \geq 0$ and for fixed f and g. If f and g are continuous functions, then so too is the function G; likewise for differentiability. We first examine the situation for $t = 0$, and it follows that

$$\begin{aligned} G(0,\mathbf{x}) = &\int \Upsilon(\mathbf{x},\lambda)^* \,\lambda\, \Upsilon(\mathbf{x},\lambda)\,d\lambda \\ &+ \int \Upsilon(\mathbf{x},\lambda)^* \,\lambda\, c(\lambda)\,d\lambda + \int c(\lambda)\,\lambda\, \Upsilon(\mathbf{x},\lambda)\,d\lambda \,. \end{aligned}$$

Here the (principal value) integral $\int \lambda\, c(\lambda)^2 d\lambda$ has not been included as follows from the hypothesis that $\langle 0|\varphi(t,\mathbf{x})|0\rangle \equiv 0$. Based on the definition of $\chi(\lambda)$, the expression for $G(0,\mathbf{x})$ is well defined for any $\gamma \geq 1/2$, as we show below. Now observe that each of the last two integrals are $O(\sqrt{\hbar})$. In fact, for $|\mathbf{x}| < R$, the Schwarz inequality leads to

$$\begin{aligned} &\left|\int c(\lambda)\,\lambda\, \Upsilon(\mathbf{x},\lambda)\,d\lambda\right|^2 \\ &\qquad \leq \int |\lambda'|^{2\gamma+2} c(\lambda')^2 \,d\lambda' \int ||\lambda|^{-\gamma}\, \Upsilon(\mathbf{x},\lambda)|^2 \,d\lambda \,, \end{aligned}$$

which is not only well defined for all $\gamma \geq 1/2$, but is also proportional to $\hbar$, simply on dimensional grounds. We now assume that $t \neq 0$, and thus consider

$$\begin{aligned} G(t,\mathbf{x}) = &\int \Upsilon(\mathbf{x},\lambda)^* \, e^{i\hbar t/\hbar} \,\lambda\, e^{-i\hbar t/\hbar}\, \Upsilon(\mathbf{x},\lambda)\,d\lambda \\ &+ \int \Upsilon(\mathbf{x},\lambda)^* \, e^{i\hbar t/\hbar} \,\lambda\, c(\lambda)\,d\lambda + \int c(\lambda)\,\lambda\, e^{-i\hbar t/\hbar}\, \Upsilon(\mathbf{x},\lambda)\,d\lambda \,. \end{aligned}$$

We wish to show that this expression is well defined. Let us focus on polynomial potentials with a discrete spectrum μ_r, $r \in \{0,1,2,3,\ldots\}$. For such potentials, it follows from standard quantum mechanics that there exists constants A and B such that $\mu_r \simeq A + B r^{2N/(N+1)}$, for large r, where $2N$ is the highest power of the polynomial potential in $\hbar$; see Exercise 10.3. Let $u_r(\lambda)$ denote the corresponding orthonormal eigenfunction of the self-adjoint operator $\hbar$. Suppressing the $\mathbf{x}$ dependence for convenience and introducing a natural inner product in the λ-space, we estimate the first term in G as follows:

$$\begin{aligned} T \equiv &\left|\int \Upsilon(\lambda)^* \, e^{i\hbar t/\hbar} \,\lambda\, e^{-i\hbar t/\hbar}\, \Upsilon(\lambda)\,d\lambda\right| \\ &= \left|\Sigma_{r,s}(\Upsilon,u_r)e^{i\mu_r t/\hbar}(u_r,\lambda u_s)e^{-i\mu_s t/\hbar}(u_s,\Upsilon)\right| \\ &\leq \Sigma_{r,s}|(\Upsilon,u_r)(u_r,\lambda u_s)(u_s,\Upsilon)| \\ &\leq \Sigma_r |(\Upsilon,u_r)| \,[\Sigma_s |(u_r,\lambda u_s)|^2]^{1/2} \,[\Sigma_q |(u_q,\Upsilon)|^2]^{1/2} \\ &= \Sigma_r |(\Upsilon,u_r)| \,(u_r,\lambda^2 u_r)^{1/2} \,. \end{aligned}$$

To proceed further we note that for some C, D, E, and F, $\lambda^2 \leq C + D\hbar$. Therefore $(u_r,\lambda^2 u_r) \leq C + D\mu_r \leq (E + F\mu_r)^2$. Thus

$$T \leq \Sigma_r |(\Upsilon,u_r)|\,(E + F\mu_r) \,.$$

It is easy to convince oneself that $\hbar^3 \Upsilon(\lambda)$ is square integrable, and therefore that $(u_r, \hbar^3 \Upsilon) = \mu_r^3 (u_r, \Upsilon)$ is square summable. Consequently, there exists constants G and H such that $|(\Upsilon, u_r)| \leq G/(1 + \mu_r^3)$, and therefore

$$T \leq \Sigma_r G(E + F\mu_r)/(1 + \mu_r^3) \leq \Sigma_r H/(1 + \mu_r^2) ,$$

which evidently converges for any $N \geq 1$. Thus the existence of the first term is established for any t including $t = 0$. Next consider

$$T' \equiv |\int c(\lambda) \lambda e^{-i\hbar t/\hbar} \Upsilon(\lambda) d\lambda|$$
$$= |\Sigma_r (\lambda c, u_r) e^{-i\mu_r t/\hbar} (u_r, \Upsilon)|$$
$$\leq \Sigma_r |(\lambda c, u_r)| |(u_r, \Upsilon)| .$$

In turn, using the fact that $|(\lambda c, u_r)| \simeq A + B\mu_r^{\gamma/2 - 3/4} r^{-1/2}$, for suitable A and B (see Exercise 10.3), and the fact that with enough differentiability for χ the function $\hbar^q \Upsilon(\lambda)$ is square integrable in λ, it follows that constants C_q exist such that $|(u_r, \Upsilon)| \leq C_q/(1 + \mu_r)^q$ for any integer $q > \gamma/2$. As a consequence, it follows that T' is well defined for any t including $t = 0$. Hence, $G(t, \mathbf{x})$ has been shown to be well defined. Moreover, a parallel analysis shows that $G(t, \mathbf{x})$ is infinitely differentiable with respect to t.

Since we are studying the classical limit we will drop the last two terms in $G(t, \mathbf{x})$ because they vanish in the limit $\hbar \to 0$; thus, we retain only the first term. Let us use the symbol "$\cong$" to denote an approximation in which terms that vanish as $\hbar \to 0$ are dropped. Thus we have

$$G(t, \mathbf{x}) \cong \int \Upsilon(\mathbf{x}, \lambda)^* e^{i\hbar t/\hbar} \lambda e^{-i\hbar t/\hbar} \Upsilon(\mathbf{x}, \lambda) d\lambda .$$

Based on the previous discussion, we next observe that

$$G(0, \mathbf{x}) \cong \int \Upsilon(\mathbf{x}, \lambda)^* \lambda \Upsilon(\mathbf{x}, \lambda) d\lambda$$
$$\cong g(\mathbf{x}) ,$$

as well as

$$\dot{G}(0, \mathbf{x}) \cong i\hbar^{-1} \int \Upsilon(\mathbf{x}, \lambda)^* [\hbar, \lambda] \Upsilon(\mathbf{x}, \lambda) d\lambda$$
$$= -i\hbar \int \Upsilon(\mathbf{x}, \lambda)^* (\partial/\partial\lambda) \Upsilon(\mathbf{x}, \lambda) d\lambda$$
$$\cong f(\mathbf{x}) .$$

In addition, if we focus on a quartic interaction as a specific example, we can also state that

$$\ddot{G}(t, \mathbf{x}) \cong -\hbar^{-2} \int \Upsilon(\mathbf{x}, \lambda)^* e^{i\hbar t/\hbar} [\hbar, [\hbar, \lambda]] e^{-i\hbar t/\hbar} \Upsilon(\mathbf{x}, \lambda) d\lambda$$
$$\cong -\int \Upsilon(\mathbf{x}, \lambda)^* e^{i\hbar t/\hbar} [m^2 \lambda + 4\kappa\lambda^3] e^{-i\hbar t/\hbar} \Upsilon(\mathbf{x}, \lambda) d\lambda$$
$$\cong -[m^2 G(t, \mathbf{x}) + 4\kappa G(t, \mathbf{x})^3] .$$

As a consequence, we have determined that

$$g(t, \mathbf{x}) \equiv \lim_{\hbar \to 0} G(t, \mathbf{x}) = \lim_{\hbar \to 0} \langle (f, g) | \varphi(t, \mathbf{x}) | (f, g) \rangle$$

satisfies the differential equation

$$\ddot{g}(t, \mathbf{x}) = -m^2 \, g(t, \mathbf{x}) - 4\kappa \, g(t, \mathbf{x})^3 \,,$$

subject to the initial conditions

$$g(0, \mathbf{x}) \equiv g(\mathbf{x}) \,, \qquad \dot{g}(0, \mathbf{x}) \equiv f(\mathbf{x}) \,.$$

These are just the initial conditions and equation of motion satisfied by a classical quartic ultralocal model! Of course, one of the terms neglected in the equation of motion is proportional to $\hbar^2/g(\mathbf{x})^3$. Such a term has the effect – so long as $\hbar \neq 0$ – of ensuring that $g(\mathbf{x}) \neq 0$ just as was the case in the preceding section.

It is in the sense indicated that we say that the nontrivial quantum solution we have found has the appropriate classical limit. The analysis of the classical limit of ultralocal models presented here has been inspired by [ZhK 94], which in turn was motivated by [He 74].

$$* \, * \, *$$

It is notable that ultralocal model field theories have been studied in quite different contexts as well. In particular, we draw the reader's attention to the work of Pilati [Pi 82, Pi 83, FrP 85] where ultralocal models of quantum gravity, which arise in the strong coupling limit, are studied. Another way to look at ultralocal models of gravity is given by [Kl 70b]. Although not the subject of this book, we may also mention that suitable ultralocal models for fermion fields are soluble as well, and the solutions exhibit many interesting properties; see [Kl 73c].

Exercises

10.1 For the conventional lattice-space functional integral of ultralocal models show (with $\hbar = 1$) that the two-point function of Section 10.1 is given by

$$C^{(2)}(t) \equiv \lim_{a \to 0} \mathcal{N} \int x(t) x(0) \exp\{-\tfrac{1}{2} \int [\dot{x}(t)^2 + m_a^2 \, x(t)^2] \, dt \, a^s$$
$$- \kappa_a \int x(t)^4 \, dt \, a^s\} \, \mathcal{D}x$$
$$= \langle 0|Q(t)Q(0)|0\rangle$$
$$= \sum_n \langle 0|Q|n\rangle \, e^{-iE_n t} \, \langle n|Q|0\rangle \,,$$

where $\{E_n\}_{n=0}^{\infty}$, $0 = E_0 < E_1 < E_2 < \cdots$, are energy levels determined as eigenvalues of the Schrödinger equation

$$[-\tfrac{1}{2}(\partial^2/\partial x^2) + \tfrac{1}{2}m_*^2 \, x^2 + \kappa_* \, x^4 - \text{const.}] \, \psi_n(x) = E_n \, \psi_n(x) \,,$$

where $\psi_n(x) = \langle x|n\rangle$, while $m_*^2 = \lim_{a \to 0} m_a^2$ and $\kappa_* = \lim_{a \to 0} (\kappa_a/a^s)$.

10.2 Let $w(-i\partial/\partial\lambda, \lambda)$ denote a self-adjoint operator on $L^2(\mathbb{R})$, and define

$$W(\mathbf{x}) = \int B(\mathbf{x}, \lambda)^\dagger \, w(-i\partial/\partial\lambda, \lambda) B(\mathbf{x}, \lambda) \, d\lambda \,,$$

where $B^\dagger$ and B are the operators introduced in Section 10.3. Determine the conditions on w so that $W(f)$ for real, bounded, compact f is a self-adjoint operator on $\mathfrak{H}$ [Kl 71].

10.3 On $L^2(\mathbb{R})$, consider the Hamiltonian operator

$$\hbar = -\frac{1}{2b}\frac{\partial^2}{\partial\lambda^2} + \frac{\gamma(\gamma+1)}{2b\lambda^2} + \frac{1}{2}bm^2\lambda^2 + \kappa b^{2N-1}\lambda^{2N} - \text{const.} \,,$$

where $\gamma \geq 1/2$ and $\kappa > 0$. Let the normalizable eigenfunctions be called $u_r(\lambda)$ and eigenvalues μ_r, $r \in \{0, 1, 2, \ldots\}$, which satisfy $\hbar u_r(\lambda) = \mu_r u_r(\lambda)$. Show the following:

(a) Determine that all eigenfunctions behave as $u_r(\lambda) \propto |\lambda|^{1+\gamma}$ or $u_r(\lambda) \propto \lambda|\lambda|^\gamma$ for $|\lambda| \ll 1$.

(b) Determine the qualitative behavior of $u_r(\lambda)$ for $|\lambda| \gg 1$.

(c) Given the model function $c(\lambda) = K|\lambda|^{-\gamma}\exp[-y(\lambda)]$, which satisfies $\hbar c(\lambda) = 0$, even though it is not a normalizable eigenfunction, show that

$$k_r \equiv \int u_r(\lambda)^* \, \lambda \, c(\lambda) \, d\lambda$$

is well defined for all $\gamma > 1/2$ and all r. Since $\lambda c(\lambda)$ is an odd function, k_r vanishes whenever u_r is even. For odd u_r, on the other hand, there is generally no reason for k_r to vanish or even be exceptionally small.

(d) Since $\hbar$ is a positive operator it follows that $\hbar^\epsilon$ can be defined for all $\epsilon > 0$. Let us accept that

$$\hbar^\epsilon \lambda c(\lambda) \approx \lambda^{1-2\epsilon} c(\lambda) \,.$$

Show that $\lambda^{1-2\epsilon}c(\lambda)$ is square integrable (the test is at the origin) provided $3 - 4\epsilon > 2\gamma$, or for $\epsilon < 3/4 - \gamma/2$. Show that, in general, square integrability is lost when $\epsilon > 3/4 - \gamma/2$.

(e) Based on Bohr–Sommerfeld quantization principles, show for large r that

$$\mu_r \propto r^{2N/(N+1)} \,.$$

(f) Consider the expression

$$\Sigma_r \, \mu_r^{2\epsilon} |k_r|^2 \,.$$

For $\epsilon < 3/4 - \gamma/2$ this expression converges, while for $\epsilon > 3/4 - \gamma/2$ it diverges. Use this fact to verify the estimate that

$$|k_r| \propto r^{-1/2-(3/2-\gamma)N/(N+1)} \,,$$

for large r. Specialize these results to a quartic interaction, i.e., when $N = 2$.

10.4 Based on the results of the previous exercise, determine the ϵ dependence, where $0 < \epsilon \ll 1$, of the expression

$$\int c(\lambda)\,\lambda\,e^{-\epsilon\hbar}\,\lambda\,c(\lambda)\,d\lambda$$

when $\gamma > 3/2$. (Hint: Use a Riemann sum approximation for an integral involved.)

10.5 Let the states $|f\rangle$ on this occasion be defined by

$$|f\rangle \equiv \exp[i\int \phi^\theta_R(\mathbf{x})f(\mathbf{x})\,d^s x]|0\rangle$$

for all real, bounded functions f of compact support. Since we have insisted that λ^θ is defined as an odd function in the definition of ϕ^θ_R, convince yourself that the set of states $\{|f\rangle\}$ defined in this way forms a total set for the Hilbert space $\mathfrak{H}$ for any value of $\theta > \gamma - 1/2$.

10.6 Given $\gamma = s/\beta + 1/2$ and $d_\theta = s - \theta/2$, determine that the operator form for the dilation operator is given by

$$D = -i\tfrac{1}{2}\int d^s x\,A(\mathbf{x},\lambda)^\dagger\{\mathbf{x}\cdot\boldsymbol{\nabla} + \boldsymbol{\nabla}\cdot\mathbf{x} + \tfrac{1}{2}\beta[\lambda(\partial/\partial\lambda) + (\partial/\partial\lambda)\lambda]\}A(\mathbf{x},\lambda)\,d\lambda\,.$$

In this regard make use of the fact, for stable ultralocal models, that

$$\langle f|\exp[i\ln(S)D]|f'\rangle = \langle f|f'_S\rangle$$
$$= \exp[-K^2\int d^s x \int (1 - \cos\{\lambda^\theta[f'_S(\mathbf{x}) - f(\mathbf{x})]\})|\lambda|^{-2\gamma}\,d\lambda]\,,$$

where $b = 1$ and the states $\{|f\rangle\}$ are as defined in Exercise 10.5.

10.7 Based on the discussion of $O(N)$-symmetric, vector-valued independent-value model theories in Section 9.6, determine the extension of scalar ultralocal quantum field theories to $O(N)$-symmetric, vector-valued ultralocal quantum field theories, e.g., to N-component fields, $N < \infty$, exhibiting an N-dimensional rotational symmetry [DeH 77]. When $N \geq 2$, the auxiliary term in the classical Hamiltonian density is of the form $\hbar^2/\vec{g}(\mathbf{x})^2$. Unlike the case $N = 1$ for which, for any solution to the classical equations of motion, g can not change sign, discuss how, for the case of a vector field, the classical solution can pass around the singular point $\vec{g} = (g_1,\ldots,g_N) = 0$ and so, effectively, $\vec{g}$ is able to change sign.

10.8 Using the freedom in the undetermined parameter K in the definition of the model function – and based on an analogous discussion in Section 9.7 – determine the form of the quantum field theory for rotationally symmetric, infinite-component ultralocal model theories [ZhK 95]. Compare and contrast the N-dependence of the several coupling constants involved in your treatment with the N-dependence of the conventional treatment for rotationally symmetric, infinite-component fields [Ya 82].

10.9 In the analysis of the classical limit for ultralocal models in Section 10.9 we treated the argument fields f and g of the states $|(f, g)\rangle$ as if, for small $\hbar$, these fields could be regarded as the classical momentum and classical field, respectively. These fields may be so identified provided that $i\hbar\langle(f, g)|d|(f, g)\rangle = -\int g(\mathbf{x})df(\mathbf{x})\, d^s x + O(\hbar)$. Show that this property does indeed hold true for the set of states $\{|(f, g)\rangle\}$.

11

Summary and Outlook

WHAT TO LOOK FOR

In this chapter we initially reiterate some basic lessons developed in earlier chapters, and, to assist the reader, we stress those particular points needed for a first appreciation of our proposal to overcome triviality. In our survey we introduce a rather simple and deductive way to uncover a great deal about the Euclidean-space lattice functional integral formulation appropriate to ultralocal models. Basing further discussion on an analog of this simple and deductive approach, we extend the concepts of earlier chapters – especially including hard-core interactions and pseudo-free theories as zero-coupling limits of interacting models – to form a conjecture on how ϕ_n^4 theories for space-time dimensions $n \geq 4$ may possibly be reformulated so as to yield nontrivial results.

11.1 Recapitulation – with commentary

The quantum theory of an infinite number of degrees of freedom is a rich and varied subject, and it is not self-evident how best to approach such a vast set of topics. One pragmatic approach has been based on the following concept. Since classical field theory can be approached through a sequence of problems with a finite number N of degrees of freedom in the limit that $N \to \infty$, it stands to reason that the problems of quantum field theory can be approached the same way, by quantizing a system of N degrees of freedom and then letting $N \to \infty$. The success of this approach for free field models and for a number of models with suitably mild nonlinear interactions, gives rise to the belief that all "physically relevant" models (whatever that means) may be approached in the same fashion. The very significant successes of the renormalization program did much to strengthen this concept, and a vast core of phenomenologically relevant formalism and calculations has been and is still being developed. From this perspective, it would be fair to state that the conceptual development of quantum field theory is essentially a completed discipline. We strongly reject this point of view.

As a many-faceted subject, it is always of value to regard one's theories in a larger context – even within an enlarged set including clearly nonphysical theories – in order to gain further insight into any particular theory. For example, we have in mind changing space-time dimensions, considering alternative interactions, etc. The study and attempted solution of a great diversity of problems is a time-honored tradition in statistical mechanics, and it deserves to be so in quantum field theory as well. And in this text we have attempted to further the understanding of general quantization procedures for fields precisely by such model studies.

Triviality

In several earlier chapters we have, among other subjects, studied a number of model quantum field theories, all of which have the common feature that when viewed within the framework of the conventional wisdom – as defined, for example, by what is put forth in the standard textbooks on quantum field theory – they are generally trivial. *With the strongest possible emphasis we wish to assert that triviality must be regarded as an unacceptable answer.* Why unacceptable? Simply because a trivial theory does not have the correct limiting behavior in the classical limit. Not too many researchers appear to concern themselves with the classical limit anymore, but the present author views the fulfillment of such a limit as an indispensable guide in selecting, among various contending quantizations, that subset that has any claim of relevance to the problem at hand. As a matter of principle, it appears impossible that the correct quantization of a *nontrivial* classical theory should be a *trivial* quantum theory; this association flies in the face of everything we believe about the correspondence principle.

Consider the famous "triviality" of the relativistic ϕ_n^4 theory [FeFS 92] which asserts that given the right starting point the quantization of this manifestly nontrivial classical theory turns out to be trivial whenever $n > 4$ and probably for $n = 4$ as well. The "right starting point" in this case – and indeed it is a very natural place to start – is that of the conventional approach which consists of regularizing a Euclidean functional integral on a lattice space-time using a naive lattice action.* If such a starting point has led to a trivial quantization for a nontrivial classical theory, then one should not necessarily conclude that the theory actually *"is"* trivial, but rather hold suspect the very set of starting assumptions that has led to this conclusion. To propose new starting assumptions that can be fully tested and eventually accepted is a daunting task; after all,

---

* A complementary operator analysis has been given by Baumann [Ba 87].

there is always more light to look for one's lost set of keys under the lamppost than anywhere else. Nevertheless, there are reasons why an alternative starting point for such theories may be relevant. One strong argument is simply the fact that such theories are known to be *not* asymptotically free, meaning that their short distance – or high momentum – behavior is *not* equivalent to that of a free theory. On the other hand, the naive lattice starting point for the quantization of such models is *exactly* one compatible with the asymptotic freedom appropriate to how the free theory itself is properly represented on the lattice. It is no wonder that triviality is the only answer that an asymptotically nonfree theory can give once it has been put into the straitjacket of an asymptotically free lattice formulation. To find a successful alternative lattice action to start with, specifically one that leads to a covariant and nontrivial quantization, is not an easy task; several workers have tried over a number of years. But the failure of this relatively modest effort to find a successful starting point does not mean that one does not exist. In point of fact, there is a certain amount of circumstantial evidence that an alternative and suitable starting point may possibly exist.

Soluble models

Being unable to solve the relativistic models of interest, we have chosen to discuss several models that share the property that when quantized within the conventional framework they also lead to triviality, but which when viewed *beyond* the confines of conventional quantization procedures do indeed have fully satisfactory and nontrivial quantizations. For the rotationally symmetric models, discussed in Chapter 7, conventional quantization methods lead to trivial results that are incompatible with the nontrivial classical results. Nevertheless, thanks to the huge symmetry built into these models, a meaningful and nontrivial solution rigorously emerged when we admitted the possibility of *reducible* representations for the canonical sharp-time field and momentum operators, thereby going beyond one of the standard tenets of conventional quantum field theory. The extension to reducible representations carried with it several special problems, the most significant of which was that uniqueness of the ground state then forced the quantum Hamiltonian not to be expressed entirely in terms of the canonical field operators alone but to involve additional operators in a fundamental way. This fact required us to find an alternative way to connect a quantum theory with a classical theory so as to justifiably claim that the quantum theory in question did indeed qualify as the quantization of the given classical theory. Fortunately, the use of the weak correspondence principle to associate a quantum and a classical theory was completely adequate in the case of the rotationally symmetric models of Chapter 7, and, along with symmetry and general principles, led to a fully satisfactory resolution of providing an acceptable solution for such models.

Chapters 9 and 10 dealt with two quite different kinds of models from those which were discussed in Chapter 7. In the case of the independent-value models, discussed in Chapter 9, we only have functional integration techniques to use. A standard lattice approach – reminiscent of the standard lattice starting point used in establishing triviality of the covariant ϕ_n^4 models (see below) – leads to triviality for any locally interacting independent-value model. To guess at an unconventional starting lattice approach would be all but impossible for such a model, just as it appears to be in the relativistic case. Fortunately, guessing is not necessary for the independent-value models, because these models enjoy such an enormous degree of symmetry that a rigorous construction becomes possible based only on that symmetry and on general principles. Given the result determined by such arguments, it becomes possible to ask and answer the question of just what would be a suitable lattice action within a functional integral in order to obtain the same nontrivial result. Thus, thanks to their rich symmetry, this class of model theories is solved in an entirely satisfactory manner. But it is also important to appreciate, from a functional integral point of view, that each of the independent-value models is actually *more* singular than the corresponding covariant model would be. This assertion is based on the absence of all space-time gradients from the continuum action, terms that heuristically should have the effect of softening the class of distributions that carry the support of the functional integration measure when dealing with a Euclidean formulation where such issues have the most meaning. In other words, we claim that not only are the independent-value models more singular than their covariant counterparts, they also admit fully satisfactory nontrivial solutions! Based on that fact, it is not unreasonable to assert that if the more singular situation is soluble and leads to nontrivial results, then so too should the less singular case.

The situation regarding the ultralocal models treated in Chapter 10 is similar to that for the independent-value models. Ultralocal models differ from independent-value models by the presence of a temporal gradient squared in the classical action. They differ from a corresponding covariant model by the absence of all spatial gradient terms in the free action. Thus from the point of view of qualitative support properties in a Euclidean functional integral formulation, the ultralocal models lie between the independent-value models and covariant models. Specifically, ultralocal models are less singular than independent-value models and more singular than covariant models. With temporal gradients present, the ultralocal models can be approached through all conventional quantization schemes, in particular, by functional integration and by operator techniques, both of which lead to triviality. Again we choose not to accept this result, but are at a loss as to what alternative starting lattice action to choose. By design, however, these models have been chosen with sufficient symmetry so that a fully satisfactory and nontrivial solution can be found beyond conventional quantization procedures. That solution again involved only symmetry and general principles, and from it we were able to determine what sort of starting lattice action within a functional integral would lead to the same nontrivial result. The alternative starting point essentially agrees with the naive lattice action apart from a special choice of an auxiliary, nonclassical (order $\hbar^2$) contribution to the potential that, without undue exaggeration, accounts for the ultimate nontriviality of the continuum limit of the regularized functional integral. For all interacting ultralocal models, it is noteworthy that a canonical momentum field does not exist as a sharp-time local operator, and so these models cannot satisfy canonical commutation relations. This fact complicates traditional efforts to take the classical limit, leading therefore to a problem that called for a new and unconventional approach. This alternative approach fortunately confirmed that the classical limit of the quantum theory gave rise to the original motivating classical theory. Once again, we have in hand a class of models that are trivial when treated conventionally but become nontrivial when given a suitable starting point, and, significantly, each of the ultralocal models in question is more singular than the corresponding covariant theory.

There are general similarities in the properties of the solutions of the independent-value models and the ultralocal models that are worth noting. Let us confine attention to the class of semi-bounded polynomial interactions for each model. Then, according to the criterion discussed in Chapter 8 for continuous and discontinuous perturbations, it is clear for both kinds of models that every (local) interacting theory is a discontinuous perturbation of the associated free theory. Although for field theories the general criterion based on finiteness of the (Euclidean) free action and the interaction action is heuristic at best, it nevertheless coincides with the facts for these two classes of models. As discussed at great length in each of the appropriate chapters, all locally interacting independent-value and ultralocal theories pass to appropriate pseudofree theories, substantially different from any of the free theories, as the coupling constant is turned off. This behavior has been attributed to the presence of a *hard-core interaction potential* that has the effect of projecting out contributions to the functional integral that would otherwise be allowed by the free theory. We have illustrated this phenomenon in the case of ordinary quantum mechanics in Chapter 8, where the consequences of a hard-core interaction can effectively be visualized and where the zero coupling limit of the interacting theory retains the effects of the hard-core giving rise to the pseudofree theory in the limit. On the other hand, in the case of a quantum field theory, the action of a hard-core interaction is nearly impossible to visualize in any meaningful way. Nevertheless, the relevance of a pseudofree limit for interacting independent-value and ultralocal models, in contrast to an ordinary free theory limit, strongly suggests that the mechanism of a hard-core interaction is at work.

Relevance for covariant theories

As discussed in Chapter 8, classical multiplicative inequalities, sometimes loosely grouped under the name of Sobolev inequalities, suggest that the interaction for a covariant ϕ_n^4 theory for $n > 4$ also satisfies the heuristic criterion for a hard-core interaction and an associated pseudofree theory as the vanishing coupling limit of an interacting theory. We cannot solve the covariant theories – this book would only be about them if we could! – but we are convinced that a solution that involves something beyond conventional quantization procedures exists; that something could very well emerge from an alternative starting lattice action that differs from the naive version by terms of order $\hbar$. Unfortunately, there does not appear to be enough symmetry within covariant models to determine a solution simply on general grounds, although this remark should not be taken as minimizing the importance of a continued search for just such a solution. Lacking any direct, symmetry-based approach to find solutions leaves only one other alternative at the present time, namely that of "guessing" what form of alternative starting lattice action to try. Of course, such an attempt may seem as likely to succeed as the proverbial search for a needle in a hay stack. That view may well be true, but nevertheless we are persuaded to have a try, and more particularly, we shall offer a proposal for a nontrivial formulation of ϕ_n^4, for $n > 4$ (and $n = 4$), motivated by the insights gleaned from a study of singular models as presented in this text, especially from a study of the ultralocal models. At present our proposal for a possible nontrivial relativistic theory is just that, namely it is no more than a proposal; it may well be entirely wrong – and yet even in that case it may have a germ of the truth, or perhaps suggest to another researcher a significant improvement for which we should all be grateful.

Certain readers may find themselves saying: So what! Who cares whether or not one can find a nontrivial quantization for a clearly unphysical theory such as ϕ_n^4 for $n > 4$. To such a reader we hasten to point out that the ϕ_n^4 models themselves need not be taken as the ultimate goal; such models may be regarded as simplified examples of asymptotically nonfree and nonrenormalizable theories, others of which may have far more physical relevance. For example, we can always point to the gravitational field, which is clearly an important physical field, the quantization of which must surely be obtained some day despite the fact that viewed perturbatively it is nonrenormalizable. Elsewhere [Kl 75b] we have discussed the fact that the gravitational field is also an example involving a potentially hard-core interaction, and therefore it would appear to be a natural candidate for a solution beyond conventional quantization.

We could hardly discuss the quantum theory of the gravitational field and not make mention of all the efforts surrounding the ideas of string theory (e.g., [Ka 93]). As string theory proponents like to say, it is the only consistent quantum theory of gravitation we have. Even granting the possibility that string theory is entirely consistent, that fact does not guarantee that it is the *correct* quantum theory of gravitation, the consistency of which may not have been established yet. In short, one should not give up alternative searches for quantum gravity based on the assertion that a consistent theory may already have been found. Should a hard-core theory of quantum gravity turn out to be consistent – as in fact the hard-core independent-value and ultralocal theories are – then it would be a far less radical way to deal with quantum gravity than via most other proposed schemes.

Despite the fact that coming to grips with quantum gravity has partially motivated the author's interest in finding soluble examples of nonrenormalizable model field theories, any talk about a hard-core theory of quantum gravity at this point in time is pure speculation. Let us return to the principal focus of our efforts, namely trying to offer some alternative and unconventional proposal for a covariant ϕ_n^4 for $n \geq 4$. But before doing so, let us dip once more into the wellspring of so many of our inspirations.

11.2 Essentials of ultralocal models

It is useful if we begin with a summary of the starting and ending statements in the case of ultralocal models, and also recall the Euclidean lattice functional integral that leads to the proper results. For convenience, we present our formulation in terms of expectation functionals. In point of fact, however, those cases where the singularity parameter $\gamma \geq 3/2$ are more properly formulated in terms of multipoint correlation functions with test functions so chosen that coincident points are excluded [OsS 73, OsS 75]; this modification makes for a modest change in the formulas which we shall illustrate below in an example. We focus our attention on a quartic self-coupled interaction.

The classical quartic ultralocal theory has a classical field $g(\mathbf{x})$ and momentum $f(\mathbf{x})$ defined for all $\mathbf{x} \in \mathbb{R}^s$. The dynamics for such models is implicitly contained in the classical Hamiltonian which is given by

$$H(f,g) = \int \{\tfrac{1}{2}[f(\mathbf{x})^2 + m_o^2 \, g(\mathbf{x})^2] + \kappa_o \, g(\mathbf{x})^4\} \, d^s x \ .$$

It is evident that the dynamical solutions of the classical theory are generally nontrivial. The formal real time functional integral for such a model has the form

$$Z\{h\} = \mathcal{M} \int \exp(i \int \{h(x)\phi(x)$$
$$+ \tfrac{1}{2}[\dot{\phi}(x)^2 - m_o^2 \, \phi(x)^2] - \kappa_o \, \phi(x)^4\} \, d^n x) \, \mathcal{D}\phi \ .$$

A formal functional integral needs to be defined – say by a lattice prescription – in order to begin its evaluation. A naive lattice prescription, as shown in Chapter 10, leads to a trivial (Gaussian) result. However, symmetry and general principles offer us another answer for what this functional integral might represent, namely

$$Z\{h\} = \exp\{-\int d^s x \int c(\lambda) \left[1 - T e^{i \int \lambda(t)\, h(t,\mathbf{x})\, dt}\right] c(\lambda)\, d\lambda\}\ .$$

In this expression

$$\lambda(t) \equiv e^{i\hbar t}\, \lambda\, e^{-i\hbar t}\ ,$$

$$\hbar \equiv -\frac{\partial^2}{2b\, \partial\lambda^2} + \frac{\gamma(\gamma+1)}{2b\, \lambda^2} + \frac{1}{2} bm^2 \lambda^2 + \kappa b^3 \lambda^4 - \text{const.}\ ,$$

$$\hbar c(\lambda) \equiv 0\ ,$$

$$c(\lambda) \equiv \frac{K e^{-y(\lambda)}}{|\lambda|^\gamma}\ , \qquad \gamma \geq 1/2\ ,$$

where b is a constant with dimensions (length)$^{-s}$, K is separated from y by the condition that $y(0) = 0$, and the relation between the coupling constants is formally given by $m_o^2 = Z m^2$, $\kappa_o = Z^3 \kappa$, where $Z = b/\delta(\mathbf{0})$, which involves an s-dimensional δ-function at the origin. Finally, K does not enter the definition of $\hbar$; it may ultimately be fixed by choosing the scale of the two-point function. So much for a characterization of the *solution* for the formal functional integral determined on the basis of symmetry and general principles. We emphasize that in this determination of the proper answer, *no functional integral has actually been performed.*

This last point is, in some sense, the most important one, and based on it one is in a position to use the symmetry-derived answer to *define* a suitable functional integral that yields the same answer. In particular, switching now to Euclidean time, the answer may be determined as the continuum limit (implicitly combined with an infinite volume limit) of the well-defined, lattice-regulated functional integral represented by

$$S\{h\} = \lim_{a \to 0} \lim_{\epsilon \to 0} M \int \exp\{\Sigma\, h_k \phi_k a^s \epsilon$$
$$- \tfrac{1}{2}(ba^s) \Sigma\, (\phi_{k^*} - \phi_k)^2 \epsilon^{-2} a^s \epsilon - \tfrac{1}{2}(ba^s) m^2 \Sigma\, \phi_k^2 a^s \epsilon - (ba^s)^3 \kappa \Sigma\, \phi_k^4 a^s \epsilon$$
$$- \tfrac{1}{2} b^2 (ba^s)^{-3} \Sigma\, [\gamma(\gamma+1)\phi_k^2 - \gamma a^{-2s}\delta^2](\phi_k^2 + a^{-2s}\delta^2)^{-2} a^s \epsilon\} \Pi\, d\phi_k\ .$$

Here a denotes the spatial lattice spacing, a^s is the volume of an elementary s-dimensional spatial lattice cell, ϵ is the temporal lattice spacing, b is the same dimensional parameter introduced above, and for $\gamma > 1/2$, $\delta = \delta(a) \equiv (GK^2 a^s)^{1/(2\gamma-1)}$ with $G = \Gamma(\tfrac{1}{2})\Gamma(\gamma - \tfrac{1}{2})/\Gamma(\gamma)$. For completeness, we note that M is a normalization chosen so that $S\{0\} = 1$, $k = (k^0, k^1, \ldots, k^s)$, $k^j \in \{0, \pm 1, \pm 2, \ldots\}$ for all j, and k^* denotes the point next to k advanced one step in time ($k^0 \to k^0 + 1$). The combination ba^s is dimensionless, vanishes as $a \to 0$, and represents the lattice regularized version of the formal expression $Z = b/\delta(\mathbf{0})$. So much for the details; what is essential to observe is that a single auxiliary potential in the lattice action accounts for convergence in the continuum limit to the proper and nontrivial continuum limit. This auxiliary potential is unavoidably proportional to $\hbar^2$, and therefore it is absent in the classical model. Its presence must be determined on general grounds, as was the case in our extensive discussion of the ultralocal models in Chapter 10 – or it must be guessed.

Of course, guessing need only be confined to plausible candidates, and we have stressed several times in this book the naturalness of an ambiguity in the coefficient of just such a term in the quantum Hamiltonian. For example, in Chapter 4 we argued that such a term was a natural and potential ambiguity in quantizing even a simple, single-degree-of-freedom system with a classical Hamiltonian of the form $H(p,q) = p^2/2 + V(q)$. Such a term also arises in the study of infinite-component scalars in the conventional treatment of such systems, e.g., in the coherent state approach well discussed by Yaffe [Ya 82]. We can find such terms in nonrelativistic particle quantum mechanics; for instance, consider the quantum treatment of a hydrogen atom with an angular momentum l. In that case the quantum Hamiltonian contains the term $\hbar^2 l(l+1)/2r^2$ in typical notation; see [Da 68]. With l fixed and finite, the classical limit of this problem reduces to the classical hydrogen atom with vanishing angular momentum. Hence, the explicit term proportional to r^{-2} does not survive in the classical limit in this class of problems as well. In yet another example, when one investigates path-integral formulations of three-dimensional spherically symmetric potentials it is generally accepted that an auxiliary potential of the form $\hbar^2 c/r^2$, c a constant, needs to be added to the classical action to obtain the correct quantization [Kle 95]. All of this discussion is put forward simply to argue that the auxiliary term $\hbar^2 \gamma(\gamma+1)/2b\,\lambda^2$ introduced into $\hbar$ is, in fact, not unusual for such Hamiltonians; what is unusual, or at least unexpected, is the fact that such a term is actually *necessary* in order for the proper and nontrivial result to emerge in the case of ultralocal models.

In general, there are many different sequences that lead to one and the same limit. Thus we can imagine that there are other forms of the lattice action which, after integration and the continuum limit, would also lead to the same result, the result favored on the basis of symmetry and general principles. However, these alternative sequences cannot deviate too much from the one indicated, or otherwise the continuum limit may differ in its result, or may not even exist at all. We do not expect that any other sequence will achieve the desired result if it does not contain the auxiliary potential in the form given or in a form that resembles it closely. We shall not pursue the question of alternative sequences that have the same limiting behavior; rather, we shall assume that the given form of the regularized auxiliary potential is sufficiently representative of such possibilities, and we shall continue to accept the given form of the auxiliary potential within the lattice functional integral, namely, as

$$\tfrac{1}{2}b^2(ba^s)^{-3}\hbar^2 \frac{[\gamma(\gamma+1)\phi_k^2 - \gamma a^{-2s}\delta^2]}{(\phi_k^2 + a^{-2s}\delta^2)^2}.$$

Design by trial and error

Now that we have recalled the correct form of the lattice-space functional integral for ultralocal models, let us see to what extent we can deduce this lattice space expression for ultralocal models based simply on certain self-consistency arguments. We suppose that the final answer $S\{h\}$ is given by a double limit of $S_{a,\epsilon}(h)$ as the temporal lattice spacing $\epsilon \to 0$ and the spatial lattice size $a \to 0$. The quantity $S_{a,\epsilon}(h)$ is the generator of correlation functions, to wit

$$S_{a,\epsilon}(h) = \langle e^{\sum h_k \, \phi_k a^s \epsilon} \rangle = \exp\{\langle [e^{\sum h_k \, \phi_k a^s \epsilon} - 1]\rangle^T\} \; .$$

Let us focus on the generator of the truncated correlation functions

$$S_{a,\epsilon}^T(h) \equiv \langle [e^{\sum h_k \, \phi_k a^s \epsilon} - 1]\rangle^T$$
$$= \Sigma_{r=1}^\infty [1/(2r)!] \, \Sigma \, h_{k_1} \cdots h_{k_{2r}} \langle \phi_{k_1} \cdots \phi_{k_{2r}} \rangle^T \, a^{2rs} \epsilon^{2r} \; ,$$

where we have used the symmetry of the model to exclude all odd-order correlation functions and the sum on r begins with $r = 1$. Just as before, we set $k \equiv (p, q)$, where $p \equiv k^0$ and $q \equiv (k^1, k^2, \ldots, k^s)$. Due to the statistical independence of the field values at distinct spatial points, it follows that

$$\langle \phi_{k_1} \cdots \phi_{k_{2r}} \rangle^T = \delta_{q_1,q_2} \delta_{q_2,q_3} \cdots \delta_{q_{2r-1},q_{2r}} \langle \phi_{p_1} \cdots \phi_{p_{2r}} \rangle^T \; ,$$

in which the spatial labels appear in the Kronecker δs and the temporal labels remain on the stochastic variable. This form means that

$$S_{a,\epsilon}^T(h) = \Sigma_r [1/(2r)!] \, \Sigma \, h_{p_1,q} \cdots h_{p_{2r},q} \langle \phi_{p_1} \cdots \phi_{p_{2r}} \rangle^T \, a^{2rs} \epsilon^{2r} \; .$$

Evidently there are too many a factors for this expression to make good sense in the continuum limit, so let us change that. We first make a rescaling transformation of the entire test sequence

$$h_{p,q} \to (ba^s)^{-1} h_{p,q} \; ,$$

where b is a parameter with units (length)$^{-s}$ chosen so that the dimensions of the test sequence remain unaffected. After this change it follows that

$$S_{a,\epsilon}^T(h) \to S_{a,\epsilon}^T((ba^s)^{-1}h)$$
$$= \Sigma_r [1/(2r)!] \, b^{-2r} \, \Sigma \, h_{p_1,q} \cdots h_{p_{2r},q} \langle \lambda_{p_1} \cdots \lambda_{p_{2r}} \rangle^T \, \epsilon^{2r} \; ,$$

where we have renamed the stochastic variable λ. Observe that now all external a factors have disappeared; and, additionally, observe that this result is also unsatisfactory. What we really need is a *single* factor a^s to go with *each* term so as to accommodate the ultimate integral over the spatial variable. Thus we are led to the notion that we should arrange the multitime correlation functions so that, for all $r \geq 1$,

$$\langle \lambda_{p_1} \cdots \lambda_{p_{2r}} \rangle^T \propto ba^s \; .$$

Obtaining this final proportionality is exactly what the auxiliary, nonclassical potential added to the naive lattice action is designed to accomplish!

The lessons of this plausibility discussion can be put into action through the following argument. Let us first introduce the expression

$$\overline{M}_a \prod_q \int \exp[\Sigma\, h_{p,q}\lambda_p a^s \epsilon - \tfrac{1}{2}b^{-1}\Sigma\,(\lambda_{p+1} - \lambda_p)^2 \epsilon^{-1}$$

$$- \tfrac{1}{2}b^{-1}m^2\,\Sigma\,\lambda_p^2\epsilon - \kappa b^{-1}\Sigma\,\lambda_p^4\epsilon$$

$$- \Sigma\,P_{a,\epsilon}(\lambda_p)\epsilon]\,\Pi d\lambda_p$$

as a product over the spatial sites q of a temporal lattice for a quartic anharmonic oscillator, where $P_{a,\epsilon}$ represents the unknown auxiliary potential. Observe that without this last term, the rest of the lattice action has been chosen to yield well-defined temporal correlation functions. As discussed above, we next make the transformation $h_{p,q} \to (ba^s)^{-1}h_{p,q}$ in this formula for all lattice sites, and at the same time we conveniently change integration variables by setting $\lambda_p \equiv (ba^s)\phi_p$. The result is

$$M_a \prod_q \int \exp[\Sigma\, h_{p,q}\phi_p a^s \epsilon - \tfrac{1}{2}(ba^s)\Sigma\,(\phi_{p+1} - \phi_p)^2 a^s \epsilon^{-1}$$

$$- \tfrac{1}{2}(ba^s)m^2\,\Sigma\,\phi_p^2 a^s \epsilon - (ba^s)^3\kappa\,\Sigma\,\phi_p^4 a^s \epsilon$$

$$- \Sigma\,P_{a,\epsilon}((ba^s)\phi_p)\epsilon]\,\Pi d\phi_p\,,$$

where M_a denotes a new normalization factor. At this point it is useful to change each variable of integration from ϕ_p to $\phi_{p,q} = \phi_k$.

As can be seen by comparison with the form of the lattice expression given previously, we have been able to derive by this plausibility argument all of the needed renormalization factors for each of the parameters in the conventional lattice action. This argument does not determine an acceptable form for the "miraculous" auxiliary potential, but it does tell us what basic property it should have, i.e., arranging that the initial (before a change of variables) temporal correlation functions on the lattice are proportional to ba^s. It is important to observe that there exists an explicit auxiliary potential – the one described above and derived in Chapter 10 – which has the effect of rendering all (truncated) temporal correlation functions on the lattice small (proportional to ba^s) *without compromising the desire that the temporal correlation functions are long ranged*, i.e., that the range is determined by fixed parameters such as m and κ, and for small a, that the range spans a large number of lattice spacings, indeed, a number which tends to infinity in the continuum limit inversely proportional to a.

11.3 Continuum limit of covariant lattice models

In this section let us start with a brief review of a conventional argument that leads to triviality for ϕ_n^4 covariant theories for $n > 4$. To that end we consider the conventional lattice-regularized Euclidean functional integral

$$S_a(h) \equiv \langle e^{\sum h_k \phi_k a^n} \rangle \equiv \exp[\langle (e^{\sum h_k \phi_k a^n} - 1)^T \rangle]$$

$$= N_a \int \exp[\sum h_k \phi_k a^n - \tfrac{1}{2} \sum (\phi_{k^*} - \phi_k)^2 a^{n-2}$$

$$- \tfrac{1}{2} m_o^2 \sum \phi_k^2 a^n - \kappa_o \sum : \phi_k^4 : a^n] \Pi d\phi_k \,,$$

possibly allowing for an a-dependent coefficient of the gradient term as well. In this expression half of all nearest neighbors contribute to the gradient term as needed. Let us focus on the generator of the truncated correlation functions,

$$\sum_{r=1}^{\infty} [1/(2r)!] \sum h_{k_1} \cdots h_{k_{2r}} \langle \phi_{k_1} \cdots \phi_{k_{2r}} \rangle^T a^{2rn} \,;$$

by symmetry all odd-order correlation functions vanish. Due to translation invariance in the continuum limit each correlation function approximately depends on a set of difference arguments. For example, $\langle \phi_{k_1} \phi_{k_2} \rangle$ is approximately a function of $k_1 - k_2$. For large values of this difference argument the correlation function falls to zero. This fact can be seen in the approximate form of the correlation function for a free field of mass m and small a, for $n \geq 3$, given (apart from irrelevant constants of proportionality) by

$$\langle \phi_0 \phi_k \rangle \approx |k|^{-(n-2)} e^{-m|ka|} \,, \qquad k \neq 0 \,.$$

Observe that the physical rate of falloff is determined by the mass m which we assume is positive. At short distances, or more precisely whenever $m|ka| \ll 1$, the two-point correlation function is given by

$$\langle \phi_0 \phi_k \rangle \approx |k|^{-(n-2)} \,, \qquad k \neq 0 \,,$$

and this functional form is characteristic of any theory that is "asymptotically free", that is, has the short-distance behavior of the free theory itself. For a test sequence which is effectively constant over a vast interval (like the size of the solar system!) that is much greater than $1/m$, and may, for example, be taken to be $h_k = \overline{h}_k \equiv A \exp[-m_S^2(ka)^2]$, $m_S \ll m$, it follows (dropping the a^n factors) that

$$\sum_{k_1, k_2} \overline{h}_{k_1} \overline{h}_{k_2} \langle \phi_{k_1} \phi_{k_2} \rangle = \sum_l (\overline{h}_l)^2 \sum_k \langle \phi_0 \phi_k \rangle \,,$$

apart from terms that are $O(m_S/m)$. Similar reasoning for the same test function sequence leads to

$$\sum \overline{h}_{k_1} \cdots \overline{h}_{k_{2r}} \langle \phi_{k_1} \cdots \phi_{k_{2r}} \rangle^T = \sum (\overline{h}_l)^{2r} \sum \langle \phi_0 \phi_{k_2} \cdots \phi_{k_{2r}} \rangle^T \,.$$

This final result suggests that we focus our attention on

$$\chi_r \equiv \sum \langle \phi_0 \phi_{k_2} \cdots \phi_{k_{2r}} \rangle^T \,.$$

For $r = 1$ it is common to set $\chi_1 \equiv \chi$, which is then known – in the context of statistical physics – as the *susceptibility*; the terms χ_r for larger values of r may be called generalized susceptibilities. Although the aggregate $\{\chi_r\}$ is but one set of all possible correlation functions, it is nonetheless an important set. In particular, for the conventional self-coupled quartic interaction, the Lebowitz inequality [GlJ 87] asserts that

$$\langle \phi_0 \phi_{k_2} \phi_{k_3} \phi_{k_4} \rangle^T \leq 0 \,.$$

Hence, if it happened that $\chi_2 = 0$, then it follows that the truncated four-point correlation function vanishes *identically*, and any distribution of this sort with a vanishing truncated four-point correlation function is necessarily Gaussian. Thus, to study the question of triviality, it suffices to consider the special correlation functions χ_r, $r \geq 1$. However, these quantities normally carry dimensions and change under a rescaling of the variables ϕ_k. Therefore they are not intrinsic. To achieve dimensionless and rescaling invariant versions of these special correlation functions, it is first useful to introduce a particular second moment expression, namely

$$\mu_2 \equiv \Sigma k^2 \langle \phi_0 \phi_k \rangle \,,$$

and the quotient

$$\xi^2 \equiv \frac{1}{6} \frac{\Sigma k^2 \langle \phi_0 \phi_k \rangle}{\Sigma \langle \phi_0 \phi_k \rangle} \,,$$

namely, $\xi^2 = \mu_2/6\chi$. The important quantity ξ is called the *correlation length* (despite the fact that it is dimensionless). For the free field correlation function given above, it follows for small a that

$$\xi^2 = \frac{1}{6} \frac{\Sigma k^2 |ka|^{-(n-2)} e^{-m|ka|}}{\Sigma |ka|^{-(n-2)} e^{-m|ka|}}$$

$$\simeq \frac{1}{6a^2} \frac{\int x^2 |x|^{-(n-2)} e^{-m|x|} \, d^n x}{\int |x|^{-(n-2)} e^{-m|x|} \, d^n x}$$

$$= \frac{1}{(ma)^2} \,.$$

This equation provides an important relation between the correlation length ξ, the lattice spacing a, and the mass m.

We next introduce dimensionless and rescaling invariant versions of the generalized susceptibilities. Let us start with the continuum theory for which we introduce $g_1 \equiv 1$ and, for $r \geq 2$,

$$g_r \equiv m^{n(r-1)} \frac{\int \langle \phi(0)\phi(x_2) \cdots \phi(x_{2r}) \rangle^T \, d^n x_2 \cdots d^n x_{2r}}{[\int \langle \phi(0)\phi(x) \rangle \, d^n x]^r} \,.$$

Observe that the factor $m^{n(r-1)}$ takes care of the dimensions of the unbalanced number of integration variables in the numerator and denominator. We now translate this continuum expression into a corresponding version on the lattice, and it readily follows (omitting the continuum limit) that $g_1 = 1$ and, for $r \geq 2$,

$$g_r = \frac{\chi_r}{\chi^r \xi^{n(r-1)}} = \frac{\Sigma \langle \phi_0 \phi_{k_2} \cdots \phi_{k_{2r}} \rangle^T}{[\Sigma \langle \phi_0 \phi_k \rangle]^r [\Sigma k^2 \langle \phi_0 \phi_k \rangle / 6 \Sigma \langle \phi_0 \phi_k \rangle]^{n(r-1)/2}} \,.$$

Note well that the expressions for g_r are *dimensionless* and *invariant* under a general field rescaling such as $\phi_k \to K\phi_k$, $K > 0$, for all k.

Mean field theory

In space-time dimensions $n \geq 5$ it is generally accepted that mean field theory applies, and as a consequence the behavior of special correlation functions such as χ_r may be estimated as $a \to 0$ [Fi 67]. In particular, it follows that

$$\chi = \Sigma\langle\phi_0\phi_k\rangle \propto a^{-2} \, ,$$
$$\xi^2 = \Sigma k^2\langle\phi_0\phi_k\rangle/6\Sigma\langle\phi_0\phi_k\rangle \propto a^{-2} \, ,$$
$$\chi_r = \Sigma\langle\phi_0\phi_{k_2}\cdots\phi_{k_{2r}}\rangle^T \propto a^{-2-6(r-1)} \, .$$

As a result, we learn that

$$g_r \propto \frac{a^{-2-6(r-1)}}{a^{-2r}\,a^{-n(r-1)}} = a^{(n-4)(r-1)} \, .$$

Here, in this final formula, are the seeds for the triviality of ϕ_n^4 for $n \geq 5$. In particular, for $n-4 > 0$ and $r > 1$, it follows that $g_r \to 0$ as $a \to 0$. Consequently, the only nonvanishing truncated correlation function of the type investigated is χ, namely the second-order one, and as noted above this fact implies that the continuum limit is Gaussian, hence trivial. Let us see how we may overcome this conclusion.

Nowhere in this analysis have we paid any attention to the general *amplitude* dependence of the correlation functions, rather only the *singular* behavior as $a \to 0$. Suppose, by the introduction of a suitable auxiliary potential into the lattice action, we could arrange that the multipoint correlation functions for two or more fields are *uniformly rescaled* so that

$$\langle\phi_0\phi_{k_2}\cdots\phi_{k_{2r}}\rangle^T \propto a^{n-4} \, .$$

In such a case, we would find that

$$\xi^2 \propto a^{-2} \, ,$$
$$\chi_r \propto a^{n-4}\,a^{-2-6(r-1)} = a^{n-6r} \, .$$

Thus, taking the amplitude factors as well as the singular behavior into account, we would find that

$$g_r = \frac{\chi_r}{\chi^r\,\xi^{n(r-1)}} \propto \frac{a^{n-6r}}{a^{(n-6)r}\,a^{-n(r-1)}} = 1 \, .$$

In other words, with this uniform rescaling of the multipoint truncated correlation functions, all dimensionless and rescaling invariant versions of the generalized susceptibilities become $O(1)$. This result is expected to hold true even if $\kappa_o = 0$ because the auxiliary potential which leads to the rescaling also ensures that the result is non-Gaussian; this is exactly the case of the pseudofree model.

What remains to deal with, however, is the *magnitude* of the several correlation functions. To rescale these functions to macroscopic values, we need only uniformly rescale the test function according to $h_k \to a^{-(n-4)/2}h_k$, for all k. Under the transformations considered we note – contrary to the situation which leads to the Lebowitz inequality – that the truncated four-point correlation function is now nonnegative, and as a consequence the vanishing of χ_2, for example, implies that the model is a degenerate Gaussian with a vanishing variance, hence trivial. Once again it is sufficient to study χ_2.

Let us summarize the present state of affairs by rewriting the Euclidean lattice expression once again, but this time including an auxiliary potential. In this case, we consider

$$\overline{S}_a(h) = \overline{N}_a \int \exp[\, \Sigma\, h_k\, \phi_k a^n - \tfrac{1}{2}\Sigma\,(\phi_{k^*} - \phi_k)^2 a^{n-2} - \tfrac{1}{2}m_o^2\,\Sigma\,\phi_k^2 a^n$$
$$- \kappa_o\,\Sigma : \phi_k^4 : a^n - \Sigma\,P_a(\phi_k)a^n\,]\,\Pi d\phi_k \;,$$

where P_a is chosen – in the same spirit as for the ultralocal models – so that, for all $r \geq 1$,

$$\langle \phi_{k_1} \phi_{k_2} \cdots \phi_{k_{2r}} \rangle^T \propto a^{n-4}$$

with coefficients that are well behaved and long ranged as $a \to 0$.

To account for the final rescaling we let $h_k \to (ma)^{-(n-4)/2}h_k$, for all k, using some mass m to preserve dimensions. After a rescaling of the integration variables, $\phi_k \to (ma)^{(n-4)/2}\phi_k$, we are led to the final expression for the Euclidean-space lattice generating function given by

$$S_a(h) = N_a \int \exp[\, \Sigma\, h_k\, \phi_k a^n - \tfrac{1}{2}(ma)^{n-4}\,\Sigma\,(\phi_{k^*} - \phi_k)^2 a^{n-2}$$
$$- \tfrac{1}{2}(ma)^{n-4}\,m_o^2\,\Sigma\phi_k^2 a^n - (ma)^{2(n-4)}\,\kappa_o\,\Sigma : \phi_k^4 : a^n$$
$$- \Sigma\,P_a((ma)^{(n-4)/2}\phi_k)a^n\,]\,\Pi d\phi_k \;.$$

Here in the final expression we have the essence of our proposal for a Euclidean-space lattice functional integral for a ϕ_n^4 theory for $n \geq 5$; unfortunately, however, we have not been able to determine a functional form for the auxiliary potential P_a by this argument. Undaunted, we *conjecture* that

$$P_a(\phi_k) = \frac{A(a)\gamma(a)[\,(\gamma(a) + 1)\phi_k^2 - \Delta^2(a)\,]}{[\,\phi_k^2 + \Delta^2(a)\,]^2}$$

for some (unknown) functions $A(a)$, $\gamma(a)$, and $\Delta(a)$. As $a \to 0$, we expect that it is necessary that $\Delta(a) \to 0$. In addition, as $a \to 0$, either $\gamma(a) \to \gamma > 0$, or it may even hold that $\gamma(a) \to 0$. In the former case, it may be possible to assume $\gamma(a) = \gamma$, and then $A(a)$ and $\Delta(a)$ become the unknowns. In the latter case, we first set $B(a) \equiv A(a)\gamma(a)$, and second it may be possible to set the remaining $\gamma(a) = 0$; then $B(a)$ and $\Delta(a)$ become the unknowns.*

Case of n=4

The previous discussion focused on $n \geq 5$, but a plausible scenario may be suggested for $n = 4$ as well. In the present case, logarithmic corrections to the given estimates typically hold, and as seems likely to be the case, let us assume, for $ma \ll 1$, that

$$g_r \propto |\ln(ma)|^{-(r-1)} \, ,$$

from which it follows, if we arrange for all $r \geq 1$ that

$$\langle \phi_{k_1} \cdots \phi_{k_{2r}} \rangle^T \propto |\ln(ma)|^{-1}$$

and rescale the test sequence according to

$$h_k \to |\ln(ma)|^{1/2} h_k \, ,$$

we should achieve a set of nonvanishing truncated correlation functions in the continuum limit.

The resultant Euclidean-space lattice generating function in this case is given by

$$S_a(h) = N_a \int \exp[\Sigma \, h_k \, \phi_k \, a^n - \tfrac{1}{2}|\ln(ma)|^{-1} \Sigma \, (\phi_{k^*} - \phi_k)^2 a^{n-2}$$
$$- \tfrac{1}{2}|\ln(ma)|^{-1} m_o^2 \Sigma \, \phi_k^2 \, a^n - |\ln(ma)|^{-2} \kappa_o \Sigma : \phi_k^4 : a^n$$
$$- \Sigma \, P_a(|\ln(ma)|^{-1/2} \phi_k) \, a^n \,] \, \Pi d\phi_k \, ,$$

where the conjectured form for the auxiliary potential is qualitatively the same as given previously.

The reader may well wonder what happens for $n = 2$ and $n = 3$. In those cases mean field theory is *not* appropriate and suitable hyperscaling relations already enforce the conditions that $g_r = O(1)$ without any need for reweighting the distribution or introducing a rescaling of the test functions [WiK 74, Zi 96].

Features of the proposal

If the proposals suggested for dealing with quartic self-interacting covariant models survive closer scrutiny, then it is worth recording some properties and immediate consequences of the auxiliary potential. First we note that the auxiliary potential is proportional to $\hbar^2$, and hence does not survive in a naive classical

---

* A slightly more general form for the auxiliary potential has been previously proposed; see [Kl 94].

limit in which $\hbar \rightarrow 0$. Nevertheless, its presence in the quantum theory leads to several profound effects: (i) the quartic interaction represents a discontinuous perturbation of the free theory; (ii) the zero-coupling limit of the quartic interacting theory is a pseudofree theory substantially different from a free (or generalized free) theory; and (iii) the quantum field operators do not satisfy canonical commutation relations. Moreover, the proposed auxiliary potential is not arbitrary; rather, the functional form chosen is a regularized form of the one ambiguity that we have claimed could in fact be anticipated. All of these properties are analogous to ones that *inescapably* hold for ultralocal models, and, by analogy, they should hold as well in the covariant case – provided, of course, that our proposal has merit.

Conclusion

With the formulas offered in this section, we have tried our best to extract lessons from the several models we have studied – and solved – in previous chapters, especially the ultralocal models, and have put forth our present best guess for a Euclidean-space lattice functional integral for ϕ_n^4 models, $n \geq 4$. Although rather more complicated by the inclusion of the nonclassical auxiliary potential, it is noteworthy that preliminary Monte Carlo studies of such models in four space-time dimensions have begun [Ga 96]. Unfortunately, additional evidence needs to be collected, and it is too soon to determine whether the ideas advanced in this section are supported or rejected by these computer studies. At any rate, should they eventually be found to be inadequate, we urge the interested reader to propose and study alternative expressions for the nonclassical, auxiliary potential in an effort to find a prescription that may ultimately prove successful.

References

Abarbanel, H.D.I., Klauder, J.R. and Taylor, J.G. (1966). Green's Functions for Rotationally Symmetric Models, *Phys. Rev.* **152**, 1198-1206.

Adler, S. and Dashen, R. (1968). *Current Algebra and Applications to Particle Physics* (W.A. Benjamin, New York).

Aizenman, M. (1982). Geometric Analysis of ϕ_4^4 Fields and Ising Models. Parts I and II, *Commun. Math. Phys.* **86**, 1-48.

Arai, A. (1992). Momentum Operators with Gauge Potentials, Local Quantization of Magnetic Flux, and Representations of Canonical Commutation Relations, *J. Math. Phys.* **33**, 3374-3378.

Araki, H. (1960*a*). *Hamiltonian Formalism and the Canonical Commutation Relations in Quantum Field Theory*, Princeton University thesis.

Araki, H. (1960*b*). Hamiltonian Formalism and the Canonical Commutation Relations in Quantum Field Theory, *J. Math. Phys.* **1**, 492-504.

Aslaksen, E.W. and Klauder, J.R. (1969). Continuous Representation Theory Using the Affine Group, *J. Math. Phys.* **10**, 2267-2275.

Bargmann, V. (1961). On a Hilbert Space of Analytic Functions and an Associated Integral Transform, *Commun. Pure Appl. Math.* **14**, 187-214.

Baumann, K. (1987). On Relativistic Irreducible Quantum Fields Fulfilling CCR, *J. Math. Phys.* **28**, 697-704.

Blank, J., Exner, P. and Havlíček, M. (1994). *Hilbert Space Operators in Quantum Physics* (AIP Press, New York).

Callaway, D.J.E. (1988). Triviality Pursuit: Can Elementary Scalar Particles Exist?, *Phys. Rep.* **167**, 241-320.

Davydov, A.S. (1968). *Quantum Mechanics* (Pergamon Press, New York).

DeFacio, B. and Hammer, C.L. (1977). Symmetries of Ultralocal Quantum Field Theories, *J. Math. Phys.* **18**, 1216-1223.

DeFinetti, B. (1975). *Theory of Probability, Vol. 2* (John Wiley & Sons, London).

Dirac, P.A.M. (1976). *The Principles of Quantum Mechanics* (Oxford Press, Oxford, Fourth Edition).

Emch, G.G. (1972). *Algebraic Methods in Statistical Mechanics and Quantum Field Theory* (Wiley-Interscience, New York).

Ezawa, H., Klauder, J.R. and Shepp, L.A. (1974). A Path Space Picture for Feynman-Kac Averages, *Annals of Physics* **88**, 588-620.

Ezawa, H., Klauder, J.R. and Shepp, L.A. (1975). Vestigial Effects of Singular Potentials in Diffusion Theory and Quantum Mechanics, *J. Math. Phys.* **16**, 783-799.

Fernández, R., Fröhlich, J. and Sokal, A. (1992). *Random Walks, Critical Phenomena, and Triviality in Quantum Field Theory* (Springer-Verlag, New York).

Fisher, M. (1967). The Theory of Equilibrium Critical Phenomena, *Rep. Prog. Phys.* **30**, 615-730.

Francisco, G. and Pilati, M. (1985). Strong-Coupling Quantum Gravity. III. Quasiclassical Approximation, *Phys. Rev. D* **31**, 241-250.

Fröhlich, J. (1982). On the Triviality of $\lambda\phi_4^4$ Theories and the Approach to the Critical Point in $d \geq 4$ Dimensions, *Nuclear Phys.* **B200**, 281-296.

Gausterer, H. (1996). (private communication).

Gel'fand, I.M. and Shilov, G.E. (1964). *Generalized Functions, Vol. 1: Properties and Operations* (Academic Press, New York).

Gel'fand, I.M. and Vilenkin, N.Ya. (1964). *Generalized Functions, Vol. 4: Applications of Harmonic Analysis* (Academic Press, New York).

Gilmore, R. (1974). *Lie Groups, Lie Algebras, and Some of Their Applications* (Wiley-Interscience, New York).

Glimm, J. and Jaffe, A. (1987). *Quantum Physics* (Springer-Verlag, New York, Second Edition).

Grimmer, H. (1970). Cell Model Field Theories, *J. Math. Phys.* **11**, 3283-3295.

Haag, R. (1955). On Quantum Field Theories, *Mat.-Fys. Medd. Kong. Danske Videns. Selskab* **29**, No. 12.

Haag, R. (1996). *Local Quantum Physics* (Springer-Verlag, Berlin, Second Edition).

Hegerfeldt, G.C. and Klauder, J.R. (1970). Metrics on Test Function Space for Canonical Field Operators, *Commun. Math. Phys.* **16**, 329-346.

Hegerfeldt, G.C. and Klauder, J.R. (1972). Fields Without Partners, *Il Nuovo Cimento* **10**, 723-738.

Hegerfeldt, G.C. and Klauder, J.R. (1974). Field Product Renormalization and Wilson-Zimmermann Expansion in a Class of Model Field Theories, *Il Nuovo Cimento* **19**, 153-172.

Hepp, K. (1974). The Classical Limit for Quantum Mechanical Correlation Functions, *Commun. Math. Phys.* **35**, 265-277.

Hida, T. (1970). *Stationary Stochastic Processes* (Princeton University Press, Princeton).

Isham, C.J. and Klauder, J.R. (1990). Affine Fields and Operator Representations for the Nonlinear σ-Model, *J. Math. Phys.* **31**, 699-711.

Itô, K. and McKean, H. (1965). *Diffusion Processes and Their Sample Paths* (Springer-Verlag, New York).

Itzykson, C. and Zuber, J.-B. (1980). *Quantum Field Theory* (McGraw-Hill, New York).

Jauch, J.M. and Rohrlich, F. (1980). *The Theory of Photons and Electrons* (Springer-Verlag, Berlin, Second Expanded Edition, Second corrected printing).

Kaku, M. (1993). *Quantum Field Theory* (Oxford University Press, New York).

Kay, B.S. (1981). On Klauder's Pseudo-Free Oscillator, *J. Phys. A: Math. Gen.* **14**, 155-164.

Klauder, J.R. and McKenna, J. (1965). Continuous-Representation Theory V. Construction of a Class of Scalar Boson Field Continuous Representations, *J. Math. Phys.* **6**, 68-87.

Klauder, J.R. (1965). Rotationally Symmetric Model Field Theories, *J. Math. Phys.* **6**, 1666-1679.

Klauder, J.R., McKenna, J. and Woods, E.J. (1966). Direct-Product Representations of the Canonical Commutation Relations, *J. Math. Phys.* **7**, 822-828.

Klauder, J.R. (1967). Weak Correspondence Principle, *J. Math. Phys.* **8**, 2392-2399.

Klauder, J.R. and Sudarshan, E.C.G. (1968). *Fundamentals of Quantum Optics* (W.A. Benjamin, New York).

Klauder, J.R. (1970a). Ultralocal Scalar Field Models, *Commun. Math. Phys.* **18**, 307-318.

Klauder, J.R. (1970b). Soluble Models of Quantum Gravitation, in *Relativity*, eds. Carmeli, M., Fickler, S. and Witten, L. (Plenum Press, New York), 1-17.

Klauder, J.R. (1971). Ultralocal Quantum Field Theory, *Acta Phys. Austr., Suppl. VIII*, 227-276.

Klauder, J.R. (1973a). Functional Techniques and their Application in Quantum Field Theory, in *Mathematical Methods in Theoretical Physics*, ed. Britten, W.E. (Colorado Assoc. Univ. Press, Boulder), 329-421.

Klauder, J.R. (1973b). Field Structure through Model Studies: Aspects of Nonrenormalizable Theories, *Acta Phys. Austr. Suppl. XI*, 341-387.

Klauder, J.R. (1973c). Ultralocal Spinor Field Models, *Annals of Physics* **79**, 111-130.

Klauder, J.R. (1975a). On Model Fields with Independent Values at Every Space-Time Point, *Acta Phys. Austr.* **41**, 237-247.

Klauder, J.R. (1975b). On the Meaning of a Nonrenormalizable Theory of Gravitation, *Gen. Rel. Grav.* **6**, 13-19.

Klauder, J.R. and Narnhofer, H. (1976a). Multi-Component, Independent-Value Quantum Field Models, *Acta Phys. Austr.* **44**, 161-171.

Klauder, J.R. and Narnhofer, H. (1976b). Large-N behavior for Independent-Value Models, *Phys. Rev.* **13**, 257-266.

Klauder, J.R. (1978). Continuous and Discontinuous Perturbations, *Science* **119**, 735-740.

Klauder, J.R. and Skagerstam, B.-S. (1985). *Coherent States: Applications in Physics and Mathematical Physics* (World Scientific, Singapore).

Klauder, J.R. (1994). Self-Interacting Scalar Fields and (Non-) Triviality, in *Mathematical Physics Toward the XXIst Century*, eds. R. Sen and A. Gersten (Ben-Gurion University Press, Beer Sheva), 87-98.

Kleinert, H. (1995). *Path Integrals in Quantum Mechanics, Statistics and Polymer Physics* (World Scientific, Singapore, Second Edition).

Kolmogorov, A.N. and Fomin, S.V. (1957). *Elements of the Theory of Functions and Functional Analysis* (Graylock Press, Rochester).

Kövesi-Domokos, S. (1976). A Strong-Coupling Approximation: Covariant Perturbation Expansion around Independent-Valued Field Theories, *Il Nuovo Cimento* **33A**, 769-785.

Ladyženskaja, O.A., Solonnikov, V. and Ural'ceva, N.N. (1968). *Linear and Quasi-linear Equations of Parabolic type* (American Math. Soc., Rhode Island, Vol. 23).

Lukacs, E. (1970). *Characteristic Functions* (Hafner, New York, Second Edition).

Naimark, M.A. (1964). *Normed Rings* (P. Noordhoff, Groningen).

Nelson, E. (1964). Feynman Integrals and the Schrödinger Equation, *J. Math. Phys.* **5**, 332-343.

Newman, C.M. (1971). *Ultralocal Quantum Field Theory in Terms of Currents*, Princeton University Ph.D. thesis.

Newman, C.M. (1972). Ultralocal Quantum Field Theory in Terms of Currents, *Commun. Math. Phys.* **26**, 169-204.

Osterwalder, K. and Schrader, R. (1973). Axioms for Euclidean Green's Functions I, *Commun. Math. Phys.* **31**, 83-112.

Osterwalder, K. and Schrader, R. (1975). Axioms for Euclidean Green's Functions II, *Commun. Math. Phys.* **42**, 281-305.

Peskin, M.E. and Schroeder, D.V. (1995). *An Introduction to Quantum Field Theory* (Addison-Wesley, Reading).

Pilati, M. (1982). Strong-Coupling Quantum Gravity. I. Solution in a Particular Gauge, *Phys. Rev. D* **26**, 2645-2663.

Pilati, M. (1983). Strong-Coupling Quantum Gravity. II. Solution without Gauge Fixing, *Phys. Rev. D* **28**, 729-744.

Reed, M. and Simon, B. (1980). *Methods of Modern Mathematical Physics I: Functional Analysis* (Academic Press, New York).

Reed, M. (1976). *Abstract Linear Wave Equations* (Springer-Verlag, Berlin).

Reeh, H. (1988). A Remark Concerning Canonical Commutation Relations, *J. Math. Phys.* **29**, 1535-1536.

Riesz, F. and Sz.-Nagy, B. (1955). *Functional Analysis* (F. Ungar, New York).

Roepstorff, G. (1996). *Path Integral Approach to Quantum Physics* (Springer-Verlag, Berlin).

Schweber, S.S. (1962). *An Introduction to Relativistic Quantum Field Theory* (Harper & Row, New York).

Segal, I. (1963). *Mathematical Problems of Relativistic Physics* (Am. Math. Soc., Providence).

Shepp, L., Ezawa, H. and Klauder, J.R. (1974). On the Divergence of Certain Integrals of the Wiener Process, *Ann. Inst. Fourier, Grenoble* **24**, 189-193.

Simon, B. (1973). Quadratic Forms and Klauder's Phenomenon: A Remark on Very Singular Perturbations, *J. Funct. Anal.* **14**, 295-298.

Skorohod, A.V. (1974). *Integration in Hilbert Space* (Springer-Verlag, Berlin).

Sudarshan, E.C.G. and Mukunda, N. (1974). *Classical Dynamics: A Modern Perspective* (Wiley-Interscience, New York).

Szego, G. (1959). *Orthogonal Polynomials* (American Mathematical Society, New York).

Thirring, W.E. (1958). *Principles of Quantum Electrodynamics* (Academic Press, New York).

Trotter, H.F. (1958). Approximation of Semi-Groups of Operators, *Pacific Journal of Mathematics* **8**, 887-919.

von Neumann, J. (1931). Die Eindeutigkeit der Schrödingerschen Operatoren, *Math. Ann.* **104**, 570-578.

von Neumann, J. (1938). On Infinite Direct Products, *Compositio Math.* **6**, 1-77.

Weinberg, S. (1995). *The Quantum Theory of Fields, I* (Cambridge University Press, New York).

Wilson, K. and Kogut, J. (1974). The Renormalization Group and the ϵ Expansion, *Phys. Rep.* **12**, 75-200.

Yaffe, L.G. (1982). Large N Limits as Classical Mechanics, *Rev. Mod. Phys.* **54**, 407-435.

Zhu, C. and Klauder, J.R. (1994). The Classical Limit of Ultralocal Scalar Fields, *J. Math. Phys.* **35**, 3400-3409.

Zhu, C. and Klauder, J.R. (1995). Operator Analysis of Nonrenormalizable Multi-Component Ultralocal Field Models, *J. Math. Phys.* **36**, 4012-4019.

Zinn-Justin, J. (1996). *Quantum Field Theory and Critical Phenomena* (Oxford University Press, New York, Third Edition).

Index

Printed in the United States
By Bookmasters